Microgrids and the Integration of Energy Storage Systems

Microgrids and the Integration of Energy Storage Systems

Editors

Alexander Micallef

Zhaoxia Xiao

MDPI

Basel • Beijing • Wuhan • Barcelona • Belgrade • Novi Sad • Cluj • Manchester

Editors
Alexander Micallef
University of Malta
Msida, Malta

Zhaoxia Xiao
Tiangong University
Tianjin, China

Editorial Office
MDPI
St. Alban-Anlage 66
4052 Basel, Switzerland

This is a reprint of articles from the Special Issue published online in the open access journal *Energies* (ISSN 1996-1073) (available at: https://www.mdpi.com/journal/energies/special_issues/microgrids_integration_energy_storage_systems).

For citation purposes, cite each article independently as indicated on the article page online and as indicated below:

Lastname, A.A.; Lastname, B.B. Article Title. *Journal Name* **Year**, *Volume Number*, Page Range.

ISBN 978-3-0365-8834-6 (Hbk)
ISBN 978-3-0365-8835-3 (PDF)
doi.org/10.3390/books978-3-0365-8835-3

Contents

About the Editors

Alexander Micallef

Alexander Micallef is currently a senior lecturer at the Department of Electrical Engineering, the University of Malta. He is an Associate Member of the Institute for Sustainable Energy at the University of Malta. Since 2015, he has also worked as a guest lecturer on a PhD course titled "Power Quality and Synchronization Techniques in Microgrids" at the Center for Research on Microgrids, Aalborg University, Denmark. Dr. Micallef has published more than 40 international peer-reviewed journal and conference papers on microgrids, electrical transportation, zero-energy buildings, and power quality. He was listed by Stanford University as being among the top 2% of the world's scientists in 2021 and 2022. Dr. Micallef is currently the technical coordinator of three projects that were granted through national funding, i.e., ESTELLE (EWA 110/20/2/001-C), RIVIERA (UM Research Excellence Grant 2021), and ATLANTES (funded through the "Setting up of transdisciplinary research and knowledge exchange (TRAKE) complex at the University of Malta (ERDF.01.124)" project, which is co-financed by the European Union through the European Regional Development Fund 2014–2020). He is an associate editor for IEEE Access and IET Smart Grids. He is a senior member of the IEEE and the vice-chair of the IEEE Malta Section.

Research interests: microgrids; maritime microgrids for electric ships, offshore applications, and seaports; integration of renewable energy sources; distributed energy storage systems; hierarchical and cooperative control; energy management systems.

Zhao-xia Xiao

Zhao-xia Xiao was born in Hunan, China, in 1981. She received a B.Sc. degree in electrical engineering and automation from Hebei University of Technology, as well as M.Sc. degree in control theory and control engineering and a Ph.D. degree in power systems and their Automation from Tianjin University, Tianjin, China, in 2002, 2005, and 2008, respectively. From 2007 to 2008, she studied at the University of Manchester, UK, as a Ph.D. Guest. In 2009, she began to work at Tiangong University. She became an associate professor and a full professor in 2011 and 2018, respectively. From September 2012 to December 2012, she worked as a visiting scholar at Fraunhofer IWES, Germany. From December 2016 to January 2018, she worked as a visiting scholar at the Department of Energy Technology, Aalborg University, Denmark. She has authored and co-authored more than 40 technical papers published in journals and at conferences.

Research interests: AC/DC Microgrids with renewable energy; maritime microgrids; planning and operation of distribution networks.

Preface

The decarbonization of the utility grid, market integration, consumer empowerment, and technical innovations will all be key objectives in international energy policy in the coming decades. Grid integration constraints are limiting the deployment potential of renewable energy sources. Therefore, while renewable energy sources are essential to reaching these key objectives, energy storage is the enabler that facilitates the integration of renewable energy sources in a cost-effective and flexible manner. The diversification of RES generation and integration of energy storage into modern power systems are also leading to the formation of the island microgrids and microgrid clusters/communities required to develop more reliable and sustainable electricity networks.

This Special Issue for MDPI Energies targeted innovations and the novelty in integration of energy storage systems. The call for papers attracted significant attention from researchers in the field ,and the Guest Editors express their gratitude to all authors who submitted manuscripts for inclusion in this Special Issue. The 10 papers that were published in this Special Issue represent a snapshot of the ongoing research concerned with the integration of energy storage systems into modern electricity distribution networks. A wide range of topics were covered in this special issue, including sizing strategies, AC/DC microgrids, renewable energy communities, and ancillary services under high penetrations of variable renewable energy sources.

A number of the published articles consider the strategies used to integrate energy storage systems into microgrids. In particular, the authors have proposed strategies to develop multiagent-based controllers, exact feedback linearization strategies, state of charge balancing, and adaptive droop coefficient algorithms. Other articles demonstrate how energy storage could be applied to specific case studies, including utility-scale deployment in islands, renewable energy communities, and green hydrogen production systems supplied by photovoltaics. Finally, the articles consider the sizing, methods of operation, and optimal operation of hybrid power plants (local generation that includes energy storage) applied to islanded microgrids, coupled power markets, and vehicle-to-grid (V2G) systems to build microgrids.

The articles in this Special Issue contribute to the body of knowledge by addressing the research gaps and challenges related to the integration of energy storage systems into microgrids and modern electricity distribution networks. Researchers, practitioners, policymakers, and other stakeholders will find that this Special Issue is a valuable resource that can enable further advancements in this field.

Alexander Micallef and Zhaoxia Xiao
Editors

energies MDPI

Review

Review on Recent Strategies for Integrating Energy Storage Systems in Microgrids

Ritu Kandari [1], Neeraj Neeraj [1] and Alexander Micallef [2,*]

[1] Department of Electronics and Communication Engineering, Indira Gandhi Delhi Technical University for Women, Delhi 110006, India
[2] Department of Industrial Electrical Power Conversion, University of Malta, MSD 2080 Msida, Malta
* Correspondence: alexander.micallef@um.edu.mt

Abstract: Energy security and the resilience of electricity networks have recently gained critical momentum as subjects of research. The challenges of meeting the increasing electrical energy demands and the decarbonisation efforts necessary to mitigate the effects of climate change have highlighted the importance of microgrids for the effective integration of renewable energy sources. Microgrids have been the focus of research for several years; however, there are still many unresolved challenges that need to be addressed. Energy storage systems are essential elements that provide reliability and stability in microgrids with high penetrations of renewable energy sources. This study provides a systematic review of the recent developments in the control and management of energy storage systems for microgrid applications. In the early sections, a summary of the microgrid topologies and architectures found in the recent literature is given. The main contributions and targeted applications by the energy storage systems in the microgrid applications is defined for each scenario. As various types of energy storage systems are currently being integrated for the reliable operation of the microgrids, the paper analyses the properties and limitations of the solutions proposed in the recent literature. The review that was carried out shows that a hybrid energy storage system performs better in terms of microgrid stability and reliability when compared to applications that use a simple battery energy storage system. Therefore, a case study for a DC microgrid with a hybrid energy storage system was modelled in MATLAB/Simulink. The presented results show the advantages of hybrid energy storage systems in DC microgrids.

Keywords: distributed energy sources; energy storage; microgrids; hybrid energy storage system; photovoltaic system

Citation: Kandari, R.; Neeraj, N.; Micallef, A. Review on Recent Strategies for Integrating Energy Storage Systems in Microgrids. *Energies* **2023**, *16*, 317. https://doi.org/10.3390/en16010317

Academic Editor: Asif Ali Tahir

Received: 10 November 2022
Revised: 19 December 2022
Accepted: 22 December 2022
Published: 27 December 2022

1. Introduction

A microgrid is an interconnected group of loads, energy storage systems (ESSs) and distributed generators that can exchange power with the main grid through a single point of common coupling (PCC) [1]. Microgrids (MGs) have the capability of working together with the main grid, and as separate entities (i.e., as islands). Therefore, MGs can be deployed to provide electricity in remote areas, thereby facilitating the generation, distribution, and regulation of the power flow to the local consumers. MGs are being considered as one of the key concepts that will enable the deployment of high penetrations of renewable energy sources (RESs) in our electricity networks [2]. Amongst present and emerging RESs, RES integration in MGs typically consists of technologies including photovoltaic (PV) modules, small-capacity hydro units, ocean energy, and wind turbines. MGs can improve the energy security and reliability of the local energy network through the integration of complimentary distributed renewable sources. MGs are the key enablers for future smart grids, which have the potential to transition the present, centralised electricity networks into fully distributed architectures. MG architectures are categorised as alternating current microgrid (ACMG), direct current microgrid (DCMG) and hybrid microgrid (HMG). The

HMG combines the advantages of the ACMG and DCMG architectures since the AC and DC buses are interlinked by a power electronic converter.

Power generation from RESs is typically intermittent and variable as the output power depends on the environmental conditions. Examples of this type of behaviour are the fluctuations in PV generation due to cloud coverage and the variable output characteristics of wind turbines. These uncertainties in RES generation can disrupt conventional planning by utilities. Any large unplanned fluctuations in non-dispatchable generators (i.e., RESs) can potentially affect the stability of the system [3–6]. Novel dispatch strategies are essential to maintaining the balance between generation and demand in scenarios with high penetrations of distributed RESs. Due to their hierarchical control architecture, MGs can maintain the voltage, phase angle, and frequency changes under permissible levels during fluctuations in RES generation. The hierarchical architecture enables the integration and management of distributed energy storage technologies that can provide the required additional reliability and energy security. High penetrations of distributed energy sources also cause power imbalances in the LV distribution network and can affect the transient stability. Local MG control strategies can maintain the voltage and frequency as stable, thus providing a reliable electricity supply to consumers during all possible modes of operation (islanded operation, grid-connected, and transitions between the two modes of operation) [7]. It is also a known phenomenon that high penetrations of distributed energy sources can also cause power imbalances, especially in cases where there are single-phase residential prosumers [8].

The authors in [9] examine the impact of integrating a PV system, an ESS, and electric vehicles into the distribution network of a campus. The authors applied an EMS to the campus microgrid while considering future integrations of RESs. A linear optimisation approach implemented in MATLAB was used to investigate the optimal PV and ESS scenarios. Without the local PV system, the utility provided for the entire energy needs of the campus under a time-of-use (ToU) tariff system. The integration of the PV and ESS, determined from the linear optimisation approach, predicted a reduction of up to 44.80% in the daily energy consumption costs. Investigations were also conducted into the effects of additional local issues such as power outages.

In [10], the authors mapped the interval forecast data from a PV system to the solution space with various weight assignment schemes for day-ahead optimisation. A thorough case study was conducted that shows a significant increase in the hybrid vessel's operational flexibility. The dispatching system that was used in this study uses a multi-objective function to schedule the ship operation in scenarios with competing objectives. The operational cost is the main objective that is addressed by the optimisation scheme while the degradation in the lifetime of the ESS is minimised.

Economic load dispatch was also proposed in [11] to meet the load demand while reducing the overall operational costs by distributing power across several ESSs. In [11], the authors propose a multi-agent consensus-distributed control strategy that was designed to achieve multiple goals simultaneously. The multi-agent consensus-distributed control strategy considers the frequency/voltage droop controllers and the BESSs' hierarchical control architecture. The presented results show that with this strategy, several BESSs tracked the time-of-use-based pricing-generated SoC reference trajectories during a 24 h period with a variable load. The multi-agent consensus strategy also distributes the load active and reactive power, and simultaneously achieves frequency and voltage management (using the leader-follower consensus approach). However, each BESS requires information from nearby BESSs, in addition to the local information. The suggested communication strategy includes plug-and-play functionality and robustness against communication-link failure and transmission delays of up to 15 ms.

A mixed-integer, nonlinear programming model for PV-battery systems was proposed by the authors in [12], which considers long-term battery deterioration. The main objective was to minimise the lifecycle cost by using a novel two-layer optimisation method that takes into account the self-consumption ratios, optimum battery capacity, and two types of tariff systems. The results showed that the battery degradation could cause an increase in operational costs. By taking the battery degradation into account in the optimisation strategy, the resulting battery capacities and lifecycle costs show a significant increase when compared to the scenario without the battery degradation effects.

In [13], a methodology is created which enables the efficient and effective management of the numerous measurements and uncertainties that come with renewable energy sources. By characterizing the measured data by representative days, clustering algorithms were employed to address these two problems. Historical data was used to reflect the specificities of the considered grid. The architecture that was created can successfully govern across various time horizons. HOMER Pro® (Boulder, CO 80301, USA) was employed for planning purposes, while other pertinent indicators were gathered through a day-ahead optimal scheduling tool. A case study shows that the proposed methodology can be used to identify the best battery technology and DOD strategy. The selection criteria for the battery storage were based on parameters such as the lifetime and annual operating costs of each battery technology. Results for the specific case study have shown that, even though NiCd batteries have the best operational costs, their profitability is limited due to their high fixed costs. From an economic standpoint, Li-ion batteries have shown similar behaviour. On the other hand, lead-acid and NaS batteries appeared to be profitable alternatives for the considered microgrid. However, further studies on various microgrid and nanogrid configurations are necessary in order to further corroborate these results.

The authors in [14] examined current research topics that are crucial for the planning, control, and operation of campus microgrid architectures. Several approaches for different types of campus microgrids were studied and compared. These campus microgrids were investigated using a variety of optimisation methodologies, modelling tools, and energy storage technology types. It was determined that different campus microgrids throughout the world lack effective energy management strategies. Most of the evaluated campus microgrids have outdated energy management methods as there have been several advancements in this field that can be deployed to further improve the energy management of their systems.

In [15], the authors identify a trade-off between minimizing energy usage and maximizing user comfort caused by the existing scheduling systems' disregard for user activities. The trade-off between user comfort and electricity cost was alleviated by directly involving user actions in a proposed load-optimisation technique. This trade-off was taken into account, and optimisation models for various home appliances were designed and implemented. An analysis of the simulation's outcomes was performed in terms of occupancy, cost, and energy-consumption reduction.

In [16], the authors discuss energy-efficient power grid technologies. A thorough analysis was performed that takes into account the numerous difficulties in smart-grid demand-side management. The authors propose that line planning and low-cost scheduling make up the first two tiers of the demand-side load management architecture. Demand response is at the third level and is a topic which has seen considerable research activity in the past decade. This study investigates the viability of reprogramming consumer goods under heavy loads and system overloads to meet distribution system requirements.

Energy storage systems (ESSs) are critical elements in MGs as these can allow the effective integration of RESs with loads while maintaining transient stability and reliability. ESSs in microgrids are flexible resources that can provide a range of functions, such as grid resiliency (e.g., seamless transitions to/from islanded operations) and power-quality mitigation (e.g., voltage regulation and harmonic compensation). There are a few review papers that have been published in recent years that mainly discuss ESS sizing strategies, hierarchical control of MGs, and the state of the art in ESS technologies. The main contribution of the present study is to highlight the latest contributions in the literature on the optimal integration of ESSs with a specific focus on microgrid applications.

The remainder of the paper is organised as follows: Section 2 gives an overview of microgrid operating modes and architectures. Energy storage methods and their applications are discussed in detail in Section 3. Section 4 provides a description and analysis of a case study that considers the integration of a hybrid energy storage system (HESS) in a DC microgrid. The conclusions of the paper are presented in Section 5.

2. Microgrid Operation Mode and Architectures

Microgrids can operate in grid-connected and islanded operation. A brief overview of these two modes of operation now follows. A detailed description of these modes of operation is widely available in the literature and is beyond the scope of this article.

2.1. Modes of Operation

2.1.1. Grid-Connected Operation

In this mode, the MG is connected to the main grid through a single PCC. The MG exchanges power with the main grid depending upon the mismatch in the load power and the power generated by the RESs. Any excess power generated by the RESs (i.e., when the load demand is low, and the generation is high) can be used to charge the ESS. In scenarios where the ESS is fully charged, any excess power generated within the microgrid can be exported to the main grid. On the other hand, in case of partial shading or cloudy conditions (i.e., when the generation is lower than the load demand), the load may be supplied by the ESS, depending on the available SoC. When the ESS reaches its lower SoC limit, the required power may be imported directly from the main grid. Hence, the RES, ESS, and main grid need to work together to maintain the reliability and stability of the microgrid. Hence, the power flow through a PCC can be bidirectional.

2.1.2. Islanded Mode of Operation

Faults occurring in the main grid may cause abnormal conditions at the PCC of the microgrid. In this scenario, the microgrid can be isolated from the main grid and continue to operate as an islanded microgrid. In this mode of operation, the local frequency and voltage are regulated by the distributed RESs (e.g., wind and solar PV) and ESSs [17]. In this mode, ESSs are critical elements of the MGs that can maintain the energy balance, minimise power fluctuations, and improve the reliability and system efficiency [18]. The ESSs absorb excess RES generation when the generation exceeds the demand. The ESSs can be used to supply power to the MG in periods where the demand exceeds the local generation. This minimises any instances where RES power curtailment and/or load shedding should be carried out. In addition, ESSs can also be used to improve the voltage and frequency regulation of the islanded microgrid.

2.2. Microgrid Architectures

As described in an earlier section, MG architectures can be categorised into ACMGs, DCMGs, and HMGs. This section provides a brief overview of the main characteristics of each architecture.

2.2.1. AC Microgrids (ACMGs)

In ACMGs, the local RESs, ESSs, and loads are all connected to a common AC bus. Any DC generating units (e.g., PV panels) and ESSs (e.g., batteries) must connect to the common AC bus through dedicated DC-to-AC inverters [19]. The control and management of ACMGs is difficult due to the presence of critical and non-critical loads that require harmonic currents [20]. In ACMGs, harmonic suppression is achieved either by the introduction of passive/active filters or through the addition of special functionality in the primary control loops of the power electronic converters in the microgrid. While AC microgrids can be easily integrated into the main grid, re-synchronisation of the ACMG with the main grid is complex. Synchronisation of the MG involves matching the voltage amplitude, frequency, and phase at the PCC with that of the main grid. In the literature, there are three main AC distribution architectures for microgrids, namely, single-phase, three-phase with neutral, and three-phase without neutral.

2.2.2. DC Microgrids (DCMGs)

In these MGs, the local RES generation, ESSs, and loads are connected to a common DC bus. Any AC sources and loads must be connected to the common bus through dedicated AC-to-DC passive/active rectifiers. In the literature, one can find three main types of DC microgrids: monopolar, bipolar, and homopolar distribution systems. A monopolar DC grid consists of a two-wire distribution system, between which the DC bus voltage is defined. On the other hand, the bipolar and homopolar DC grids are three-wire DC distribution systems. In addition to the ground return conductor, the bipolar DC grid has two low-voltage conductors with different polarities, while the homopolar DC grid has two low-voltage conductors with the same voltage polarity. DCMGs have several advantages over their AC counterparts. These include greater reliability, higher efficiency (fewer power electronic converters), and improved stability. DC microgrids have been recently employed in special applications, such as shipboard microgrids, EVs, and telecommunication systems. However, the main limitation of DC microgrids is the complexity and high cost of the protection system when compared to AC microgrids [21].

2.2.3. Hybrid Microgrids (HMGs)

In HMGs, the AC sources and loads are connected to the AC bus, while DC sources and DC loads are connected to the DC bus. HMGs have the advantages of both ACMGs and DCMGs, and they result in fewer power conversion stages since these can simultaneously support both AC and DC sources/loads. The AC and DC sub-grids are interconnected via a bidirectional interlinking converter. This converter is the most important part of the HMG as it manages and coordinates the power flow between and within the sub-grids. The HMG uses a transformer to convert voltage on the AC side and a DC–DC converter for voltage conversion on the DC side [21,22]. Depending upon the load requirement and the condition of main grid, HMGs can also be made to work in grid-connected or islanded mode.

3. Recent Microgrid Architectures and Applications

Table 1 summarises the applications and main contributions described in the recent literature on microgrids that include ESSs. The remainder of this section provides a detailed review of the work carried out in these articles.

The authors in [23] performed an extensive review on the integration of RES with the main grid, while focusing on commercial applications. This study has shown that the role of ESSs in microgrids becomes crucial to ensure the continuous power supply to the loads due of the intermittent nature of the RES. In addition, the authors also provide a detailed analysis for the design of microgrids for commercial applications, with a specific focus on reliability criteria. Hybrid energy storage systems (HESSs) were proposed by the authors so as to benefit from the advantages of the combined types of ESSs. In [24], a HESS was used to overcome the inherent limitations of individual ESSs and thus achieve reliable storage solutions. A centralised control strategy is used with the conventional method for managing the HESS, whereby a filter-based technique is typically used. Both centralised and distributed control strategies were evaluated by the authors in order to improve the reliability of the ESS. Independent SoC recovery was performed to maintain the SoC of the HESS within the permissible limits.

An intelligent-control strategy aimed at scheduling the power purchasing in a microgrid consisting of wind turbines and BESSs was described in [25]. A carbon tax was also proposed by the authors to complement the RES generation since this would result in additional reductions in the dependency on fossil fuels. Quantum-based particle swarm optimisation was applied to a number of case studies to improve performance. The optimisation strategy was aimed at minimizing the generation cost while ensuring that the RES generation was used to the maximum and was increased using the optimisation and control technique. The authors in [26] proposed a self-sustainable microgrid architecture that can work with minimal dependence on the utility grid. BESSs were used to support the microgrid in the absence of the main grid. The sizing strategy for the BESS was based on minimizing the tie line flow from the utility grid. The test microgrid and control scheme were successfully implemented in a region in Thailand, and measures regarding the ways this strategy can be applied to other similar regions were also provided.

Table 1. Categorisation of recent microgrid publications, including ESSs, available in the literature.

MG Configuration	Regulatory Control	Mode of Operation	Application	Contribution	Year	Reference
DC	Distributed	Grid connected	Commercial	Reliability	2020	[23]
DC	Centralised	Grid connected	Industrial	Enhances system reliability, voltage regulation, and SoC recovery	2015	[24]
DC	Distributed	Grid connected	Commercial	Reduces the generating cost and enhances the power capacity	2015	[25]
DC	Distributed	Autonomous	Commercial	BESS sizing for system reliability	2015	[26]
AC	Distributed	Grid connected	Industrial	Power losses and battery sizing are strategies for economic benefits of the microgrid	2017	[27]
DC	Distributed	Grid connected	Residential	Battery life and reduction of the voltage fluctuations	2017	[28]
AC	Distributed	Grid connected	Commercial	Optimal sizing of the ESS, enhances the battery life	2017	[29]
DC	Distributed	Autonomous	Residential	Cost of BESS is highlighted for reducing overvoltages, energy loss, and emissions	2018	[30]
DC	Distributed	Autonomous	Commercial	System reliability, reduction of energy costs using intelligent techniques	2018	[31]
DC	Distributed	Grid connected	Commercial	Energy cost is analysed for efficient BESS operation	2017	[32]
DC, AC	May be applied to al	May be applied to all	May be applied to all	Cost and capacity of the ESS to reduce the peak-load demand	2019	[33]
DC, AC	Centralised	Autonomous	Commercial	Power loss and cost minimisation of the ESS	2017	[34]
DC	Decentralised	Autonomous	Industrial	Cost and sizing of the storage system	2019	[35]
DC, AC	Decentralised	Grid connected	Industrial	Cost, capacity, and sizing of ESS	2017	[36]
AC	Centralised	Grid Connected	Commercial	Operation and sizing of an ESS for a windfarm	2020	[37]
Hybrid	Decentralised	Both	Household	Cost	2020	[38]
DC, AC	Decentralised	Autonomous	Residential	Hybrid power system; limit the power and energy for efficient ESS operation	2021	[39]
DC, AC	Grid connected	Autonomous	Commercial	Two control techniques are proposed for the charging and discharging of ESSs	2021	[40]
DC	Decentralised	Both	Commercial	An ESS is integrated with a microgrid for reliability in normal and abnormal conditions	2021	[41]
DC, AC	Centralised	Both	Industrial	Hybrid ESS for resilient microgrid operation	2022	[42]
DC	Centralised	Grid connected	Residential	Intelligent method for estimating the battery SoC	2021	[43]
DC	Centralised	Autonomous	Commercial	Hybrid energy storage approach is used to minimise the operating cost in the microgrid system and minimise waste energy	2021	[44]

The authors in [27] propose an optimisation strategy that serves two purposes: minimising the cost and minimising the power losses. Traditionally, there is a trade-off between these two objectives. Hence, a fuzzy logic strategy was proposed by the authors in order to reach the optimal solution. A cost-efficient model was proposed to determine the optimal size and location of the ESS units. Demand response was used for the load management, where the peak loads were shifted to off-peak periods. The optimal siting of the BESS for voltage-balancing applications in grid-connected microgrids with high penetrations of RESs was explored in [28]. A genetic algorithm-based optimisation technique was utilised for voltage-deviation mitigation while considering the maximisation of the battery lifetime. A qualitative cost model was designed to minimise the overall cost of the system. In addition, a sensitivity analysis was performed for varying costs.

Intelligent methods were applied in [29] for the optimal sizing of the ESS while maximising the battery life by avoiding overuse. The CAPEX of BESSs is highlighted in [30] in applications that target the reduction of overvoltages, energy losses, and emissions. Intelligent techniques were applied in [31] to reduce the cost and improve the reliability of the system. The authors in [32] analyses the energy costs by effectively utilising the BESS. Cost and capacity of the ESS are highlighted, and a reduced peak-load demand is achieved in [33]. Power loss and the cost of the ESS is discussed in [34] for hybrid microgrids using a centralised control scheme. The cost and capacity of the ESS is analysed in [35] for a DC microgrid.

Optimal sizing and cost of the ESS are considered in [36] for a hybrid microgrid. Charging and discharging strategies for a storage system used in a microgrid that includes a wind farm are discussed in [37]. The proposed system explores the most economical ESS for a wind-based microgrid. The cost and charging and discharging patterns of the ESS are discussed in [38] for a hybrid microgrid. The system was found to be environmentally friendly. The hybrid power system based on different scenarios was compared in [39]. The limits of the power and energy for efficient ESSs are also discussed and analysed. Two control strategies were applied to regulate the unbalanced power in the microgrid and to control the charging/ discharging of the storage system in [40]. The authors in [41] propose an ESS integrated in a microgrid for reliable system operation by balancing demand and supply in normal and abnormal conditions. The authors in [42] propose a hybrid ESS for improving the resiliency of microgrids. The authors in [43] use an intelligent method for estimating the SoC of a battery in a microgrid system. A hybrid energy storage approach is used to minimise the operating cost in the microgrid system and minimise the waste energy in [44].

3.1. Energy Storage Systems for Microgrid Prosumers

With the integration of renewable energy generation into the existing power system, many users are becoming prosumers by playing the combined roles of producer and consumer. The main grid may be negatively impacted by the huge amount of PV and WT feed-in; the feed-in pricing is substantially lower than the retail pricing to entice prosumers to use RES. Prosumers must also deal with inconsistency in the load demand and the unpredictability of the RES output. All of these requirements have led to a pressing demand for prosumer energy storage resources. The broad use of personal energy storage systems is hampered by the high investment cost, and in many cases, it is still not economical for the prosumers to install their own storage systems. Community ESSs for the shared use of prosumers is a workable solution to the issue [45]. Consumers who use community-scale ESSs have a distinct financial advantage over all other consumers who use personal energy storage systems. Many studies have recently concentrated on the energy exchange between prosumers and community ESSs. For instance, the authors in [46] developed a two-stage paradigm for CESSs and prosumers to share energy storage, in which community ESSs determine the price of virtual storage capacity in the first stage, and prosumers determine capacity and charging/discharging power in the second stage. In [47], a revised two-stage approach for shared energy storage is introduced. A distributed energy capacity trade and operation game proposed in [48] allows consumers to choose capacity bidding actions and day-ahead charging and discharging routines.

3.2. Machine Learning in the Energy Management System of Microgrids

The electric power system is undergoing a drastic modernisation process, which is being driven by the most recent developments and implementations of smart-grid technologies. Microgrids are a crucial component of the modernisation of electricity networks because they offer a flexible means of integrating distributed RESs into the electrical grid. Distributed RESs, like solar and wind, can, nevertheless, be quite unpredictable and intermittent. It is challenging to properly operate an MG because of these erratic resources, load demand, and the random variations on the supply and demand sides. Many studies have been performed to address this issue and provide energy management strategies for the real-time scheduling of an MG, taking into account the adequacy of renewable energy, electricity prices, and load demand uncertainty [49]. In [50], a learning-based solution is provided which does not require an explicit model of the uncertainty, in contrast to traditional model-based approaches that call for a predictor to estimate the uncertainty. The goal of energy management is to reduce the daily running costs using a Markov decision process [51]. A deep reinforcement learning strategy was devised to resolve the Markov decision process. The deep Q-network algorithm was used to train the neural network in the deep reinforcement learning approach, which uses a deep feedforward neural network to approximate the ideal action-value function [52,53]. The necessity of an explicit system model and a predictor to manage the uncertainty can be relaxed by learning-based approaches. The MG is treated as a mysterious black box, and by interacting with it, they develop an almost ideal strategy. Deep reinforcement learning techniques were suggested as potential solutions to the issues a few years ago by the machine learning community. DRL methods use deep neural networks' end-to-end learning capabilities to circumvent the difficulty of learning from high-dimensional state inputs. A deep learning architecture based on a convolutional neural network (CNN) was developed to extract knowledge from historical time series of energy consumption and PV generation.

4. Energy Storage Methods

ESSs play a crucial role in microgrids, and various types of ESS technologies have been considered in the literature. Figure 1 shows abroad classification of ESSs that encompasses many technologies into six main categories. Table 2 lists the different energy storage methods and outlines their main benefits and their disadvantages.

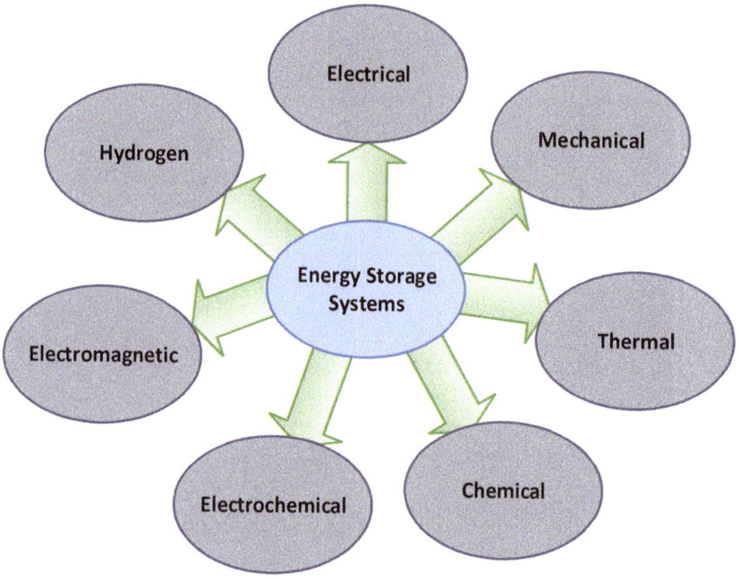

Figure 1. Classification of energy storage systems.

Table 2. Advantages and Disadvantages of available energy storage technologies.

Type of ESS	Advantages	Disadvantages	Reference
Flywheel-based	Environmentally friendly, efficient system, power density is high, low maintenance cost, longer lifespan	Large capital investment, high self-discharge rate, low energy density	[54–56]
Compressed-air-based	Peak shaving performance, provides better control, better quality of air, more stable, smoothened power	Implementation is difficult as appropriate geographical regions needs to be chosen, large capital investment, water loss	[57,58]
Gravity-based	Difference in elevation is not an issue, can be coupled with high-voltage transmission system easily	Capacity of storage system needs to be high, bigger in size, shorter lifetimes	[59,60]
Electrochemical	Low losses, different sizes are available	Uneconomical, Low energy density, shorter lifespan, requires maintenance	[61,62]
Battery	Enables better utilisation of DGs, reliable in grid-connected as well as islanded mode, Provides better control	Shorter lifespan, high maintenance required, SoC limits needs to be maintained	[63–65]
Thermal	Economical, environmentally friendly, rate of self-discharge is low	Temperature needs to be varied for the energy to be stored, unpredictable lifespan, capacity of storage system needs to be high	[66–68]
Chemical	High duration of energy storage, high storage capability	High energy losses, high cost high, low energy density, maintenance is required	[69,70]
Electrical	Better power quality, better response during peak hours, high power density	Uneconomical, high self-discharge rate	[63,71]
Hybrid	HESS has high energy density and power density because of the presence of both BESS and SC both, their energy storage capability is also high, making the system more reliable and stable	HESS has high energy density and power density because of the presence of both BESS and SC, their energy storage capability is also high, making the system more reliable and stable	[72–74]
Hydrogen-based	High energy density, independent charge/discharge rate	relatively low round-trip efficiency	[75,76]

4.1. Mechanical Energy Storage Systems

Mechanical ESSs encompass technologies such as pumped hydro, compressed air and gravity storage devices. The flywheel-based ESS is a robust and efficient method to store electrical energy as kinetic energy in a rotating mass with low frictional losses. Flywheel-based ESSs have low maintenance costs that result in long lifecycles but have very high capital costs. These storage systems are mainly suitable for enhancing the transient stability of microgrids due to their fast repost times. These storage systems typically have high power densities and low energy densities, which limits their flexibility. Thus, this ESS is not ideal for microgrid energy storage applications [54–56].

4.2. Compressed-Air Energy Storage Systems

The compressed-air ESS (CAESS) is a technically mature storage technology in which electricity is stored as compressed air that can be used when required. CAESSs can be built for small-scale and large-scale applications as backup power systems. CAESSs can be used for black start applications and peak shaving; however, their moderate-to-long time constants and spatial constraints (due to the storage tank) limits their widespread application in microgrid systems. Large-scale deployments also require significant capital investments, and the siting of a CAESS is not a trivial task [57,58].

4.3. Gravity-Based Energy Storage Systems

A gravity-based ESS (GESS) is a new solution for large-scale energy storage in which energy is stored as potential energy by the hydraulic lifting of a large mass. It can be coupled with a high-voltage transmission system easily but has a shorter lifespan. The authors in [59,60] examine and survey current gravity-based energy storage technologies. This storage method provides the means for high-capacity energy storage as well as a pollution-free, cost-effective, and extended lifespan. GESSs have quite a significant potential and could be employed in the future to reserve sustainable power to complement existing large-scale energy storage solutions. However, a significant number of studies and innovations must be performed to put these systems into practise and make them interoperable with other ESSs.

4.4. Electrochemical Energy Storage Systems

Secondary batteries, supercapacitors (SC), and fuel cells (FC) fall into the category of electrochemical energy storage. Batteries use an electrochemical reduction reaction to transform the chemical energy found in their active components into electric energy. SCs are high power density devices with time ranges from seconds to minutes as only the electrode surface of material is used, unlike in BESSs and FCs. Electrochemical ESSs have low losses and are available in different sizes [61,62].

4.5. Battery Energy Storage Systems

The battery-based energy storage system (BESS) is a reliable type of electrochemical energy storage which enables better utilisation of distributed generation. If necessary, the battery discharges to supply local loads. The most common technologies that are available on the market for BESSs include lithium-ion, lead-acid, nickel–cadmium, and nickel–metal hydride. Each battery type has unique performance details that define BESS implementations and impact the efficiency of the BESS. The SoC of a battery-based ESS needs to be maintained, otherwise it directly affects the battery's lifetime. It provides better control but requires more maintenance [63–65]. Peak shaving, flexibility, and load shifting are the main advantages of BESSs.

4.6. Thermal Energy Storage Systems

Each year, renewable energy makes up more of the total energy produced throughout the world. Apart from hydroelectric energy, renewable energy sources, such as solar energy and wind energy are among the most used. Due to being intermittent energy sources, they cannot be utilised to their full potential yet. The sun does not always shine, and the wind is not always blowing. If this intermittent energy can be stored, it can be employed even during times when the sources are not actively providing energy. Thermal energy storage is a viable solution. It can be described as storing energy as heat or cold in a storage medium to be used later. Thermal energy storage's main use is to overcome the imbalance between energy generation and energy use. A few essential requirements for an effective thermal energy storage system are a storage material with a high energy density, very good insulation to ensure minimal heat loss, a storage material that is chemically stable, and a completely reversible process that can be repeated numerous times. There are three main types of thermal energy storage systems: sensible heat storage, dealing with a mild increase or decrease in the temperature of a storage material; latent heat storage, which involves the phase change of a storage material; and thermochemical energy storage, where a reversible chemical reaction with high energy involved is used to store energy. The main limitation of these systems is that the change in temperature required for the energy to be stored. A thermal ESS is an economical and environmentally friendly way to store energy that has a low self-discharge rate and a high storage capacity [66–68].

4.7. Chemical Energy Storage Systems

A chemical ESS has high storage capability but low energy density. Their cost is usually high, and the losses are even greater. More maintenance is required for such systems [69,70]. Electrical ESSs provides high power density and better power quality. They show a better response during peak hours, but their self-discharge rate is very high [63,71]. Hybrid ESSs area reliable and stable solution to these problems as they have high energy density as well as high power density due to the presence of both BESSs and supercapacitors. Their energy storage capability is also high. The drawback is that their control is comparatively complex [72–74]. Figure 2 summarises the main applications of ESSs from small-scale systems to large-scale systems.

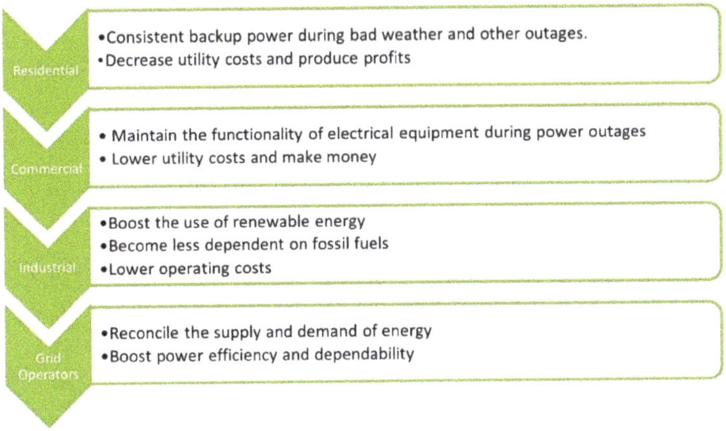

Figure 2. Applications of energy storage systems.

4.8. ESS Integration in Microgrids: Research Gaps

Table 3 summarises the main contributions to the field of ESS integration in microgrids by recent studies, together with the identified research gaps. A thorough review of energy storage methods and their respective analyses are discussed in [77]. However, the shortcomings of battery ESSs and their optimisation approaches were not considered in that work. In [78], a comprehensive review of energy-storage-based applications was performed. The drawbacks with respect to the degradation in the battery lifetime were discussed, but no solutions were provided to maximise the expected lifetime of the ESS. In [79], PID- and ANN-based control methods for frequency control in an ESS are discussed. Their results showed that the ANN-based methods show better performance, and PID controllers have certain limitations. The stability of a microgrid with high penetrations of renewable energy sources was improved in [40] by the introduction of complex controllers designed to regulate the SoC of the ESSs. Different sizing methods for battery ESSs are discussed in [80]. However, only a few of the considered optimisation methods are considered in sufficient detail to benefit academic and industrial applications. A battery's lifetime depends on its operational conditions [81], with the developed model identifying and including the main variables in the degradation process of lithium-ion batteries in V2G applications.

Table 3. Energy Storage Methods and research gaps.

Proposed Research	Research Gaps	Year	Reference
Control techniques are used to reduce the instability of a microgrid during the huge integration of renewable energy sources in load consumption	The numbers of controllers are designed to regulate the SoC conditions.	2021	[40]
Microgrid network, which predicts the uncertainties using deep learning techniques for efficient energy management	The sample batch size is reduced, which impacts the performance of the system.	2019	[49]

Table 3. *Cont.*

Proposed Research	Research Gaps	Year	Reference
Detailed techniques for battery life and SoC condition	The bus voltage is not considered, and battery SoC shows minor fluctuations.	2017	[74]
A thorough explanation of ES methods and their applications	For BESS techniques, the difficulties and problems are not discussed. The optimisation approaches are also not provided.	2016	[77]
Energy-storage-based applications are discussed in detail	The issues related to batteries' life cycles are identified, but solutions are not suggested.	2017	[78]
PID- and ANN-based control methods are discussed for frequency control	The ANN-based methods show excellent results, but the PID methods have some limitations	2020	[79]
Different sizing methods of BESSs for renewable energy systems	Details of the optimisation approach for sizing BESSs is limited to a few approaches.	2017	[80]
BESSs are delivered based on a confined forecast horizon of uncertainties and bound by the SoC constraints	The present BESS dispatch decisions can be appropriate for the current period but not for day-ahead planning.	2016	[81]
A two-stage coordinated technique is introduced to reduce operational costs	Energy losses, uncertainties, and real-time electricity prices are not considered.	2020	[82]
Charging and discharging among different microgrid networks	The real-time connections failed in some microgrid networks.	2015	[83]
A rule-based controller is used to reduce the annual cost and save the operational cost of ESSs	This method is limited to one ESS.	2020	[84]
Performance optimisation of residents in multi-microgrid networks	All four MPC techniques have their drawbacks in terms of size and communication channels.	2015	[85]
Deep learning techniques for energy management, cost reduction, and energy savings	This work lacks accuracy. So, more feasibility is required.	2021	[86]
PV-battery system for residential loads. Battery scheduling and electricity cost reduction are considered	This research is restricted to one consumer.	2021	[87]
Various decision-making approaches in microgrids are discussed	The related works do not involve the sizing of batteries and optimisation methods.	2019	[88]
Deep learning techniques are implemented for microgrid network hybrids	Time-based(day/hour) consumption does not distinguish between day and night.	2021	[89]
Application and principles of lead-acid batteries are discussed for different countries	The battery sizing and optimisation approaches are not included. It is only focused on lead-acid batteries.	2018	[90]
Smart homes have reduced energy costs and temperature fluctuations for energy management	Energy cost and temperature variations are effectively discussed but show the variations in grid parameters' performance.	2019	[91]
Energy sizing techniques for decarbonisation	The attributes are not discussed clearly.	2020	[92]
Categorised into four parts: electrical, mechanical, thermal, and chemical	Sizing and optimisation methods are not included.	2021	[93]

A two-stage coordinated technique was introduced in [82] to reduce the operational costs in microgrids. However, energy losses, uncertainties, and real-time electricity prices were not considered in the two-stage coordination technique, thereby limiting its practicality. Control techniques were proposed in [74] to control the SoC and improve the battery lifetime. However, minor fluctuations can still be observed in SoC conditions using the proposed technique. In [83], model predictive charging and discharging of energy storage was used to schedule the power exchanges between microgrid clusters. The proposed cooperation among grids shares similar limitations with centralised architectures when it comes to the failure of real-time communication. A rule-based controller was used in [84] to reduce the annual costs and reduce the operational cost of the ESS. The main drawback of this controller is that it is limited to only one ESS. The performance of residential ESSs in multi-microgrid networks was also considered by the authors in [85]. All four MPC techniques have their drawbacks in terms of size and communication channel. Deep learning techniques were proposed in [86] to deal with network complexity for the better energy management, cost reduction, and energy saving of a TESS. However, there are still numerous challenges that need to be addressed before deep learning techniques can find their way into the consumer market. The authors in [87] deal with battery scheduling and electricity cost-reduction for a hybrid PV-battery system in a residential setting. However, this research was restricted to one consumer, thereby limiting its applicability. The prediction of uncertainties using deep learning techniques for the efficient energy management of microgrids is considered in [49]. The sample batch size is small, which impacts the performance of the system. Various decision-making approaches are discussed in [88] for the integration of BESSs within microgrids. The related works do not involve the sizing of batteries or optimisation methods. Deep learning techniques are also implemented in [89] for hybrid microgrid networks. The time-based (day/hour) consumption, however, does not distinguish between day and night. May et al. [90] discuss the properties and applications of lead-acid batteries for utility-scale applications based on the operational experiences of different countries. However, the study does not describe battery sizing and optimisation approaches. Smart homes have reduced energy costs and temperature fluctuations for energy management [91]. The energy cost and temperature variations are effectively discussed but show the variations in grid parameters' performance. The grid-connected and standalone modes of operation are discussed in [92] for energy sizing techniques for decarbonisation. It is categorised into four parts, such as electrical, mechanical, thermal, and chemical [93]. The sizing and optimisation methods are not included.

Table 4 shows the comparative study of different approaches for the integration of ESSs into microgrids. The authors in [23] discuss a microgrid consisting of a PV and a BESS for smart grid applications. The exergy principle was applied in [94] for multiple energy sources with different energy levels and qualities. Using a multi-objective optimisation technique, the overall efficiency of the system was found to be improved, while the energy cost was reduced. Branch-and-cut was utilised to combat the trade-off between the cost and exergy objective functions. The proposed technique ensures minimum economic cost and improved efficiency. A comprehensive review of optimal energy management techniques was performed in [95]. Power flow stability was discussed for a complex power system consisting of utility grid, DER, ESS, DC loads, etc. Intelligent energy management technologies utilising optimisation approaches were discussed for coordination in the power flow of the smart grid. The smart grid ensures resilient and reliable demand and supply management. The energy internet was found to be a novel approach for a multi-energy system in a microgrid. A multi-level HESS topology and novel energy management scheme was proposed in [96] to improve battery life. A comparison of various HESS approaches was also performed by the authors. The system was found to be technically and financially viable when compared to existing ESSs. The life of an ESS is improved while the operational cost is minimised using the proposed technique. A novel coordination control algorithm is introduced in [97] for the control of voltage and frequency within the permissible limits. A BESS is connected to the microgrid to improve the efficiency and power quality of the system. The BESS is controlled by the main control centre, which controls the charging and discharging of the BESS. BESSs are divided into master BESSs and slave BESSs, depending upon their capacities. The BESS with a larger capacity, i.e., the master BESS, contributes to the charging/discharging process first. When the SoC limits of the master BESS are reached, the control centre sends a signal to the slave BESS to serve the purpose. Lesser deviations in voltage and frequency are obtained as compared to the on-load tap-changer technique.

Table 4. Comparative study of different approaches for energy storage system integration with the microgrid.

Storage Type	Renewable Energy Type	Highlights	Year	Reference
BESS	PV	Smart grid storage application,	2020	[23]
BESS	Hybrid	Energy cost is reduced, overall energy efficiency is improved	2015	[94]
ESS	Hybrid	Optimal energy management, reducing energy cost	2019	[95]
HESS	PV	Energy management, minimising the energy cost, comparison of HESS approaches	2017	[96]
BESS	RES is not included	Voltage and frequency control, central control of BESS, novel coordination control algorithm is introduced	2015	[97]
BESS	PV	Voltage and SoC control, local ESS controller is used.	2014	[98]
HESS	PV	ESS is centralised control, active distribution network is considered, IEEE-34 test feeder	2014	[99]
BESS	PV	Coordination control to manage the charging and discharging conditions, implemented on are al dataset	2018	[100]
BESS	PV	Predictive control methods used for managing the energy storage condition	2018	[101]
HESS	PV	Droop control and LPF is used to control the battery conditions, communication traffic is minimised	2021	[102]
HESS	PV	Designed an energy management system to increase the performance of the optimisation approaches used in the control scheme	2021	[103]
HESS	PV	Augmented filters used to increase the life of the battery; PI controller is used to control the reference current of the battery	2021	[104]
HESS	PV	Hybrid optimisation approach is used for energy management and battery sizing, predictive control method is implemented	2021	[105]
HESS	Hybrid	Reviews the energy storage approaches and applications for hybrid renewable power system	2022	[106]

The PV feed-in tariffs are lower than the utility grid electricity consumption rates in countries such as Germany. Hence, PV plants with ESSs are emerging in such locations. With high penetrations of PVs into the power system, the voltage tends to increase since the peak of the PV generation and the peak electricity demand do not coincide. Self-consumption does not necessarily imply an advantage for distribution networks with significant PV penetration. Voltage control techniques also need to be applied in such microgrids in addition to self-consumption strategies. The local PV storage control achieved in [98] ensures reactive power compensation and PV power curtailment, while managing the BESS charging, depending on the voltage. Grid simulations and an economic evaluation are used to determine their capacity to facilitate PV–grid integration while boosting self-consumption.

A coordinated control scheme was designed in [99] to regulate the charging/discharging of BESSs. By combining the local droop-based control method with a distributed control scheme, the voltage can be maintained within the permissible levels. Two distinct consensus algorithms were used. The first algorithm is employed to assess the BESSs' ability to participate in terms of their installed capacity, while the second algorithm basically performs voltage control. The latter adjusts the SoC of the BESSs to avoid over-charging/discharging of the BESS, thereby ensuring that the BESS is utilised effectively for various scenarios. The predictive control method was used to manage the energy storage units in [100] for a microgrid with a PV and a BESS. Droop control is used to control the battery SoC in [101] in a microgrid with PV and HESS. The advantage of this proposed algorithm is that the communication traffic is also minimised. In [102], an augmented filter was used to increase the lifetime of the battery, while a PI controller was used to control the reference current of the battery. A hybrid optimisation approach was used for the energy management and the battery sizing in [103]. The predictive control method was implemented for proper energy management. An extensive review of the energy storage approaches and applications for hybrid renewable power systems was performed in [104].

5. Case Study

A standalone microgrid with an energy storage system is an attractive alternative for remote electrification. Since BESSs have a short lifespan, they are unreliable and expensive to operate as the predominant solution for sustainable microgrids. In an effort to extend the lifetime of a BESS, the hybridisation of the ESS by combining elements with complimentary characteristics and their associated energy management strategies was investigated.

A MATLAB/Simulink model of a test microgrid was developed to evaluate the effectiveness of a HESS for accurate power sharing and voltage-deviation mitigation in dynamic power-exchange scenarios. The schematic diagram of the small-scale test standalone DC microgrid simulated in MATLAB/Simulink is presented in Figure 3. The test microgrid is composed of a PV source, a battery, a supercapacitor, and a load. A 120 W PV source is interfaced with the DC bus using a DC/DC boost converter, which implements a perturb-and observe-based MPPT technique. A 24 V, 14 Ah lithium-ion battery pack having an initial SoC of 50% and a 32 V, 29 F supercapacitor were connected to the DC bus through their respective bidirectional DC/DC converters. Tables 5–7 show the parameters of the PV, SC, and BESS, respectively. It can be observed that, for the varying power scenarios, the HESS supports the system such that the load is served all the time. It can further be observed that the DC-link voltage is maintained at a constant value, except for the negligible fluctuations during a change in PV power. These fluctuations during transients are minimised by the action of the SC, while the BESS maintains this DC-link voltage during a steady state.

Energy management in the microgrid is achieved by ensuring adequate coordination between the BESS and the SC. The PV was implemented with maximum power point tracking (MPPT) mode, using the well-known perturb and observe technique. The duty cycle is maintained depending on the SoC of the BESS and the deviation of the voltage from the reference value (50 V). Switching signals are sent to the boost converter interfacing the PV with the DC bus.

The DC-link voltage and reference voltage are compared, and the error signal is fed to the PI controller. This PI controller generates the reference current for the BESS. Now, the battery current reference is compared with the measured battery current, and their difference is fed to the PI controller. The output of the PI controller is sent to the PWM generator, which sends switching signals to switches S1 andS2 of the DC/DC bidirectional buck–boost controller, which in turn controls the charging and discharging process of the BESS. If the generation is more than the demand, a rise in voltage is observed, and the BESS works in buck mode. On the other hand, when a dip in voltage is observed, the BESS works on boost mode and supports the load. Hence, energy management is achieved while maintaining the DC-link voltage at its reference value [107].

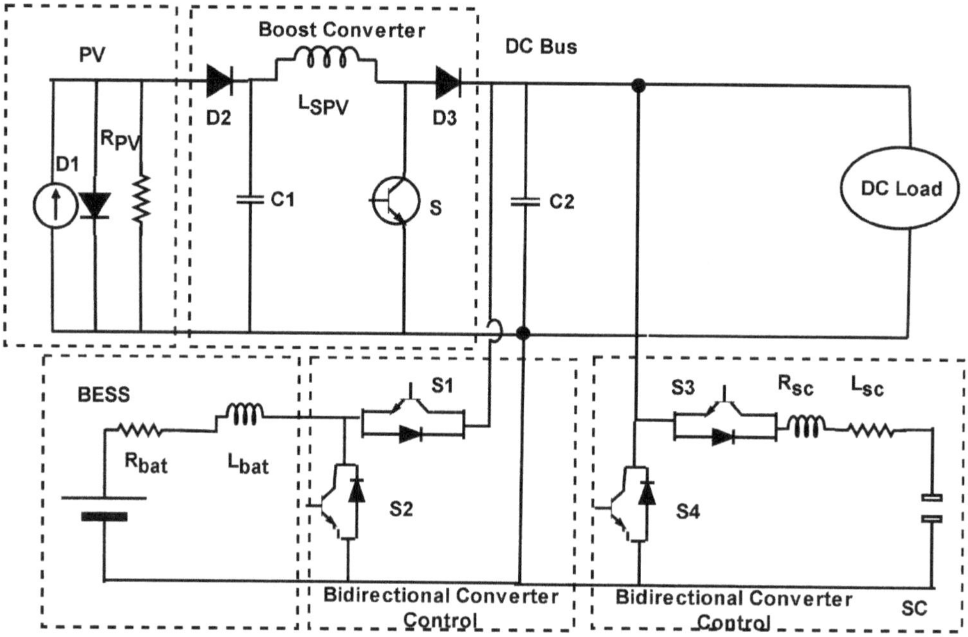

Figure 3. Simplified Schematic of the Test Microgrid structure.

Table 5. PV parameters used.

S. No.	Parameter	Value
1	Maximum power (W)	120.7
2	Cells per module (Ncell)	72
3	Open circuit voltage (V)	21
4	Short-circuit current (A)	8
5	Voltage at maximum power point (V)	17
6	Parallel strings	4
7	Series-connected modules per string	2
8	Operating temperature (Celsius)	25

Table 6. Supercapacitor parameters used.

S. No.	Parameter	Value
1	Rated capacitance (F)	29
2	Equivalent DC series resistance (ohms)	0.003
3	Rated voltage (V)	32
4	Number of series capacitors	1
5	Number of parallel capacitors	1
6	Initial voltage (V)	32
7	Operating temperature (Celsius)	25

The battery reference current and error signal of the voltage generates the SC current reference signal. The SC current reference signal is compared with the SC current, and the error signal is sent to the PI controller. The output of the PI controller is converted to switching signals by the PWM generator. The switching signals are sent to switches S3 andS4 of the DC/DC bidirectional converter connecting the SC and the DC bus. The BESS controls the DC-link voltage during the steady state, while the SC participates in the process during transients. A moving average and low-pass filter are utilised for the power allocation so as to remove the low frequency element and reduce BESS stresses.

Table 7. Battery parameters used.

S. No.	Parameter	Value
1	Type	Lithium-ion
2	Nominal voltage (V)	24
3	Rated capacity (Ah)	14
4	Initial SoC (%)	50
5	Battery response time (s)	0.1
6	Maximum capacity (Ah)	14
7	Cut-off voltage (V)	18
8	Fully charged voltage (V)	27.93
9	Nominal discharge current(A)	6.087
10	Internal resistance (ohms)	0.0171
11	Capacity at nominal voltage (V)	12.66

Figure 4 shows the output PV power for the considered case scenario. Initially, the PV power is increased to 900 W as per the irradiance. At 1.2 s, the PV power is dropped to 400 W. The microgrid serves a constant load of 500 W. Since the PV generation is reduced to less than the load at 1.2 s, PV power is not sufficient to serve the load. Then, the transient supercapacitor comes into action and makes it stable. The supercapacitor power can be observed in Figure 5. The SoC of the SC is shown in Figure 6. It can be observed that, since the supercapacitor works only during the transient, the supercapacitor SoC drops considerably at 1.2 s, which further remains constant.

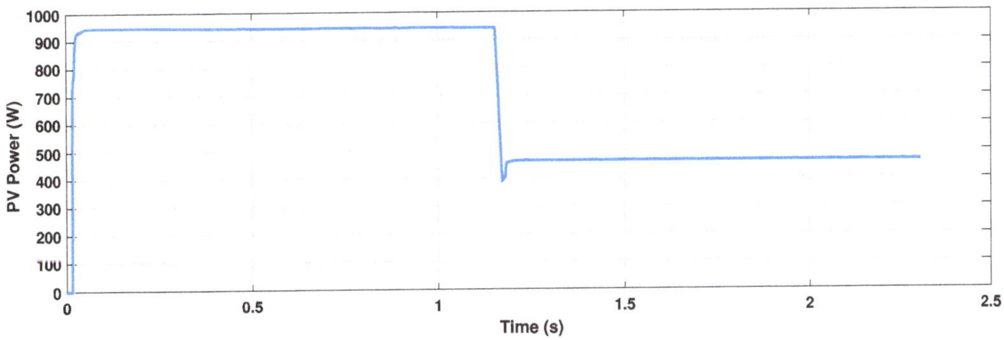

Figure 4. Variations in PV output power with time. The PV power is initially set to 900 W while at 1.2 s the PV power reduces to 400 W.

Figure 7 shows the battery charging and discharging power characteristics for the considered case study. It can be observed that the battery is charging initially until the time that the PV power is more than the load power. However, at 1.2s, the battery starts discharging itself to support the load by providing the deficit amount of power to the load. Variations in battery SoC can be observed in Figure 8. Since the battery is initially charged by the excess amount of power generated by the PV, the battery SoC can be seen to increase, while, after 1.2s, the battery starts being discharged, so the battery SoC also decreases. The DC-link voltage variation is shown in Figure 9. It can be observed from the figure that the DC-link voltage is maintained as constant throughout due to the action of the HESS. Fluctuations in the DC-link voltage can be observed at 1.2s. Due to the action of the HESS, the deviation is mitigated, and the voltage is maintained at the same constant value. Figure 10 shows the variation in battery power and the supercapacitor power with the variation in PV power. It can be observed that a constant load power of 500W is required throughout. Initially, the generated power is sufficient to serve the load; hence, the SC and BESS do not need to discharge. The BESS takes the excess amount of power by charging itself while the supercapacitor acts only during transients. A drop in the generated power can be observed at 1.2s. Now, during this transient condition, the supercapacitor consumes the power to maintain the DC-link voltage. After 1.2s, the generated power is less than that required by the load. The battery comes into action to compensate for the rest of the required power by discharging itself. The DC-link voltage is maintained by the prompt action of the HESS.

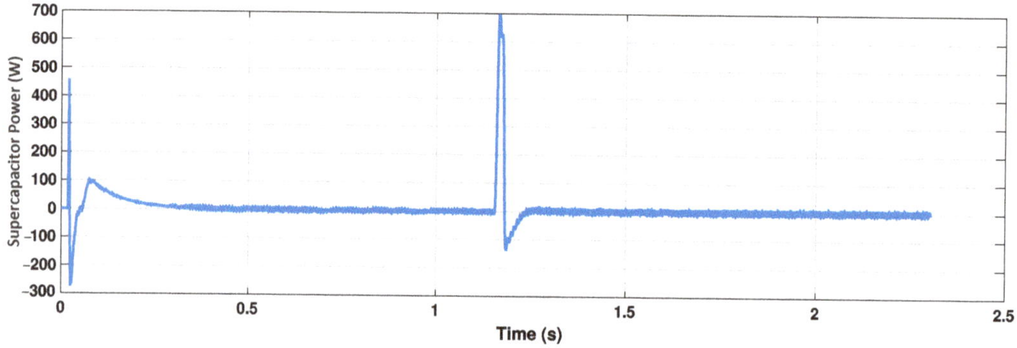

Figure 5. Variation in supercapacitor power with time for a step reduction in the PV output power from 900 W to 400 W at t = 1.2 s.

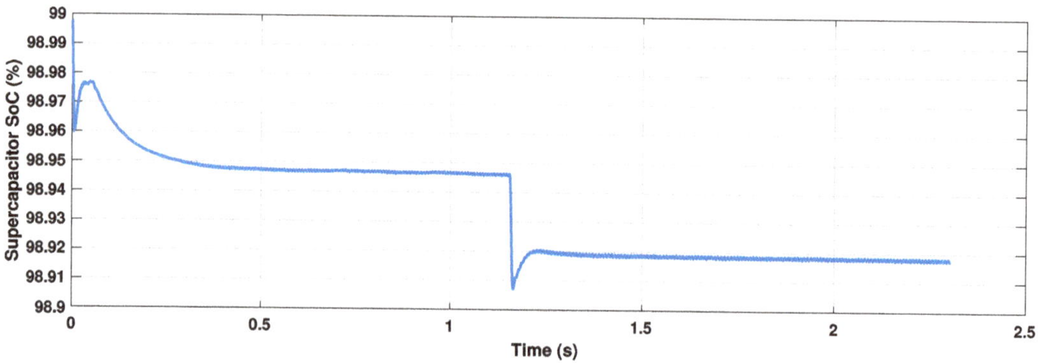

Figure 6. Variation in supercapacitor SoC with time for a step reduction in the PV output power from 900 W to 400 W at t = 1.2 s.

In order to provide a fairly stable DC bus, the battery bank is positioned between the PV and the load. This basic design balances the mismatch between PV generation and load by charging and discharging the BESS. This conventional approach operates effectively when generation and load requirements are steady, but PV microgrids frequently experience fluctuations in generation and load requirements. Hence, the battery is constantly under stress from dynamic supply and absorption, which could possibly shorten its lifecycle.

In the proposed scheme, the SC and BESS are connected in parallel to the DC-link through their respective bidirectional DC/DC converters. It can be observed that, by combining the SC and the BESS, the stress on the battery can be reduced as transient power sharing is assigned to the SC. To minimise DC-link voltage imbalance, the HESS arrangements are meticulously planned because the SC and BESS are on the same DC bus. The SC responds to the high-frequency power exchange or to mitigate the DC-link voltage fluctuations during transients, whereas the BESS, which has a high energy density, is designed to accommodate the low-frequency power variation during steady state conditions.

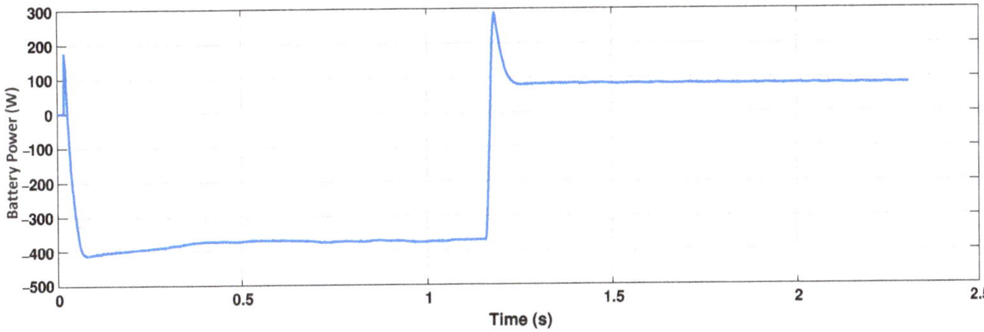

Figure 7. Variation in battery power with time for a step reduction in the PV output power from 900 W to 400 W at t = 1.2 s.

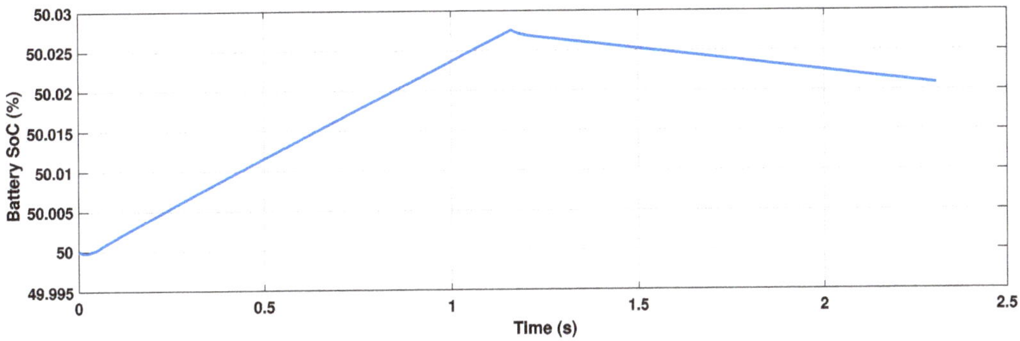

Figure 8. Variation in battery SoC with time for a step reduction in the PV output power from 900 W to 400 W at t = 1.2 s.

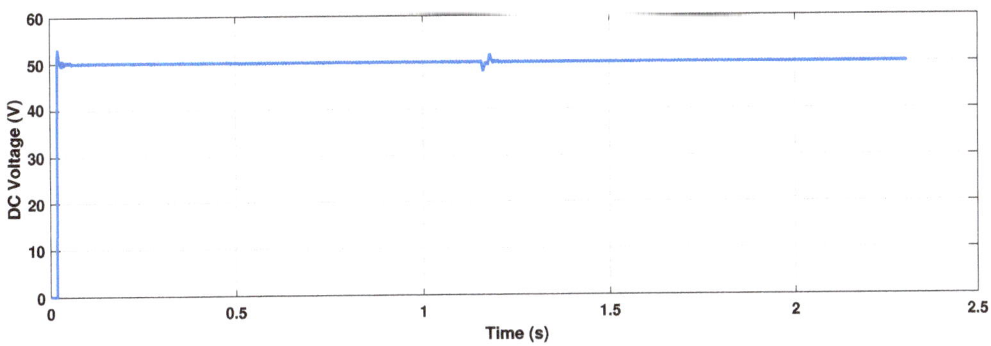

Figure 9. Variation in DC-link voltage with time for a step reduction in the PV output power from 900 W to 400 W at t = 1.2 s.

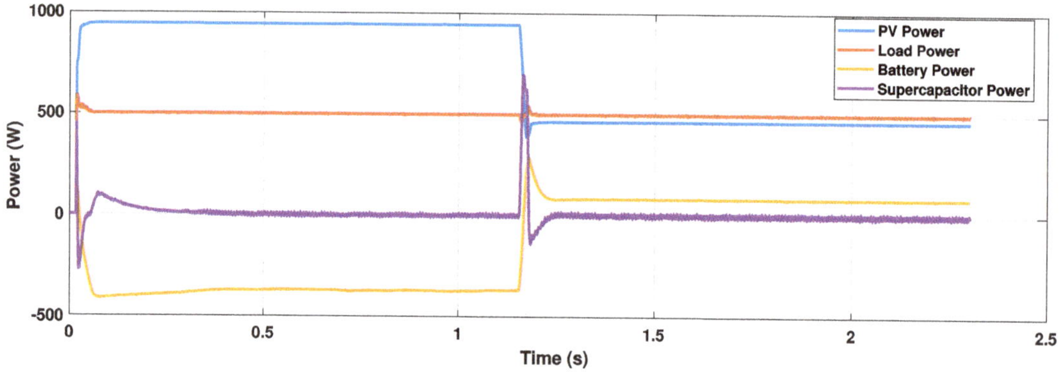

Figure 10. Variation in supercapacitor power/battery power with the variation in demand and supply for a step reduction in the PV output power from 900 W to 400 W at t = 1.2 s.

6. Conclusions

This study provides a systematic review of the recent developments in the control and management of energy storage systems for microgrid applications. One can observe that there is an apparent trend towards shifting from research on AC microgrid topologies to DC microgrid architectures. This can be attributed to the advantages with respect to reducing the number of power electronic converters when integrating storage systems and renewable energy sources in DC microgrids. While battery storage is the predominant system for microgrid applications in the evaluated literature, there was an increase in the studies involving alternative storage systems. The present trends have shifted towards hybrid energy storage systems, combining multiple complementary storage technologies to exploit their advantages. Hybrid energy storage systems perform better in terms of energy security and reliability when compared to applications that use a simple battery energy storage system. Finally, intelligent control strategies have also become prominent in the literature, with a specific focus on machine learning and artificial intelligence starting to become integrated into the management strategies of the storage devices in microgrids.

Author Contributions: Conceptualisation, R.K., N.N. and A.M.; methodology, R.K., N.N. and A.M.; validation, R.K., N.N. and A.M.; formal analysis, R.K., N.N. and A.M.; investigation, R.K., N.N. and A.M.; resources, R.K., N.N. and A.M.; data curation, R.K., N.N. and A.M.; writing—original draft preparation, R.K., N.N. and A.M.; writing—review and editing, R.K., N.N. and A.M.; visualisation, R.K., N.N. and A.M. All authors have read and agreed to the published version of the manuscript.

Funding: This research received no external funding.

Data Availability Statement: The original data used for the simulations are not publicly available.

Conflicts of Interest: The authors declare no conflict of interest.

References

1. Rajesh, K.S.; Dash, S.S.; Rajagopal, R.; Sridhar, R. A review on control of ac microgrid. *Renew. Sustain. Energy Rev.* **2017**, *71*, 814–819. [CrossRef]
2. Mohammed, A.; Refaat, S.S.; Bayhan, S.; Abu-Rub, H. AC microgrid control and management strategies: Evaluation and review. *IEEE Power Electron. Mag.* **2019**, *6*, 18–31. [CrossRef]
3. Roslan, M.; Hannan, M.; Ker, P.J.; Uddin, M. Microgrid control methods toward achieving sustainable energy management. *Appl. Energy* **2019**, *240*, 583–607. [CrossRef]
4. Chandak, S.; Rout, P.K. The implementation framework of a microgrid: A review. *Int. J. Energy Res.* **2021**, *45*, 3523–3547. [CrossRef]
5. Hartono, B.S.; Budiyanto, Y.; Setiabudy, R. Review of microgrid technology. In Proceedings of the 2013 International Conference on QiR, Yogyakarta, Indonesia, 25–28 June 2013; pp. 25–28.

6. Falahi, M.; Lotfifard, S.; Ehsani, M.; Butler-Purry, K. Dynamic model predictive-based energy management of DG integrated distribution systems. *IEEE Trans. Power Deliv.* **2013**, *28*, 2217–2227. [CrossRef]
7. Liang, H.; Choi, B.J.; Abdrabou, A.; Zhuang, W.; Shen, X.S. Decentralized Economic Dispatch in Microgrids via Heterogeneous Wireless Networks. *IEEE J. Sel. Areas Commun.* **2012**, *30*, 1061–1074. [CrossRef]
8. Tenti, P.; Caldognetto, T. Master/Slave Power-Based Control of Low-Voltage Microgrids. In *Microgrid: Advanced Control Methods and Renewable Energy System Integration*; Mahmoud, M.S., Ed.; Butterworth-Heinemann, Elsevier: Oxford, UK, 2017; pp. 101–135. [CrossRef]
9. Nasir, T.; Raza, S.; Abrar, M.; Muqeet, H.A.; Jamil, H.; Qayyum, F.; Cheikhrouhou, O.; Alassery, F.; Hamam, H. Optimal Scheduling of Campus Microgrid Considering the Electric Vehicle Integration in Smart Grid. *Sensors* **2021**, *21*, 7133. [CrossRef]
10. Hein, K.; Yan, X.; Wilson, G. Multi-Objective Optimal Scheduling of a Hybrid Ferry with Shore-to-Ship Power Supply Considering Energy Storage Degradation. *Electronics* **2020**, *9*, 849. [CrossRef]
11. Ullah, S.; Khan, L.; Badar, R.; Ullah, A.; Karam, F.W.; Khan, Z.A.; Rehman, A.U. Consensus based SoC trajectory tracking control design for economic-dispatched distributed battery energy storage system. *PLoS ONE* **2020**, *15*, e0232638. [CrossRef]
12. Wu, Y.; Liu, Z.; Liu, J.; Xiao, H.; Liu, R.; Zhang, L. Optimal battery capacity of grid-connected PV-battery systems considering battery degradation. *Renew. Energy* **2021**, *181*, 10–23. [CrossRef]
13. Arévalo, P.; Tostado-Véliz, M.; Jurado, F. A novel methodology for comprehensive planning of battery storage systems. *J. Energy Storage* **2021**, *37*, 102456. [CrossRef]
14. Muqeet, H.A.; Munir, H.M.; Javed, H.; Shahzad, M.; Jamil, M.; Guerrero, J.M. An Energy Management System of Campus Microgrids: State-of-the-Art and Future Challenges. *Energies* **2021**, *14*, 6525. [CrossRef]
15. Rasheed, M.B.; Javaid, N.; Ahmad, A.; Jamil, M.; Khan, Z.A.; Qasim, U.; Alrajeh, N. Energy Optimization in Smart Homes Using Customer Preference and Dynamic Pricing. *Energies* **2016**, *9*, 593. [CrossRef]
16. Balouch, S.; Abrar, M.; Muqeet, H.A.; Shahzad, M.; Jamil, H.; Hamdi, M.; Malik, A.S.; Hamam, H. Optimal Scheduling of Demand Side Load Management of Smart Grid Considering Energy Efficiency. *Front. Energy Res.* **2022**, *10*, 861571. [CrossRef]
17. Mogaka, L.O.; Nyakoe, G.N.; Saulo, M.J. Islanded and Grid-Connected Control in a Microgrid with Wind-PV Hybrid. *Int. J. Appl. Eng. Res.* **2020**, *15*, 352–357.
18. Talapur, G.G.; Suryawanshi, H.M.; Xu, L.; Shitole, A.B. A Reliable Microgrid with Seamless Transition Between Grid Connected and Islanded Mode for Residential Community with Enhanced Power Quality. *IEEE Trans. Ind. Appl.* **2018**, *54*, 5246–5255. [CrossRef]
19. Lotfi, H.; Khodaei, A. AC versus DC microgrid planning. *IEEE Trans. Smart Grid* **2015**, *8*, 296–304. [CrossRef]
20. Planas, E.; Andreu, J.; Gárate, J.I.; de Alegria, I.M.; Ibarra, E. AC and DC technology in microgrids: A review. *Renew. Sustain. Energy Rev.* **2015**, *43*, 726–749. [CrossRef]
21. Justo, J.J.; Mwasilu, F.; Lee, J.; Jung, J.-W. AC-microgrids versus DC-microgrids with distributed energy resources: A review. *Renew. Sustain. Energy Rev.* **2013**, *24*, 387–405. [CrossRef]
22. Unamuno, E.; Barrena, J.A. Hybrid ac/dc microgrids—Part I: Review and classification of topologies. *Renew. Sustain. Energy Rev.* **2015**, *52*, 1251–1259. [CrossRef]
23. Mbungu, N.T.; Naidoo, R.M.; Bansal, R.C.; Siti, M.W.; Tungadio, D.H. An overview of renewable energy resources and grid integration for commercial building applications. *J. Energy Storage* **2020**, *29*, 101385. [CrossRef]
24. Xiao, J.; Wang, P.; Setyawan, L. Hierarchical Control of Hybrid Energy Storage System in DC Microgrids. *IEEE Trans. Ind. Electron.* **2015**, *62*, 4915–4924. [CrossRef]
25. Zhang, Y.; Iu, H.H.-C.; Fernando, T.; Yao, F.; Emami, K. Cooperative Dispatch of BESS and Wind Power Generation Considering Carbon Emission Limitation in Australia. *IEEE Trans. Ind. Informatics* **2015**, *11*, 1313–1323. [CrossRef]
26. Keskamol, K.; Hoonchareon, N. Sizing of battery energy storage system for sustainable energy in a remote area. In Proceedings of the 2015 IEEE Innovative Smart Grid Technologies—Asia (ISGT ASIA), Bangkok, Thailand, 3–6 November 2015; pp. 1–4.
27. Nojavan, S.; Majidi, M.; Esfetanaj, N.N. An efficient cost-reliability optimization model for optimal siting and sizing of energy storage system in a microgrid in the presence of responsible load management. *Energy* **2017**, *139*, 89–97. [CrossRef]
28. Babacan, O.; Torre, W.; Kleissl, J. Siting and sizing of distributed energy storage to mitigate voltage impact by solar PV in distribution systems. *Sol. Energy* **2017**, *146*, 199–208. [CrossRef]
29. Moghaddam, I.N.; Chowdhury, B.H.; Mohajeryami, S. Predictive Operation and Optimal Sizing of Battery Energy Storage with High Wind Energy Penetration. *IEEE Trans. Ind. Electron.* **2017**, *65*, 6686–6695. [CrossRef]
30. Jannesar, M.R.; Sedighi, A.; Savaghebi, M.; Guerrero, J.M. Optimal placement, sizing, and daily charge/discharge of battery energy storage in low voltage distribution network with high photovoltaic penetration. *Appl. Energy* **2018**, *226*, 957–966. [CrossRef]
31. Zheng, Y.; Zhao, J.; Song, Y.; Luo, F.; Meng, K.; Qiu, J.; Hill, D.J. Optimal Operation of Battery Energy Storage System Considering Distribution System Uncertainty. *IEEE Trans. Sustain. Energy* **2017**, *9*, 1051–1060. [CrossRef]
32. Alharbi, H.; Bhattacharya, K. Stochastic optimal planning of battery energy storage systems for isolated microgrids. *IEEE Trans. Sustain. Energy* **2017**, *9*, 211–227. [CrossRef]
33. Kim, M.; Kim, K.; Choi, H.; Lee, S.; Kim, H. Practical Operation Strategies for Energy Storage System under Uncertainty. *Energies* **2019**, *12*, 1098. [CrossRef]
34. Zhao, T.; Ding, Z. Distributed Agent Consensus-Based Optimal Resource Management for Microgrids. *IEEE Trans. Sustain. Energy* **2017**, *9*, 443–452. [CrossRef]

35. Caro-Ruiz, C.; Lombardi, P.; Richter, M.; Pelzer, A.; Komarnicki, P.; Pavas, A.; Mojica-Nava, E. Coordination of optimal sizing of energy storage systems and production buffer stocks in a net zero energy factory. *Appl. Energy* **2019**, *238*, 851–862. [CrossRef]
36. Lei, J.; Gong, Q. Operating strategy and optimal allocation of large-scale VRB energy storage system in active distribution networks for solar/wind power applications. *IET Gener. Transm. Distrib.* **2017**, *11*, 2403–2411. [CrossRef]
37. Cao, M.; Xu, Q.; Qin, X.; Cai, J. Battery energy storage sizing based on a model predictive control strategy with operational constraints to smooth the wind power. *Int. J. Electr. Power Energy Syst.* **2019**, *115*, 105471. [CrossRef]
38. Mayer, M.J.; Szilágyi, A.; Gróf, G. Environmental and economic multi-objective optimization of a household level hybrid renewable energy system by genetic algorithm. *Appl. Energy* **2020**, *269*, 115058. [CrossRef]
39. Javed, M.S.; Ma, T.; Jurasz, J.; Canales, F.A.; Lin, S.; Ahmed, S.; Zhang, Y. Economic analysis and optimization of a renewable energy based power supply system with different energy storages for a remote island. *Renew. Energy* **2020**, *164*, 1376–1394. [CrossRef]
40. Jan, M.U.; Xin, A.; Rehman, H.U.; Abdelbaky, M.A.; Iqbal, S.; Aurangzeb, M. Frequency Regulation of an Isolated Microgrid with Electric Vehicles and Energy Storage System Integration Using Adaptive and Model Predictive Controllers. *IEEE Access* **2021**, *9*, 14958–14970. [CrossRef]
41. Emara, D.; Ezzat, M.; Abdelaziz, A.Y.; Mahmoud, K.; Lehtonen, M.; Darwish, M.M.F. Novel Control Strategy for Enhancing Microgrid Operation Connected to Photovoltaic Generation and Energy Storage Systems. *Electronics* **2021**, *10*, 1261. [CrossRef]
42. Tobajas, J.; Garcia-Torres, F.; Roncero-Sánchez, P.; Vázquez, J.; Bellatreche, L.; Nieto, E. Resilience-oriented schedule of microgrids with hybrid energy storage system using model predictive control. *Appl. Energy* **2022**, *306*, 118092. [CrossRef]
43. Boujoudar, Y.; Azeroual, M.; Elmoussaoui, H.; Lamhamdi, T. Intelligent control of battery energy storage for microgrid energy management using ANN. *Int. J. Electr. Comput. Eng.* **2021**, *11*, 2760–2767. [CrossRef]
44. Jani, A.; Karimi, H.; Jadid, S. Hybrid energy management for islanded networked microgrids considering battery energy storage and wasted energy. *J. Energy Storage* **2021**, *40*, 102700. [CrossRef]
45. Wu, C.; Zhou, D.; Lin, X.; Wei, F.; Chen, C.; Ma, Y.; Huang, Y.; Li, Z.; Dawoud, S.M. A novel energy cooperation framework for community energy storage systems and prosumers. *Int. J. Electr. Power Energy Syst.* **2021**, *134*, 107428. [CrossRef]
46. Zhao, D.; Wang, H.; Huang, J.; Lin, X. Pricing-based energy storage sharing and virtual capacity allocation. In Proceedings of the 2017 IEEE International Conference on Communications (ICC), Paris, France, 21–25 May 2017; pp. 1–6.
47. Jo, J.; Park, J. Demand-Side Management with Shared Energy Storage System in Smart Grid. *IEEE Trans. Smart Grid* **2020**, *11*, 4466–4476. [CrossRef]
48. Zhang, W.; Wei, W.; Chen, L.; Zheng, B.; Mei, S. Service pricing and load dispatch of residential shared energy storage unit. *Energy* **2020**, *202*, 117543. [CrossRef]
49. Ji, Y.; Wang, J.; Xu, J.; Fang, X.; Zhang, H. Real-Time Energy Management of a Microgrid Using Deep Reinforcement Learning. *Energies* **2019**, *12*, 2291. [CrossRef]
50. Mbuwir, B.V.; Ruelens, F.; Spiessens, F.; Deconinck, G. Battery Energy Management in a Microgrid Using Batch Reinforcement Learning. *Energies* **2017**, *10*, 1846. [CrossRef]
51. Kim, S.; Lim, H. Reinforcement Learning Based Energy Management Algorithm for Smart Energy Buildings. *Energies* **2018**, *11*, 2010. [CrossRef]
52. Mnih, V.; Kavukcuoglu, K.; Silver, D.; Rusu, A.A.; Veness, J.; Bellemare, M.G.; Graves, A.; Riedmiller, M.; Fidjeland, A.K.; Ostrovski, G.; et al. Human-level control through deep reinforcement learning. *Nature* **2015**, *518*, 529–533. [CrossRef]
53. Silver, D.; Huang, A.; Maddison, C.J.; Guez, A.; Sifre, L.; van den Driessche, G.; Schrittwieser, J.; Antonoglou, I.; Panneershelvam, V.; Lanctot, M.; et al. Mastering the game of Go with deep neural networks and tree search. *Nature* **2016**, *529*, 484–489. [CrossRef]
54. Arani, A.K.; Karami, H.; Gharehpetian, G.; Hejazi, M. Review of Flywheel Energy Storage Systems structures and applications in power systems and microgrids. *Renew. Sustain. Energy Rev.* **2017**, *69*, 9–18. [CrossRef]
55. Olabi, A.G.; Wilberforce, T.; Abdelkareem, M.A.; Ramadan, M. Critical Review of Flywheel Energy Storage System. *Energies* **2021**, *14*, 2159. [CrossRef]
56. Xu, Y.; Pi, H.; Ren, T.; Yang, Y.; Ding, H.; Peng, T.; Li, L. Design of a multipulse high-magnetic-field system based on flywheel energy storage. *IEEE Trans. Appl. Supercond.* **2016**, *26*, 1–5. [CrossRef]
57. Tong, Z.; Cheng, Z.; Tong, S. A review on the development of compressed air energy storage in China: Technical and economic challenges to commercialization. *Renew. Sustain. Energy Rev.* **2020**, *135*, 110178. [CrossRef]
58. Olabi, A.G.; Wilberforce, T.; Ramadan, M.; Abdelkareem, M.A.; Alami, A.H. Compressed air energy storage systems: Components and operating parameters—A review. *J. Energy Storage* **2021**, *34*, 102000. [CrossRef]
59. Berrada, A.; Loudiyi, K.; Zorkani, I. System design and economic performance of gravity energy storage. *J. Clean. Prod.* **2017**, *156*, 317–326. [CrossRef]
60. Rohit, A.K.; Devi, K.P.; Rangnekar, S. An overview of energy storage and its importance in Indian renewable energy sector. *J. Energy Storage* **2017**, *13*, 10–23. [CrossRef]
61. Zou, K.; Deng, W.; Cai, P.; Deng, X.; Wang, B.; Liu, C.; Li, J.; Hou, H.; Zou, G.; Ji, X. Prelithiation/Presodiation Techniques for Advanced Electrochemical Energy Storage Systems: Concepts, Applications, and Perspectives. *Adv. Funct. Mater.* **2020**, *31*, 2005581. [CrossRef]
62. Xu, X.; Bishop, M.; Oikarinen, D.G.; Hao, C. Application and modeling of battery energy storage in power systems. *CSEE J. Power Energy Syst.* **2016**, *2*, 82–90. [CrossRef]

63. Hadjipaschalis, I.; Poullikkas, A.; Efthimiou, V. Overview of current and future energy storage technologies for electric power applications. *Renew. Sustain. Energy Rev.* **2009**, *13*, 1513–1522. [CrossRef]
64. RezaeeJordehi, A. An improved particle swarm optimisation for unit commitment in microgrids with battery energy storage systems considering battery degradation and uncertainties. *Int. J. Energy Res.* **2021**, *45*, 727–744. [CrossRef]
65. Pham, T.T.; Kuo, T.-C.; Bui, D.M. Reliability evaluation of an aggregate battery energy storage system in microgrids under dynamic operation. *Int. J. Electr. Power Energy Syst.* **2019**, *118*, 105786. [CrossRef]
66. Sarbu, I. Thermal energy storage. In *Advances in Building Services Engineering*; Springer: Cham, Germany, 2021; pp. 559–627.
67. Tooryan, F.; HassanzadehFard, H.; Collins, E.R.; Jin, S.; Ramezani, B. Smart integration of renewable energy resources, electrical, and thermal energy storage in microgrid applications. *Energy* **2020**, *212*, 118716. [CrossRef]
68. Pelay, U.; Luo, L.; Fan, Y.; Stitou, D.; Rood, M. Thermal energy storage systems for concentrated solar power plants. *Renew. Sustain. Energy Rev.* **2017**, *79*, 82–100. [CrossRef]
69. Konstantinopoulos, S.A.; Anastasiadis, A.G.; Vokas, G.A.; Kondylis, G.P.; Polyzakis, A. Optimal management of hydrogen storage in stochastic smart microgrid operation. *Int. J. Hydrogen Energy* **2018**, *43*, 490–499. [CrossRef]
70. Farhadi, M.; Mohammed, O. Energy Storage Technologies for High-Power Applications. *IEEE Trans. Ind. Appl.* **2015**, *52*, 1953–1961. [CrossRef]
71. Kawakami, N.; Iijima, Y.; Fukuhara, M.; Bando, M.; Sakanaka, Y.; Ogawa, K.; Matsuda, T. Development and field experiences of stabilization system using 34MW NAS batteries for a 51MW wind farm. In Proceedings of the 2010 IEEE International Symposium on Industrial Electronics, Bari, Italy, 4–7 July 2010; pp. 2371–2376.
72. Singh, P.; Lather, J.S. Power management and control of a grid-independent DC microgrid with hybrid energy storage system. *Sustain. Energy Technol. Assess.* **2020**, *43*, 100924. [CrossRef]
73. San Martín, I.; Ursua, A.; Sanchis, P. Integration of fuel cells and supercapacitors in electrical microgrids: Analysis, modelling and experimental validation. *Int. J. Hydrogen Energy* **2013**, *38*, 11655–11671. [CrossRef]
74. Li, J.; Xiong, R.; Yang, Q.; Liang, F.; Zhang, M.; Yuan, W. Design/test of a hybrid energy storage system for primary frequency control using a dynamic droop method in an isolated microgrid power system. *Appl. Energy* **2017**, *201*, 257–269. [CrossRef]
75. Møller, K.T.; Sheppard, D.; Ravnsbæk, D.B.; Buckley, C.E.; Akiba, E.; Li, H.-W.; Jensen, T.R. Complex Metal Hydrides for Hydrogen, Thermal and Electrochemical Energy Storage. *Energies* **2017**, *10*, 1645. [CrossRef]
76. Alzahrani, A.; Ramu, S.K.; Devarajan, G.; Vairavasundaram, I.; Vairavasundaram, S. A Review on Hydrogen-Based Hybrid Microgrid System: Topologies for Hydrogen Energy Storage, Integration, and Energy Management with Solar and Wind Energy. *Energies* **2022**, *15*, 7979. [CrossRef]
77. Gallo, A.B.; Simõões-Moreira, J.R.; Costa, H.K.M.; Santos, M.M.; Moutinho dos Santos, E.M. Energy storage in the energy transition context: A technology review. *Renew. Sustain. Energy Rev.* **2016**, *65*, 800–822. [CrossRef]
78. Katsanevakis, M.; Stewart, R.A.; Lu, J. Aggregated applications and benefits of energy storage systems with application-specific control methods: A review. *Renew. Sustain. Energy Rev.* **2017**, *75*, 719–741. [CrossRef]
79. Iqbal, S.; Xin, A.; Jan, M.U.; Abdelbaky, M.A.; Rehman, H.U.; Salman, S.; Rizvi, S.A.A.; Aurangzeb, M. Aggregation of EVs for primary frequency control of an industrial microgrid by implementing grid regulation & charger controller. *IEEE Access* **2020**, *8*, 141977–141989.
80. Yang, Y.; Bremner, S.; Menictas, C.; Kay, M. Battery energy storage system size determination in renewable energy systems: A review. *Renew. Sustain. Energy Rev.* **2018**, *91*, 109–125. [CrossRef]
81. Farzin, H.; Fotuhi-Firuzabad, M.; Moeini-Aghtaie, M. A Practical Scheme to Involve Degradation Cost of Lithium-Ion Batteries in Vehicle-to-Grid Applications. *IEEE Trans. Sustain. Energy* **2016**, *7*, 1730–1738. [CrossRef]
82. Wang, B.; Zhang, C.; Dong, Z.Y. Interval Optimization Based Coordination of Demand Response and Battery Energy Storage System Considering SOC Management in a Microgrid. *IEEE Trans. Sustain. Energy* **2020**, *11*, 2922–2931. [CrossRef]
83. Ouammi, A.; Dagdougui, H.; Dessaint, L.; Sacile, R. Coordinated Model Predictive-Based Power Flows Control in a Cooperative Network of Smart Microgrids. *IEEE Trans. Smart Grid* **2015**, *6*, 2233–2244. [CrossRef]
84. Banfield, B.; Robinson, D.A.; Agalgaonkar, A.P. Comparison of economic model predictive control and rule-based control for residential energy storage systems. *IET Smart Grid* **2020**, *3*, 722–729. [CrossRef]
85. Worthmann, K.; Kellett, C.M.; Braun, P.; Grüne, L.; Weller, S.R. Distributed and decentralized control of residential energy systems incorporating battery storage. *IEEE Trans. Smart Grid* **2015**, *6*, 1914–1923. [CrossRef]
86. Zsembinszki, G.; Fernández, C.; Vérez, D.; Cabeza, L.F. Deep learning optimal control for a complex hybrid energy storage system. *Buildings* **2021**, *11*, 194. [CrossRef]
87. Joshi, A.; Tipaldi, M.; Glielmo, L. An actor-critic approach for control of residential photovoltaic-battery systems. *IFAC-Pap. OnLine* **2021**, *54*, 222–227. [CrossRef]
88. Baumann, M.; Weil, M.; Peters, J.F.; Chibeles-Martins, N.; Moniz, A.B. A review of multi-criteria decision-making approaches for evaluating energy storage systems for grid applications. *Renew. Sustain. Energy Rev.* **2019**, *107*, 516–534. [CrossRef]
89. Desportes, L.; Fijalkow, I.; Andry, P. Deep Reinforcement Learning for Hybrid Energy Storage Systems: Balancing Lead and Hydrogen Storage. *Energies* **2021**, *14*, 4706. [CrossRef]
90. May, G.J.; Davidson, A.; Monahov, B. Lead batteries for utility energy storage: A review. *J. Energy Storage* **2018**, *15*, 145–157. [CrossRef]

91. Yu, L.; Xie, W.; Xie, D.; Zou, Y.; Zhang, D.; Sun, Z.; Zhang, L.; Zhang, Y.; Jiang, T. Deep Reinforcement Learning for Smart Home Energy Management. *IEEE Internet Things J.* **2019**, *7*, 2751–2762. [CrossRef]
92. Hannan, M.; Faisal, M.; Ker, P.J.; Begum, R.; Dong, Z.; Zhang, C. Review of optimal methods and algorithms for sizing energy storage systems to achieve decarbonization in microgrid applications. *Renew. Sustain. Energy Rev.* **2020**, *131*, 110022. [CrossRef]
93. Olabi, A.; Onumaegbu, C.; Wilberforce, T.; Ramadan, M.; Abdelkareem, M.A.; Al-Alami, A.H. Critical review of energy storage systems. *Energy* **2020**, *214*, 118987. [CrossRef]
94. Di Somma, M.; Yan, B.; Bianco, N.; Graditi, G.; Luh, P.B.; Mongibello, L.; Naso, V. Operation optimization of a distributed energy system considering energy costs and exergy efficiency. *Energy Convers. Manag.* **2015**, *103*, 739–751. [CrossRef]
95. Mbungu, N.T.; Naidoo, R.M.; Bansal, R.C.; Vahidinasab, V. Overview of the optimal smart energy coordination for microgrid applications. *IEEE Access* **2019**, *7*, 163063–163084. [CrossRef]
96. Jing, W.; Lai, C.H.; Wong, W.S.H.; Wong, M.L.D. Dynamic power allocation of battery-supercapacitor hybrid energy storage for standalone PV microgrid applications. *Sustain. Energy Technol. Assess.* **2017**, *22*, 55–64. [CrossRef]
97. Lee, S.-J.; Kim, J.-H.; Kim, C.-H.; Kim, S.-K.; Kim, E.-S.; Kim, D.-U.; Mehmood, K.K.; Khan, S.U. Coordinated Control Algorithm for Distributed Battery Energy Storage Systems for Mitigating Voltage and Frequency Deviations. *IEEE Trans. Smart Grid* **2015**, *7*, 1713–1722. [CrossRef]
98. Von Appen, J.; Stetz, T.; Braun, M.; Schmiegel, A. Local voltage control strategies for PV storage systems in distribution grids. *IEEE Trans. Smart Grid* **2014**, *5*, 1002–1009. [CrossRef]
99. Zeraati, M.; Golshan, M.E.H.; Guerrero, J.M. Distributed Control of Battery Energy Storage Systems for Voltage Regulation in Distribution Networks with High PV Penetration. *IEEE Trans. Smart Grid* **2018**, *9*, 3582–3593. [CrossRef]
100. Gao, X.; Sossan, F.; Christakou, K.; Paolone, M.; Liserre, M. Concurrent Voltage Control and Dispatch of Active Distribution Networks by Means of Smart Transformer and Storage. *IEEE Trans. Ind. Electron.* **2017**, *65*, 6657–6666. [CrossRef]
101. Zhao, H.; Guo, W. Coordinated control method of multiple hybrid energy storage systems based on distributed event-triggered mechanism. *Int. J. Electr. Power Energy Syst.* **2020**, *127*, 106637. [CrossRef]
102. Cabrane, Z.; Kim, J.; Yoo, K.; Ouassaid, M. HESS-based photovoltaic/batteries/supercapacitors: Energy management strategy and DC bus voltage stabilization. *Sol. Energy* **2021**, *216*, 551–563. [CrossRef]
103. Hernández, J.C.; Gomez-Gonzalez, M.; Sanchez-Sutil, F.; Jurado, F. Optimization of battery/supercapacitor-based photovoltaic household-prosumers providing self-consumption and frequency containment reserve as influenced by temporal data granularity. *J. Energy Storage* **2021**, *36*, 102366. [CrossRef]
104. Qi, X.; Wang, J.; Królczyk, G.; Gardoni, P.; Li, Z. Sustainability analysis of a hybrid renewable power system with battery storage for islands application. *J. Energy Storage* **2022**, *50*, 104682. [CrossRef]
105. Rullo, P.; Braccia, L.; Luppi, P.; Zumoffen, D.; Feroldi, D. Integration of sizing and energy management based on economic predictive control for standalone hybrid renewable energy systems. *Renew. Energy* **2019**, *140*, 436–451. [CrossRef]
106. Gan, J.; Li, J.; Qi, W.; Kurban, A.; He, Y.; Guo, S. A Review on Capacity Optimization of Hybrid Renewable Power System with Energy Storage. *E3S Web Conf.* **2019**, *118*, 02055. [CrossRef]
107. Kandari, R.; Gupta, P.; Kumar, A. Coordination Control and Energy Management of Standalone Hybrid AC/DC Microgrid. *J. Eng. Res.* **2021**, *1*, 58–69. [CrossRef]

energies

MDPI

Article

Optimal Sizing and Operation of Hybrid Renewable Power Plants Participating in Coupled Power Markets with Different Execution Times

Carlos García-Santacruz *, Andrés Alcántara, Juan M. Carrasco * and Eduardo Galván

Electronical Engineering Department, University of Seville, 41092 Seville, Spain
* Correspondence: cgarcia20@us.es (C.G.-S.); jmcarrasco@us.es (J.M.C.)

Abstract: The increasing limitations in the use of fossil fuels due to their limited availability and pollution have increased the use of renewable energies and storage systems for electricity generation. To achieve the goals of the integration of renewable energy, sizing and management methods for hybrid plants are needed to make investments profitable and attractive in these resources. This work presents an optimization method for the sizing and operation of hybrid plants with storage, choosing the best combination of technologies based on resource availability, installation costs and market prices, maximizing an economic index such as the net present value. One of the main contributions of this work is to reduce the oversizing that occurs in traditional methods through a penalty term for lost energy, encouraging investment in batteries to store excess energy above the point of interconnection (POI). In addition, it is intended to cover gaps such as the operation in coupled markets with different execution periods to maximize the benefits of the investment made and to contemplate different generation alternatives together with storage. The presented method is tested through sizing and operation simulations to demonstrate its potential. The presented method is tested through sizing and operation simulations to demonstrate its potential. In scenario A, the best combination of solar energy, photovoltaic energy and storage, is chosen. In scenario B, it is shown how the curtailment of the oversizing is reduced in some months by more than 5%. In scenario C, for daily operation in coupled markets, it is possible to improve the benefits from 0.7% to 37.04% in the days of the year.

Keywords: batteries; energy storage; optimal sizing; power system management; electricity markets

Citation: García-Santacruz, C.; Alcántara, A.; Carrasco, J.M.; Galván, E. Optimal Sizing and Operation of Hybrid Renewable Power Plants Participating in Coupled Power Markets with Different Execution Times. *Energies* **2023**, *16*, 3432. https://doi.org/10.3390/en16083432

Academic Editors: Alexander Micallef and Zhaoxia Xiao

Received: 29 March 2023
Revised: 10 April 2023
Accepted: 12 April 2023
Published: 13 April 2023

1. Introduction

Renewable energy resources are clean and increasingly competitive energy sources, so their growth seems to have no limit. According to the latest projections of the International Energy Agency (IEA), the contribution of renewable energies to the global electricity supply will increase from 26% in 2018 to 44% in 2040. These sources will provide enough energy to power two thirds of the increase in electricity demand, mainly through wind and photovoltaic energy technologies [1]. This, together with various directives such as [2,3] at the European level, which set targets for clean energy, makes it essential to consider investing in renewable energies.

The International Renewable Energy Agency (IRENA), in [4], has pointed out how investments in renewable energy have grown from USD 50 to USD 300 billion in the last two decades. Together with renewable energy generation, energy storage systems (ESS) should be considered [5], especially batteries, as cost-effective and beneficial investment options, since it is a critical element in the transition to a sustainable electricity system, able to provide a wide range of services. In addition, ref. [6] noted that clean hydrogen currently has unprecedented political and business momentum, encouraging hydrogen to be widely used.

There are works that analyze investments in renewable energies, such as [7], where the financing, risk and environmental and financial connection of these investments in renewable technologies were analyzed. Similarly, in [8], investments in renewable energy projects were analyzed, pointing out the importance of technological innovation and R&D investments, which also includes sizing and management methods. The development of projects with renewable technologies and the improvement of the efficiency of their use is key [9].

This highlights the need for research in the study and analysis of investments in renewable energies, trying to obtain the maximum economic return from the renewable energy and storage mix, considering all possible markets, in order to make these investments attractive and their development even faster. The use of storage is crucial, obtaining benefits by arbitraging or reducing the possible curtailment of the plants [10], for the subsequent sale of energy in markets.

For this, it is essential to consider the correct sizing and operation of the assets [11]. It is important to base the sizing methods on obtaining benefits for the investor, since, as indicated in [12], only taking into account the technology and installation cost criteria does not provide an optimal result from a cost-effectiveness perspective.

In [13], mixed integer non-linear programming (MINLP) optimization models are used to compare the economic performance of hybrid systems with PV generation as the only alternative of energy source together with batteries. In [14], a multi-objective optimization method is used to size a storage system to maximize revenue and minimize the daily cost, in order to adjust to realistic sizes, but without taking into account investment costs or measuring profit with an economic index. In [15], an MILP problem is proposed for the sizing of storage systems participating in the frequency reserve market, without considering the installation costs to be incurred by the investor. In the work developed in [16], an optimization method for the sizing and operation of photovoltaic energy generation and storage system based on price control is proposed, with the disadvantage of oversizing the hybrid plant in some cases to make it profitable. In [17], particle swarm optimization (PSO) is used to size hybrid energy sources with the objective of minimizing the levelized cost of energy (LCOE). Minimizing the LCOE is interesting from an investor's point of view, but it is also essential to consider market prices to determine the benefit. In [18], different methods are presented for sizing batteries only in photovoltaic energy plants to maximize the total annual revenue and try to find cost-effective storage sizes. In [19], the maximization of economic indexes are evaluated to obtain a hybrid plant, but with PV generation and storage, which is the only asset to be sized. In [20], the problem of optimal storage operation together with wind generation to maximize profit is investigated. It can be observed that it is usual to consider a unique generation source together with storage. This is an important gap, since there are not several generation sources from which to choose the best option together with storage.

In [21], a sizing is proposed through a multi-objective optimization, reducing the cost of the system but also seeking to minimize the emissions of the generation, nor considering different generation sources. The stochastic formulations [22] are also applied to the problem of sizing and profit maximization for the owner participating in the daily and real-time markets, but without specifying different periods as in this paper and not contemplating investment costs.

The problem of sizing is also approached from a technical point of view. In [23], it is presented how to determine the optimal size of wind-solar photovoltaic hybrid energy plant (without storage and its management) using heuristic optimization, with an iterative algorithm to minimize fluctuating production. Studies such as [24] evaluate the sizing of storage systems to compensate for fluctuating wind and solar power generation through optimal economic dispatch. Another approach for sizing can be to determine the optimal generation mix, as in [25], where a flexible fuzzy programming approach is proposed for it.

In [26], a storage system sizing technique is proposed, taking into account the possible errors made in the temporal prediction. The impact of hybrid power plants, with photovoltaic energy and solar thermal energy, from a technical point of view, is analyzed in [27]. Following this point of view, in [28], a sizing of distributed generation and storage is proposed for the improvement of the system and maximizing benefits from a company's point of view, but not measuring with any economic index the investment over the years. The work developed in [29] focuses on solving the problem of grid inertia, without considering investment costs, through frequency control as well as a focus on the use of storage systems. A method for optimal sizing of a BESS, not a full hybrid plant, to provide different services to the power system is included in the work developed in [30]. In [31], the focus is on reducing system losses and voltage unbalance, leaving the economic criterion as an added benefit, not as the focus of the solution. In addition, only one type of generation is considered, without considering if there are more beneficial options.

The sizing problem has also been studied for island systems. In [32], the optimization of the size of an existing plant is studied. There are also studies for isolated networks, as can be seen in [33]. With regards to other, less common technologies, we refer to a review covering all types of CHP optimization problems using meta-heuristic algorithms, including operation and sizing [34].

Most methods in the literature focus on sizing with a single source of generation and storage. For a correct evaluation and maximization of the investment, it is necessary to evaluate all options with predictions of generation and installation costs. Issues of interest to investors, such as economic performance and installation costs must be evaluated in the method, to obtain real and feasible sizing in its application. Another gap in the existing methods is to perform the operation of the plant considering only one market, or several in some works, but which are executed in the same time intervals.

Thus, this article presents an optimal sizing and operation method that attempts to integrate the various renewable resources together with storage. Thanks to the model presented, it is possible to participate in different markets at the same time, including the hydrogen market, in order to make investments attractive.

As the main contributions of this work, the following are highlighted:

- Optimization model to size resources, maximizing the net present value and adjusting to the maximum investment as a constraint (even not spending the entire budget in some cases), to find highly attractive and realistic investments;
- Reduction of curtailment that occurs due to oversizing with a penalty term for lost energy, encouraging investment in storage;
- Evaluate all resources at the same level to maximize the benefit, considering availability, market prices and installation costs, choosing the best option based on the input parameters;
- Modeling and integration of participation in different time-coupled markets (hourly and fifteen minutes), with the possibility of incorporating ancillary services market;
- Consideration of the number of daily charge and discharge cycles of storage in the optimization model, taking care of the valuable life for real results.

2. Method

This section presents the proposed method for sizing and optimal operation of hybrid plants in coupled markets. It is an MINLP optimization model, with three technologies (extendable to more) as alternatives: wind energy, photovoltaic energy and storage systems. The plant can operate in two coupled markets, *M*1, with hourly execution, and *M*2, every 15 min.

An overview of the proposed method is shown in Figure 1, indicating the inputs and outputs to the system. The aim is to evaluate which is the best investment to size a hybrid plant and manage it, participating in several markets coupled at the same time, with the objective of maximizing profits.

The inputs of the algorithm are the annual (or years of analysis) unit profile of wind and PV generation, installation costs of all technologies, characteristics of the assets and annual price profiles (or years of analysis) of the *M*1 and *M*2 markets. As outputs, we obtain the size of the generation and storage, the scheduling of charge and discharge and the participation in each moment on the markets.

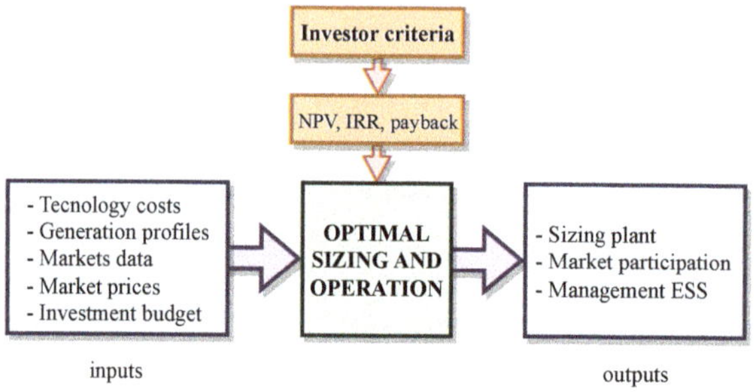

Figure 1. Flowchart of the proposed sizing and operation method.

As a result of the conceptualization and model presented, it is possible to extend it to more markets and generation sources. Table 1 shows the parameters and variables used to model the optimization problem for daily operation and the optimal sizing problem.

To simplify the nomenclature, the model is presented for a single year, which can be extrapolated to as many years as desired for the analysis of optimal sizing.

Table 1. Nomenclature.

Sets		$\pi^{M1}_{d,h}$	Selling price on the market hourly market *M1* (EUR/MWh).
D	Set of D days of a year, $d \in D$.	$\pi^{M2}_{d,h,t}$	Selling price on the 15-min market *M2* (EUR/MWh).
H	Set of H hours of a day, $h \in H$.	α	Weight factor related to the to penalization of curtailment.
T	Set of T intervals of an hour, $t \in T$.	E^{POI}	Limit of the point of connection (MWh).
Parameters		**Variables**	
C^{PV}	Installation cost for photovoltaic energy generation (EUR/MW).	PV^{size}	Maximum installed capacity for photovoltaic energy resource (MW).
C^{Wind}	Installation cost for wind generation (EUR/MW).	$Wind^{size}$	Maximum installed capacity for wind resource (MW).
C^{ESS}	Installation cost for ESS generation (EUR/MWh).	ESS^{size}	Maximum installed capacity for ESS (MWh).
$E^{load}_{d,h,t}$	Internal load in day d, hour h and period t (MW).	$E^{PV}_{d,h,t}$	Power injection by photovoltaic energy generation in interval t of hour h and day d (MWh).
$M^{PV}_{d,h,t}$	Generation profile for photovoltaic energy in day d, hour h and period t (pu).	$E^{Wind}_{d,h,t}$	Power injection by wind generation in interval t of hour h and day d (MWh).
$M^{Wind}_{d,h,t}$	Generation profile for wind in day d, hour h and period t (pu).	$E^{M1}_{d,h,t}$	Energy sold on market *M1* in the interval t of hour h and day d (MWh).
N	Number of charge and discharge cycles.	$E^{M2}_{d,h,t}$	Energy sold on market *M2* in the interval t of hour h and day d (MWh).
$E^{max,ch}_{d,h,t}$	Maximum energy charged by battery in interval t of hour h and day d (MWh).	$E^{ch}_{d,h,t}$	Energy charged by battery in interval t of hour h and day d (MWh).
$E^{max,ch}_{d,h,t}$	Maximum energy discharged by battery in interval t of hour h and day d (MWh).	$E^{dis}_{d,h,t}$	Energy discharged by battery in interval t of hour h and day d (MWh).
SOC^{min}, SOC^{max}	Minimum and maximum state of charge of storage.	$E^{Bat}_{d,h,t}$	Energy stored in the battery in interval t of hour h and day d (MWh).
η^{ch}	Battery charge efficiency.	**Binary**	
η^{dis}	Battery discharge efficiency.	$B^{Ms}_{d,h,t}$	Binary variable to indicate the sale of energy on markets *M1* and *M2*.
Inv	Total investment in assets in millions of euros (MEUR).	$B^{GridBuy}_{d,h,t}$	Binary variable to indicate the purchase of energy.
		$B^{ch}_{d,h,t}$	Binary variable to indicate the charge of battery.
		$B^{dis}_{d,h,t}$	Binary variable to indicate the discharge of battery.

2.1. Objective Functions

The two objective functions proposed in this work are presented below. The first function is utilized for the daily operation of the hybrid plant, with the goal of maximizing profits through participation in different markets. In this case, the operation is performed for a specific size of the hybrid plant. The second objective function is used for sizing the resources of the hybrid plant, seeking the maximum benefit for the investor based on an economic index. In this case, the sizes of the generation and storage components are variables to be calculated in the optimization.

2.1.1. Operation

In the daily operation of the power plant, the maximum economic benefit is sought from the sale of energy to the market. To achieve this, it is necessary to consider the largest possible number of market types, whether they have a short execution interval (minutes) or an hourly one, in order to have more alternatives.

Thus, the objective function of the optimal operation, which takes into account all types of assets, as well as two markets with different execution times, is defined as follows:

$$\max z_1 = \underbrace{\sum_{h=1}^{h=H} E^{M1}_{d,h,1} \cdot \pi^{M1}_{d,h}}_{\text{Market } M1} + \underbrace{\sum_{h=1}^{h=H} \sum_{t=1}^{t=T} E^{M2}_{d,h,t} \cdot \pi^{M2}_{d,h,t}}_{\text{Market } M2} + \underbrace{\sum_{h=1}^{h=H} \left[\sum_{t=1}^{t=T} \left(E^{ch}_{d,h,t} + E^{load}_{d,h,t} \right) \right] \cdot \pi^{M1}_{d,h}}_{\text{Operation cost}} \tag{1}$$

where the profit is determined by participating simultaneously in the $M1$ and $M2$ markets, with different execution intervals. As the operation of the power plant is limited to a short horizon of hours or days, the value of D is set to 1 and the set of hours in the desired horizon, such as $H = 24$, is defined. Additionally, the cost of providing power to internal loads and charging the battery from the grid is taken into consideration.

The MINLP optimization model for the operation is composed of the objective function (1) and the associated constraints (6)–(13), (15)–(20) defined in Section 2.2.

2.1.2. Sizing

To achieve optimal sizing of the hybrid plant, it is necessary to consider the objective of daily operation, which is to maximize daily profit through participation in various markets and to calculate cash flows accordingly. Since the sizing is done based on a given investment budget, it is important to use an economic index to evaluate the investment.

Typically, sizing algorithms based solely on investment costs and operating benefits tend to oversize the plants. This is because, numerically, it is often more profitable to install excess generation capacity and curtail the excess energy produced. To avoid this, a penalty term is proposed to be included in the objective function. This term penalizes the oversizing of the plant and seeks to produce realistic and practical results in a real-world environment.

This work proposes maximizing the net present value (NPV) to determine if asset selection and management, participation in various markets and discounting the investment result in benefits. The objective function for sizing is defined as follows:

$$\max \quad NPV \quad \rightarrow \quad \max \quad -I_0 + \underbrace{\sum_{y=1}^{y=Y} \frac{CF_y}{(1+k)^y}}_{\text{Profit markets}} - \underbrace{\sum_{y=1}^{y=Y} \frac{NCFG_y}{(1+k)^y}}_{\text{Penalty curtailments}} \tag{2}$$

where I_0 represents the total investment made in the generation and energy storage system (ESS), k is the discount rate, Y denotes the number of years of analysis or project and, finally, CF refers to the annual cash flow generated by participating in all markets. This annual cash flow is defined as:

$$\begin{aligned} CF_d = & \sum_{d=1}^{d=D} \sum_{h=1}^{h=H} \left(E^{M1}_{d,h,1} \cdot \pi^{M1}_{d,h} \right) + \sum_{d=1}^{d=D} \sum_{h=1}^{h=H} \sum_{t=1}^{t=T} \left(E^{M2}_{d,h,t} \cdot \pi^{M2}_{d,h,t} \right) + \\ & \sum_{d=1}^{d=D} \sum_{h=1}^{h=H} \left[\sum_{t=1}^{t=T} \left(E^{ch}_{d,h,t} + E^{load}_{d,h,t} \right) \right] \cdot \pi^{M1}_{d,h} \end{aligned} \tag{3}$$

The penalty term, referred to as non-cash flow generated (NCFG), is introduced to penalize excessive oversizing of the generation resources of the hybrid plant. Depending on the generation costs, resource availability and price forecasts, it is possible for the plant to be oversized beyond the point of connection (POI). The NCFG is defined as:

$$NCFG_d = \alpha \cdot \sum_{d=1}^{d=D} \sum_{h=1}^{h=H} \sum_{t=1}^{t=T} \left[(E_{d,h,t}^{PV,max} - E_{d,h,t}^{PV}) \cdot \pi_{d,h}^{M1} + (E_{d,h,t}^{Wind,max} - E_{d,h,t}^{wind}) \cdot \pi_{d,h}^{M1} \right] \quad (4)$$

The NCFG penalizes the excessive oversizing of the generation resources of the hybrid plant. It is calculated as the energy not served due to curtailment in each type of generation. This is determined as the difference between the maximum possible injection according to the sizing, $E_{d,h,t}^{PV,max}$ and $E_{d,h,t}^{Wind,max}$, and the actual energy injected, $E_{d,h,t}^{PV}$ and $E_{d,h,t}^{wind}$, during each interval t. The unserved energy is then economically valued by multiplying it with the market price. It is further weighted by the coefficient α to determine the amount of penalty.

This penalty term is intended to promote the investment and use of storage together with the generation source. The α value can be tuned from 0 to a saturation value specific to each case (from which it does not penalize more). Depending on the value it takes, it will penalize to a different degree and affect the sizing. A small α will allow a lot of curtailment, while a big value will decrease curtailment and will further adjust the sizing of the resources.

Although initially a higher NPV or other index may be obtained, oversizing the plant excessively over the connection point, a large curtailment makes that energy unusable, not allowing it to be stored and used in other markets or ancillary services. This can even be very detrimental to the investment if the prices for which the plant was sized change a lot, not having the flexibility of storage to minimize the negative effects of this change.

The MINLP optimization model for the sizing of the investment in resources is composed of the objective function (2) and constraints (5)–(13), (15)–(20), presented below.

2.2. Problem Constraints

All constraints associated with the operation and size problems are defined in this section. First, the maximum investment that can be made in assets is limited in Equation (5), where the investment is divided between the installed power of the generation sources as well as the storage systems. This constraint allows optimal sizing according to the maximum investment that can be realized.

$$Inv \geq C^{PV} \cdot PV^{size} + C^{Wind} \cdot Wind^{size} + C^{ESS} \cdot ESS^{size} \quad (5)$$

In Equation (6), the balance to be met at the POI of the hybrid plant to the grid is represented.

$$E_{d,h,t}^{PV} + E_{d,h,t}^{wind} + E_{d,h,t}^{dis} + E_{d,h,t}^{GridBuy} = E_{d,h,t}^{load} + E_{d,h,t}^{ch} + E_{d,h,t}^{M1} + E_{d,h,t}^{M2}, \quad \forall d \in D, \forall h \in H, \forall t \in T, \quad (6)$$

where this balance is applied for each period t for the entire horizon of days (in operation case) or years (in sizing) analyzed. It must be fulfilled that the available photovoltaic energy ($E_{d,h,t}^{PV}$), wind ($E_{d,h,t}^{wind}$) or discharge of the storage systems ($E_{d,h,t}^{dis}$) is equal to the load of the plant, sold energy in the different markets ($E_{d,h,t}^{M1}$, $E_{d,h,t}^{M2}$) or the charge of the storage ($E_{d,h,t}^{ch}$).

Equation (7) represents the maximum energy that can be injected into the grid at the connection point. This is determined by the net available energy of the hybrid plant.

$$E^{POI} \geq E_{d,h,t}^{M1} + E_{d,h,t}^{M2}, \quad \forall d \in D, \forall h \in H, \forall t \in T. \quad (7)$$

In order to avoid simultaneous participation in the energy purchase and sale market, it is necessary to define the restrictions (8)–(10), also limiting the maximum amount of energy to be injected or consumed:

$$B^{Ms}_{d,h,t} + B^{GridBuy}_{d,h,t} \leq 1, \quad \forall d \in D, \forall h \in H, \forall t \in T. \tag{8}$$

$$E^{M1}_{d,h,t} + E^{M2}_{d,h,t} \leq B^{Ms}_{d,h,t} \cdot E^{POI}, \quad \forall d \in D, \forall h \in H, \forall t \in T. \tag{9}$$

$$E^{load}_{d,h,t} + E^{ch}_{d,h,t} \leq B^{GridBuy}_{d,h,t} \cdot E^{POI}, \quad \forall d \in D, \forall h \in H, \forall t \in T. \tag{10}$$

Equations (11) and (12) limit the maximum PV and wind generation injections as a function of the total available energy:

$$E^{PV}_{d,h,t} \leq M^{PV}_{d,h,t} \cdot PV^{size}, \quad \forall d \in D, \forall h \in H, \forall t \in T. \tag{11}$$

$$E^{wind}_{d,h,t} \leq M^{wind}_{d,h,t} \cdot Wind^{size}, \quad \forall d \in D, \forall h \in H, \forall t \in T. \tag{12}$$

Another contribution of this work is the possibility of participating in two coupled markets simultaneously, $M1$ and $M2$, with different execution times. This could be extrapolated to more than two markets due to the modeling of the problem. The market with the shorter execution time ($M2$) will be the one that determines into how many intervals, t, an hour, h, should be split.

Instead of dividing the whole year directly into the total number of intervals, we work on the hours of the day, and these are divided into intervals determined by the market with the shortest execution time. This allows to establish a better relationship between markets and execution times. This is shown in Figure 2.

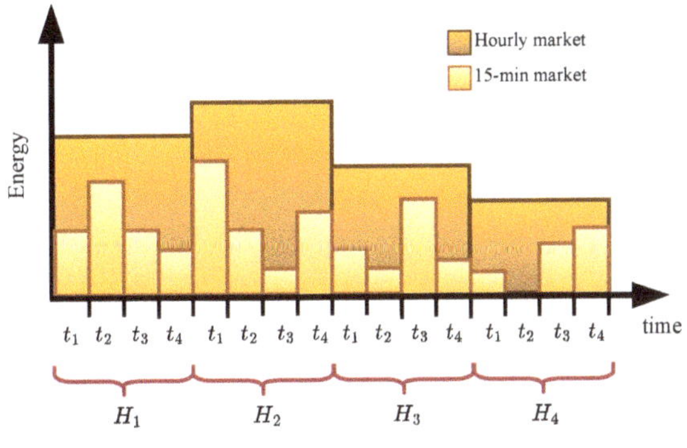

Figure 2. Diagram of coupled markets.

To relate the two markets, a set must be defined that relates the execution intervals of the smaller market to the other market. Thus, for example, if market $M2$ is executed every 15 min and market $M1$ is executed every hour, the values to be taken by t must be in the set $\{1, 2, 3, 4\}$. The value in hourly market $M1$ must be maintained during the four intervals that define an hour. That is, for each period t, the value in the first period must be maintained to complete the hour with the same energy value. The equation which models the coupled simultaneity of participation is defined as:

$$E^{M1}_{d,h,1} = E^{M1}_{d,h,t} \quad \forall d \in D, \forall h \in H, \forall t \in T. \tag{13}$$

This modeling of the problem allows adding additional markets at different times. For example, to add another market $M3$ that closes every half hour, it is sufficient to define a set $m \in \{1,3\}$, and include the following constraint equation:

$$E^{M3}_{d,h,1} = E^{M3}_{d,h,m+1}, \quad ,\forall y \in Y, \forall d \in D, \forall h \in H, \forall m \in M, \tag{14}$$

where with this set, it is possible to relate t_1, the first fifteen minutes, with t_2, which is the second fifteen minutes, making the half-hour constant. This would be the same in the second half hour.

These markets can represent the case of a main market and other markets for ancillary services or other markets. Considering participation in deviation or short-time markets is interesting, due to the variability of generation, with storage offering flexibility to the system. This can result in economic benefits, the greater possibility of renewable energy integration and greater system reliability.

Storage system is modeled by the following set of equations. Equation (15) represents the update of the stored energy in each period. In (16)–(18), simultaneous charging and discharging of the battery is constrained. Finally, the maximum and minimum energy stored in each interval is defined in (19):

$$E^{Bat}_{d,h,t} = E^{Bat}_{d,h,t-1} + E^{ch}_{d,h,t} \cdot \eta^{ch} - E^{dis}_{d,h,t} \cdot \frac{1}{\eta^{dis}}, \quad \forall d \in D, \forall h \in H, \forall t \in T. \tag{15}$$

$$B^{ch}_{d,h,t} + B^{dis}_{d,h,t} \leq 1, \quad \forall d \in D, \forall h \in H, \forall t \in T. \tag{16}$$

$$E^{ch}_{d,h,t} \leq E^{max,ch}_{d,h,t} \cdot B^{ch}_{d,h,t}, \quad \forall d \in D, \forall h \in H, \forall t \in T. \tag{17}$$

$$E^{dis}_{d,h,t} \leq E^{max,dis}_{d,h,t} \cdot B^{dis}_{d,h,t}, \quad \forall d \in D, \forall h \in H, \forall t \in T. \tag{18}$$

$$Bat^{size} \cdot SOC^{min} \leq E^{bat}_{d,h,t} \leq Bat^{size} \cdot SOC^{max}, \quad \forall d \in D, \forall h \in H, \forall t \in T. \tag{19}$$

Finally, the daily energy that can be charged or discharged by the battery is defined in (20). This is essential to protect the battery and maintain its lifetime throughout the project, and to provide realism to the solution of the problem.

$$\sum_{h=1}^{h=H} \sum_{t=1}^{t=T} \left[E^{dis}_{d,h,t} + E^{ch}_{d,h,t} \right] \leq N \cdot Bat^{size}, \quad \forall d \in D. \tag{20}$$

3. Test Case

To test the presented method, two different generation scenarios will be used for the analysis. For simplicity and to observe the differences, the same energy prices will be used for both cases. In Figure 3, the prices used for the analysis are shown, with the main market data, $M1$, with a resolution of one hour, and the secondary market data, $M2$, with a resolution of 15 min.

Table 2 shows the investment data, with generation and storage costs used in the scenarios. The cost of land or areas to place the generation is not taken into account.

Table 2. Resource costs.

Cost	Value
PV energy generation costs	550,000 EUR/MW
Wind energy generation costs	1,200,000 EUR/MW
Battery cost	300,000 EUR/MWh

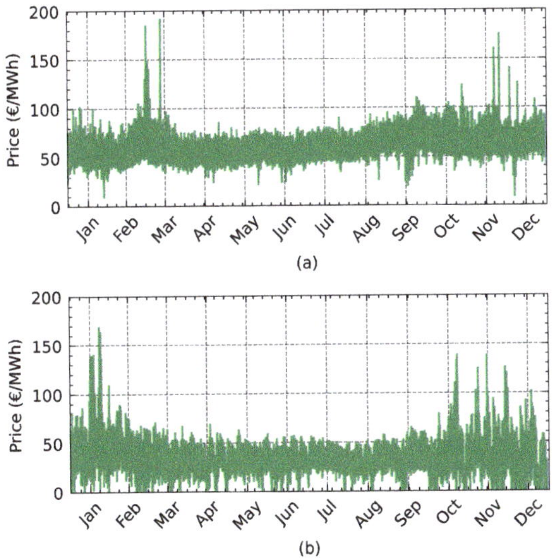

Figure 3. Day-ahead market price (**a**) and 15-min market price (**b**).

3.1. Scenario A

This scenario will use data collected in [35], specifically from Great Britain. Unit profiles of wind energy and PV energy generation, shown in Figure 4, will be used.

For this scenario, the plant has installation limitations of 60 MW of wind energy generation and 170 MW of photovoltaic energy, with no restrictions on the use of batteries, while the limit of the connection point is 100 MW. The limit for purchase at the connection point is 30 MW.

A 15-year period is analyzed, with a discount rate of 2% and a maximum investment of EUR 100 million. A 1-year simulation is performed, and its result is considered approximately constant for the entire 15-year period.

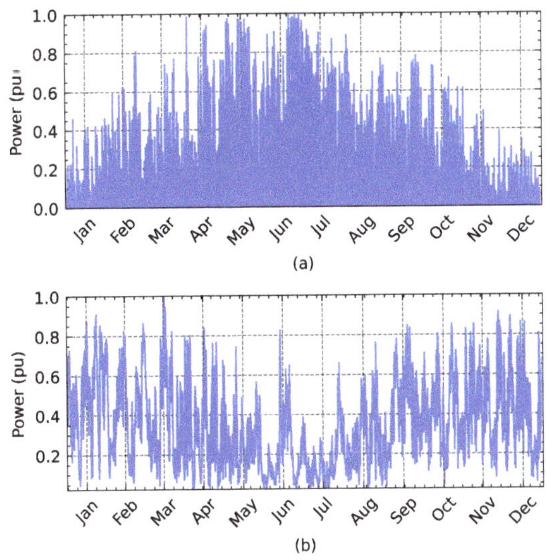

Figure 4. Photovoltaic energy (**a**) and wind energy (**b**) generation in Great Britain.

3.2. Scenario B

In this scenario, only PV energy generation is considered. An estimate of the energy generation in Seville, Spain, has been chosen using the PVGIS tool [36]. This profile is shown in Figure 5.

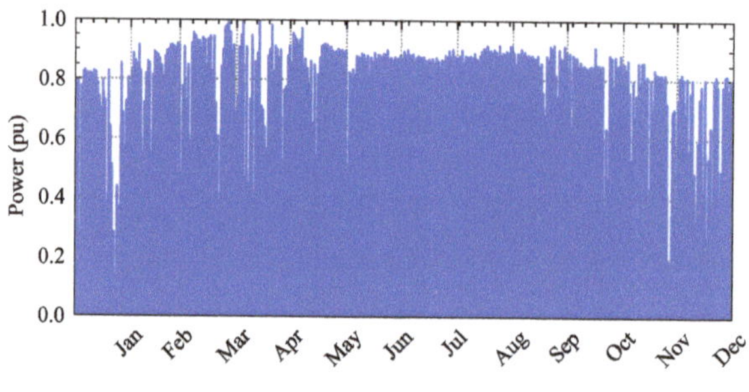

Figure 5. Photovoltaic energy generation in Seville, Spain.

The plant has installation limitations of 170 MW of photovoltaic energy generation, with no restrictions on the use of batteries, while the connection point limit is 100 MW. A period of 30 years is analyzed, with a discount rate of 2% and a maximum investment of 100 million euros.

3.3. Scenario C

Scenario C is used to test participation in two coupled markets with different closing times. For this purpose, the prices from Figure 3 are shown together with the evolution of PV energy generation from Scenario B, for a specific size of hybrid plant. To be more restrictive and observe the influence of participation in the two markets, the limit for purchase at the connection point is lowered to 10 MW.

4. Simulations and Results

This section presents the results of the application of the proposed method for sizing and operation in different cases. In scenario A, the plant is sized, taking into account storage, PV energy generation and wind energy generation for the profiles shown above. In scenario B, the sizing is performed only with PV energy generation and battery to observe the influence of the NCFG and α term. For sizing, only the hourly market will be considered, due to the complexity of predicting markets at shorter execution times. Finally, in scenario C, one of the combinations of energy generation and battery size is chosen to operate the hybrid plant in one day, participating in two coupled markets simultaneously.

4.1. Scenario A

In this scenario, the most optimal investment option according to maximizing the NPV is the installation of 60 MW of wind generation. This translates into a total investment of 72 MEUR, which is 28 MEUR less than the maximum investment limit. Despite having more budget for investment, the best option is to invest only in wind generation, not investing in other resources, because it will not increase the net present value of the investment.

It is demonstrated how the method seeks the best option for the investor without the need to spend the entire budget, saving money for the investor, which is a key advantage. In addition, this result coincides with the values collected in LCOE analysis, where onshore wind energy presents a smaller value than photovoltaic energy or batteries, making it a better investment when there is an appropriate wind profile.

Table 3 shows a sizing comparison and economic performance data if the annual profile were the same for each month of the year, to see how the method performs according to the availability of the generation resource, which changes for each month of the year.

Table 3. Sizing comparison for each month of the year participating in the day ahead market.

Month	PV (MW)	Wind (MW)	ESS (MWh)	NPV (EUR)	IRR (%)	Investment
January	0	60	0	87,888,550	15.2	72 MEUR
February	0	60	0	61,097,001	11.6	72 MEUR
March	50.91	60	0	128,773,250	15.8	100 MEUR
April	50.91	60	0	61,048,578	9.2	100 MEUR
May	141.41	0	74.04	118,245,445	14.6	100 MEUR
June	167.50	0	26.24	121,558,100	15.2	100 MEUR
July	151.59	0	55.43	160,512,939	18.7	100 MEUR
August	166.43	6.08	3.91	109,994,630	14.1	100 MEUR
September	50.91	60	0	138,465,872	16.7	100 MEUR
October	50.91	60	0	103,139,340	13.4	100 MEUR
November	0	60	0	125,544,887	19.9	72 MEUR
December	0	60	0	129,307,840	20.4	72 MEUR

It is observed how in only four months of spring and summer, photovoltaic energy generation would be installed as the main source, but in the remaining eight months, the main source is wind energy generation alone or accompanied by photovoltaic energy generation. The month-by-month analysis is consistent with the results obtained for the sizing of the entire year.

Furthermore, in the month of August, it can be seen that due to the photovoltaic energy and wind energy generation profiles and prices, the sizing model presented chooses all technologies and storage, not excluding any of the alternatives, as all of them are considered profitable.

4.2. Scenario B

The objective in this scenario is to size a hybrid plant using photovoltaic energy generation and storage in a location where there is a large amount of solar radiation, reducing curtailment and promoting the installation of storage.

For this purpose, a comparison is also made between the model without penalty, very similar to other sizing methods present in the literature, and the method proposed contemplating the penalty, with a non-zero α. This will allow showing the influence of the proposed method.

Applying the proposed sizing method, the most optimal sizing solution is formed by 131.15 MW of photovoltaic energy generation and 92.88 MWh of storage. These values are those that obtain the maximum NPV: 291,430,088 EUR, investing the maximum possible budget of 100 MEUR.

Table 4 shows a detailed analysis of the sizing according to the profiles of the different seasons of the year. In this case, the optimal sizing is very similar in all cases, due to the uniformity of the generation profile. In months with lower radiation, it can be seen that less battery size would be installed, since there would not be as much surplus photovoltaic energy for storage and subsequent sale. All this has repercussions in lower economic indexes for the months with less generation.

Table 4. Sizing comparison for each season of the year participating in the day ahead market with $\alpha = 1$.

Season	PV Size (MW)	ESS Size (MWh)	NPV (EUR)	IRR (%)
Winter	133.60	88.39	233,110,096	14.6
Spring	130.17	94.70	315,173,776	18.4
Summer	131.15	92.88	335,398,839	19.3
Autumn	133.04	89.34	166,560,782	11.4

To evaluate the influence that the term of non cash flow generated (NCFG), the sizing is carried out by defining $\alpha = 0$, obtaining results of 140.09 MW of photovoltaic energy generation and 76.49 MWh of storage with an NPV of 374,590,511 EUR. This implies that if α is not considered, allowing a great curtailment, about 8.94 MW more generation is installed, but more importantly, 16.39 MWh less of storage are installed. In Figure 6, the energy lost due to curtailment is shown, penalizing it with $\alpha = 1$, and not penalizing it ($\alpha = 0$). This is the total energy that cannot be injected, discounting that which is stored.

This oversizing results in a loss of energy over several months of 1000 MWh (an average of more than 32 MWh per day) that cannot be stored. To use this energy, a new investment in storage would have to be made, which is more expensive for the investor. By defining $\alpha = 1$, the curtailment is considerably reduced, and the storage sizing is adequate to store a large part of the energy produced.

Although with $\alpha = 0$, a slightly higher NPV is obtained in this sizing, it is interesting to size prioritizing that not too much energy is left over and stored. This makes the owner's investment more flexible in terms of profitability and changes in new scenarios, electricity prices or new markets, making the investment more interesting and less rigid throughout the years of the project.

Finally, Figure 7 shows the influence of Equation (20), which limits the maximum number of daily charge and discharge cycles for the same case as the previous figure. It is observed that, if no limit is defined, the maximum limit is exceeded on some of the days shown for January and August. If this number is not limited, the battery could be used in an uncontrolled way, maximizing the benefits, but not being a real result because the battery would degrade more, not reaching the useful life of the project.

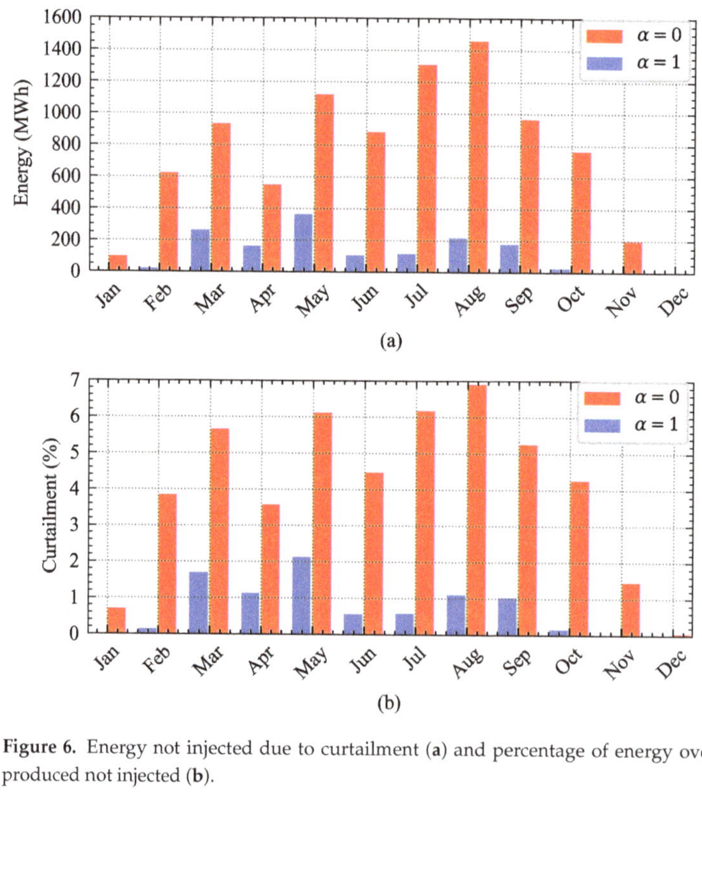

Figure 6. Energy not injected due to curtailment (**a**) and percentage of energy over total energy produced not injected (**b**).

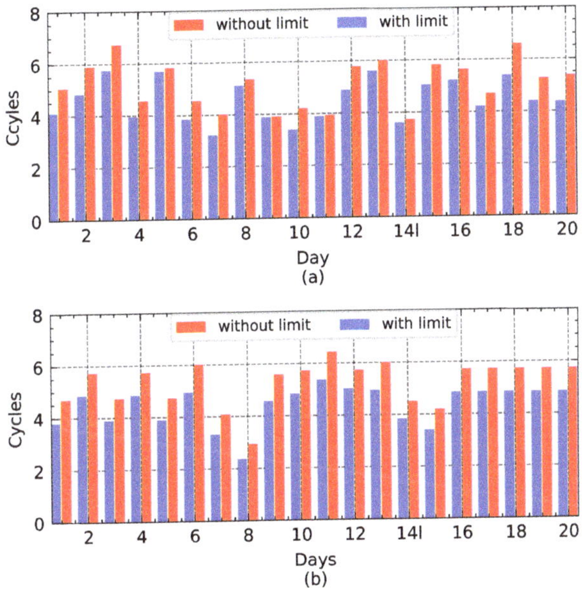

Figure 7. Number of daily charge and discharge cycles for the first 20 days of January (**a**) and July (**b**).

4.3. Scenario C

In this scenario, the participation in two coupled markets of the hybrid plant with battery is tested. The optimal plant size from the previous scenario of 131.15 MW of photovoltaic energy generation and 92.88 MWh of storage are chosen to demonstrate the operation defined in Section 2.1.1. For the analysis, an hourly execution market, *M*1, and another fifteen-minute market, *M*2, are considered. For greater clarity in the results, the day will be divided into 96 intervals of 15 min.

Figure 8 shows the evolution of prices and the results obtained in the operation of the hybrid plant for day 2 of the year with the sizing obtained. It is observed how the constraint (13) for the participation in the markets is satisfied, maintaining during four periods of 15 min the same value for the market that is executed hour by hour, while the one that is executed every 15 min remains free.

For this day, contemplating the two markets and with the price evolution shown, a daily profit of 18,853.13 EUR is achieved, while if only participating in the hourly market, *M*1, the profit is 18,359.97 EUR. This represents an increase of 2.61% profit for only one day in January, one of the worst photovoltaic energy generation seasons.

Table 5 shows some days on which the method of operation is applied to compare benefits.

Table 5. Comparison of profits due to market participation on various days of the year.

Day	Without *M*2 (EUR)	With *M*2 (EUR)	Improve (%)
1	18,359.97	18,853.13	2.61
181	47,283.94	47,622.34	0.71
295	21,544.82	34,219.59	37.04

It can be seen that since the price of the *M*2 market is more variable over time, the benefits oscillate, with reasonable increases on day 1, small increases on other days, such as day 181, or a significant increase on day 295, where it is more convenient to participate on the *M*2 market instead of the *M*1 market.

Therefore, it is shown that a method that considers two markets coupled in time is necessary and can improve the profitability of the hybrid plant, always choosing the best option at each instant of time.

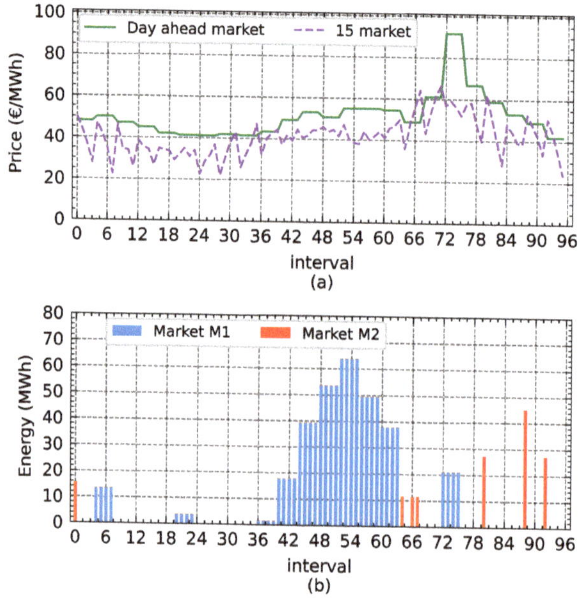

Figure 8. Electricity prices of markets (**a**) and participation (energy sale) in markets (**b**).

5. Conclusions

In this work, a method of sizing and managing hybrid plants has been proposed to make them attractive to investors, with participation in several coupled markets. All types of generation technology are put on the same level to choose which is the best decision to invest in based on a budget, while maximizing the NPV, to see which is the most profitable sizing combination based on resource availability and storage.

It is proposed to reduce the curtailment that occurs when the plant is oversized, penalizing the energy that cannot be injected or stored, in order to better adjust the generation and storage sizes. In addition, the participation in coupled markets with different execution time is defined, in order to participate in different markets seeking to increase the profitability of the investment.

The results show that an optimal sizing is produced depending on the location and budget, choosing the most optimal combination. Thanks to the proposed definition, the curtailment of the plant is reduced, reducing the oversizing that occurs in many cases. The operation also shows how profitability is improved as a result of the participation in two coupled markets with different execution intervals.

Future work will focus on improving the storage management system to improve investor profitability, as well as integrating the method into multi-objective optimizations that meet investor criteria.

Author Contributions: Conceptualization, all authors; methodology, C.G.-S. and A.A.; software, C.G.-S. and A.A.; validation, all authors; formal analysis, all authors; investigation, C.G.-S. and A.A.; resources, J.M.C. and E.G.; data curation, C.G.-S.; writing—original draft preparation, C.G.-S. and A.A.; writing—review and editing, J.M.C. and E.G.; visualization, all authors; supervision, J.M.C. and E.G.; project administration, J.M.C. and E.G.; funding acquisition, J.M.C. and E.G. All authors have read and agreed to the published version of the manuscript.

Funding: This research was supported by the Spanish State Research Agency (AEI) through the Project PDC2021-121278-I00 funded by MCIN/AEI/10.13039/501100011033 and by the European Union Next GenerationEU/PRTR and by the Spanish State Research Agency (AEI) through the Project PYC20 RE 075 US funded by Junta de Andalucía (Ministry of Economy, Knowledge, Business and University).

Data Availability Statement: Not applicable.

Conflicts of Interest: The authors declare no conflict of interest. The funders had no role in the design of the study; in the collection, analyses, or interpretation of data; in the writing of the manuscript, or in the decision to publish the results.

References

1. IEA. World Energy Outlook 2019. 2019. Available online: https://www.iea.org/reports/world-energy-outlook-2019 (accessed on 15 December 2022).
2. Directive (EU) 2018/2001 of the European Parliament and of the Council of 11 December 2018 on the Promotion of the Use of Energy from reneWable Sources. *Off. J. Eur. Union*. Available online: https://eur-lex.europa.eu/legal-content/EN/TXT/PDF/?uri=CELEX:32018L2001&from=EN (accessed on 15 December 2022).
3. Directive 2012/27/EU of the European Parliament and of the Council of 25 October 2012 on Energy Efficiency. *Off. J. Eur. Union*. Available online: https://eur-lex.europa.eu/LexUriServ/LexUriServ.do?uri=OJ:L:2012:315:0001:0056:en:PDF (accessed on 15 December 2022).
4. IRENA. *Renewable Energy Finance: Institutional Capital*; Renewable Energy Finance Brief 02; IRENA: Abu Dhabi, United Arab Emirates, 2020; pp. 1–11.
5. Ralon, P.; Taylor, M.; Ilas, A.; Diaz-Bone, H.; Kairies, K. Electricity Storage and Renewables: Costs and Markets to 2030. International Renewable Energy Agency (Issue October). 2017. Available online: http://irena.org/publications/2017/Oct/Electricity-storage-and-renewables-costs-and-markets (accessed on 15 December 2022).
6. International Energy Agency. *The Future of Hydrogen: Seizing Today's Opportunities*; IEA Publications: Paris, France, 2019.
7. Abbas, J.; Wang, L.; Ben Belgacem, S.; Pawar, P.S.; Najam, H.; Abbas, J. Investment in renewable energy and electricity output: Role of green finance, environmental tax, and geopolitical risk: Empirical evidence from China. *Energy* **2023**, *269*, 126683. [CrossRef]
8. Siddik, A.B.; Khan, S.; Khan, U.; Yong, L.; Murshed, M. The role of renewable energy finance in achieving low-carbon growth: Contextual evidence from leading renewable energy-investing countries. *Energy* **2023**, *270*, 126864. [CrossRef]
9. He, X.; Khan, S.; Ozturk, I.; Murshed, M. The role of renewable energy investment in tackling climate change concerns: Environmental policies for achieving SDG -13. *Sustain. Dev.* **2023**. [CrossRef]
10. Root, C.; Presume, H.; Proudfoot, D.; Willis, L.; Masiello, R. Using battery energy storage to reduce renewable resource curtailment. In Proceedings of the 2017 IEEE Power and Energy Society Innovative Smart Grid Technologies Conference, ISGT 2017, Arlington, VA, USA, 23–26 April 2017; pp. 1–5. [CrossRef]
11. Dimopoulou, S.; Oppermann, A.; Boggasch, E.; Rausch, A. A Markov Decision Process for managing a Hybrid Energy Storage System. *J. Energy Storage* **2018**, *19*, 160–169. [CrossRef]
12. Dykes, K.; King, J.; Diorio, N.; King, R.; Gevorgian, V.; Corbus, D.; Blair, N.; Anderson, K.; Stark, G.; Turchi, C.; et al. Opportunities for Research and Development of Hybrid Power Plants. May 2020. Available online: https://www.nrel.gov/docs/fy20osti/75026.pdf (accessed on 15 December 2022).
13. Terlouw, T.; AlSkaif, T.; Bauer, C. Techno-Economic Assessment of PV-Coupled Battery Energy Storage Systems with Different Sizing Methodologies. In Proceedings of the 2021 IEEE Madrid PowerTech, Madrid, Spain, 28 June–2 July 2021; pp. 1–6. [CrossRef]
14. Kelly, J.J.; Leahy, P.G. Sizing Battery Energy Storage Systems: Using Multi-Objective Optimization to Overcome the Investment Scale Problem of Annual Worth. *IEEE Trans. Sustain. Energy* **2019**, *11*, 2305–2314. [CrossRef]
15. Martinez-Rico, J.; de Argandona, I.R.; Zulueta, E.; Fernandez-Gamiz, U.; Armendia, M. Energy Storage Sizing Based on Automatic Frequency Restoration Reserve Market Participation of Hybrid Renewable Power Plants. In Proceedings of the 2021 International Conference on Smart Energy Systems and Technologies (SEST), Vaasa, Finland, 6–8 September 2021; pp. 1–6. [CrossRef]
16. García-Santacruz, C.; Galván, L.; Carrasco, J.M.; Galván, E. Sizing and Management of Energy Storage Systems in Large-Scale Power Plants Using Price Control and Artificial Intelligence. *Energies* **2021**, *14*, 3296. [CrossRef]
17. Nuvvula, R.S.S.; Elangovan, D.; Teegala, K.S.; Madurai Elavarasan, R.; Islam, M.R.; Inapakurthi, R. Optimal Sizing of Battery-Integrated Hybrid Renewable Energy Sources with Ramp Rate Limitations on a Grid Using ALA-QPSO. *Energies* **2021**, *14*, 5368. [CrossRef]
18. Xia, Q.; Debnath, S.; Saeedifard, M.; Marthi, P.R.V.; Arifujjaman, M. Energy Storage Sizing and Operation of an Integrated Utility-Scale PV+ESS Power Plant. In Proceedings of the 2020 IEEE Power & Energy Society Innovative Smart Grid Technologies Conference (ISGT), Washington, DC, USA, 17–20 February 2020; pp. 1–5. [CrossRef]
19. Yao, M.; Cai, X. Energy Storage Sizing Optimization for Large-Scale PV Power Plant. *IEEE Access* **2021**, *9*, 75599–75607. [CrossRef]
20. Shu, Z.; Jirutitijaroen, P. Optimal Operation Strategy of Energy Storage System for Grid-Connected Wind Power Plants. *IEEE Trans. Sustain. Energy* **2013**, *5*, 190–199. [CrossRef]

21. Parashar, S.; Swarnkar, A.; Niazi, K.R.; Gupta, N. Multiobjective optimal sizing of battery energy storage in grid-connected microgrid. *J. Eng.* **2019**, *2019*, 5280–5283. [CrossRef]
22. Krishnamurthy, D.; Uckun, C.; Zhou, Z.; Thimmapuram, P.R.; Botterud, A. Energy Storage Arbitrage Under Day-Ahead and Real-Time Price Uncertainty. *IEEE Trans. Power Syst.* **2018**, *33*, 84–93. [CrossRef]
23. Al-Shereiqi, A.; Al-Hinai, A.; Albadi, M.; Al-Abri, R. Optimal Sizing of Hybrid Wind-Solar Power Systems to Suppress Output Fluctuation. *Energies* **2021**, *14*, 5377. [CrossRef]
24. Klansupar, C.; Chaitusaney, S. Optimal Sizing of Utility-scaled Battery with Consideration of Battery Installtion Cost and System Power Generation Cost. In Proceedings of the 2020 17th International Conference on Electrical Engineering/Electronics, Computer, Telecommunications and Information Technology (ECTI-CON), Phuket, Thailand, 24–27 June 2020; pp. 498–501. [CrossRef]
25. Khan, M.F.; Pervez, A.; Modibbo, U.M.; Chauhan, J.; Ali, I. Flexible Fuzzy Goal Programming Approach in Optimal Mix of Power Generation for Socio-Economic Sustainability: A Case Study. *Sustainability* **2021**, *13*, 8256. [CrossRef]
26. Zhao, L.; Zhang, T.; Peng, X.; Zhang, X. A novel long-term power forecasting based smart grid hybrid energy storage system optimal sizing method considering uncertainties. *Inf. Sci.* **2022**, *610*, 326–344. [CrossRef]
27. Remon, D.; Cantarellas, A.M.; Martinez-Garcia, J.; Escano, J.M.; Rodriguez, P. Hybrid solar plant with synchronous power controllers contribution to power system stability. In Proceedings of the 2017 IEEE Energy Conversion Congress and Exposition (ECCE), Cincinnati, OH, USA, 1–5 October 2017; pp. 4069–4076. [CrossRef]
28. Abdeltawab, H.; Mohamed, Y.A.-R.I. Energy Storage Planning for Profitability Maximization by Power Trading and Ancillary Services Participation. *IEEE Syst. J.* **2021**, *16*, 1909–1920. [CrossRef]
29. Bera, A.; Chalamala, B.R.; Byrne, R.H.; Mitra, J. Sizing of Energy Storage for Grid Inertial Support in Presence of Renewable Energy. *IEEE Trans. Power Syst.* **2021**, *37*, 3769–3778. [CrossRef]
30. Siface, D. Optimal Economical and Technical Sizing Tool for Battery Energy Storage Systems Supplying Simultaneous Services to the Power System. In Proceedings of the 2019 16th International Conference on the European Energy Market (EEM), Ljubljana, Slovenia, 18–20 September 2019; pp. 1–6. [CrossRef]
31. Chiang, M.-Y.; Huang, S.-C.; Hsiao, T.-C.; Zhan, T.-S.; Hou, J.-C. Optimal Sizing and Location of Photovoltaic Generation and Energy Storage Systems in an Unbalanced Distribution System. *Energies* **2022**, *15*, 6682. [CrossRef]
32. Chowdhury, T.; Hasan, S.; Chowdhury, H.; Hasnat, A.; Rashedi, A.; Asyraf, M.R.M.; Hassan, M.Z.; Sait, S.M. Sizing of an Island Standalone Hybrid System Considering Economic and Environmental Parameters: A Case Study. *Energies* **2022**, *15*, 5940. [CrossRef]
33. Li, X.; Jones, G. Optimal Sizing, Location, and Assignment of Photovoltaic Distributed Generators with an Energy Storage System for Islanded Microgrids. *Energies* **2022**, *15*, 6630. [CrossRef]
34. Alsagri, A.S.; Alrobaian, A.A. Optimization of Combined Heat and Power Systems by Meta-Heuristic Algorithms: An Overview. *Energies* **2022**, *15*, 5977. [CrossRef]
35. Open Power System Data. A Free Open Data Platform for Power System Modeling. Available online: https://data.open-power-system-data.org/ (accessed on 20 January 2023).
36. The European Commission's Science and Knowledge Service. Photovoltaic Geographical Information System (PVGIS). Available online: ec.europa.eu/jrc/en/pvgis (accessed on 20 January 2023).

energies

Article

Multi-Agent-Based Controller for Microgrids: An Overview and Case Study

Necmi Altin [1], Süleyman Emre Eyimaya [2] and Adel Nasiri [3,*]

1 Department of Electrical-Electronics Engineering, Faculty of Technology, Gazi University, Ankara 06560, Turkey
2 Department of Electronics and Automation, TUSAS-Kazan Vocational School, Gazi University, Ankara 06560, Turkey
3 Electrical Engineering Department, College of Engineering and Computing, University of South Carolina (USC), Columbia, SC 29208, USA
* Correspondence: nasiri@sc.edu; Tel.: +1-803-576-7796

Abstract: A microgrid can be defined as a grid of interconnected distributed energy resources, loads and energy storage systems. In microgrid systems containing renewable energy resources, the coordinated operation of distributed generation units is important to ensure the stability of the microgrid. A microgrid needs a successful control scheme to achieve its design goals. Undesirable situations such as distorted voltage profile and frequency fluctuations are significantly reduced by installing the appropriate hardware such as energy storage systems, and control strategies. The multi-agent system is one of the approaches used to control microgrids. The application of multi-agent systems in electric power systems is becoming popular because of their inherent benefits such as autonomy, responsiveness, and social ability. This study provides an overview of the agent concept and multi-agent systems, as well as reviews of recent research studies on multi-agent systems' application in microgrid control systems. In addition, a multi-agent-based controller and energy management system design is proposed for the DC microgrid in the study. The designed microgrid is composed of a photovoltaic system consisting of 30 series-connected PV modules, a wind turbine, a synchronous generator, a battery-based energy storage system, critical and non-critical DC loads, the grid and the control system. The microgrid is controlled by the designed multi-agent-based controller. The proposed multi-agent-based controller has a distributed generation agent, battery agent, load agent and grid agent. The roles of each agent and communication among the agents are designed properly and coordinated to achieve control goals, which basically are the DC bus voltage quality and system stability. The designed microgrid and proposed multi-agent-based controller are tested for two different scenarios, and the performance of the controller has been verified with MATLAB/Simulink simulations. The simulation results show that the proposed controller provides constant DC voltage for any operation condition. Additionally, the system stability is ensured with the proposed controller for variable renewable generation and variable load conditions.

Keywords: control; distributed control; microgrid; multi-agent systems; renewable energy systems

Citation: Altin, N.; Eyimaya, S.E.; Nasiri, A. Multi-Agent-Based Controller for Microgrids: An Overview and Case Study. *Energies* **2023**, *16*, 2445. https://doi.org/10.3390/en16052445

Academic Editors: Alexander Micallef and Zhaoxia Xiao

Received: 7 February 2023
Revised: 25 February 2023
Accepted: 27 February 2023
Published: 3 March 2023

1. Introduction

In the electricity sector, distributed generation is becoming more common day by day, with the increasing energy demand and technological developments [1]. Distributed generation units, which have become widespread, have also demonstrated the microgrid concept. Microgrids are small-scale energy networks that can be operated independently or connected to the grid. They have their own energy resources and loads with certain limits [2]. Microgrids offer various advantages, such as providing energy supply in remote areas with on-site generation, reducing transmission losses between regions, increasing service quality by detecting faults instantly, using resources efficiently by supporting demand management, commissioning more domestic resources, and having a more reliable network. Microgrids can consist of renewable energy resources such as photovoltaic (PV) modules and wind turbines, energy storage systems, and controllable loads, which are among the distributed power generation tools. They can be installed at points close to the user. Additionally, they can operate in grid-connected or off-grid mode and can be independently controlled. These superior features are making them popular in small-scale grid systems [3]. This also shows a pathway for the future grid structure. The inclusion of different types of technological integrations such as distributed generation, electrical energy storage units, power system management applications, microgrid structures, and information and automation technologies in the system allows conventional power systems to operate more efficiently and flexibly [4]. A microgrid consisting of renewable energy resources, conventional generation resources, electric vehicles, energy storage systems, conventional grid connection, and loads is depicted in Figure 1.

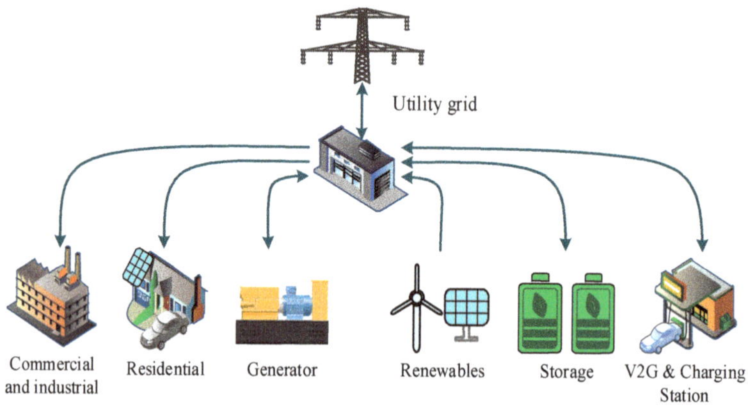

Figure 1. Architecture of a typical microgrid system.

Although AC microgrids are more common, DC microgrids have become popular in recent years with the significant increase in DC loads in applications such as LED lighting, computers and electric vehicles. The following are a few advantages of DC microgrids over AC microgrids:

- The DC microgrid is simpler due to the absence of reactive power flow control;
- Integrated distributed generations can be coordinated more easily because their control is based on DC voltage without the need for synchronization;
- Due to the prevalence of DC electronic domestic loads, the majority of DGs today produce DC outputs; this prevents the need for unnecessary AC/DC power conversions. This has a direct impact on the system's cost and losses, further reducing the size and cost of the system due to the fact that the majority of the converters used for the DC micro-source interface do not use transformers.
- In DC systems, issues such as reactive power and frequency-synchronized power management become unimportant. Additionally, skin effect, harmonics, proximity effect, and inrush current problems are absent from the DC system because it has no frequency. DC systems are thought to be safer than AC systems because they have a lower electromagnetic field.
- Compared to an AC microgrid, voltage regulation is superior.

As a result of energy generation using distributed generation units in microgrids, many economic, political, and environmental benefits are obtained. In order to obtain these benefits and to obtain the most effective use of microgrids, it is important to provide optimal design with the proper control of all components within the microgrid [3]. In microgrids, control strategies are used to control voltage and frequency, balance supply and demand, and improve the power quality by using communication between microgrid components.

Different control approaches termed centralized, decentralized, and distributed control are used in microgrids. The decentralized and distributed control strategies provide more flexible and effective control features [3]. Multi-agent systems, with their inherent features, are suitable for implementing these control strategies to achieve a more effective and flexible power system and higher power quality.

Computer systems placed in a particular environment that can take autonomous action to meet design objectives are called agents. Artificial intelligence and agent concepts emerge by transferring human characteristics, such as learning by experience and making logical decisions, into the computer environment [5]. The concept of intelligence, which is the ability to perceive and interpret an environment and situation, make decisions, and control behaviors, forms the basis of intelligent agents, which are defined as "anything that can notice and affect its environment through sensors". Agents produce output actions with the inputs they receive from their environment, as shown in Figure 2. The main purpose of using agents is to create autonomous systems that give automatic and appropriate responses to events detected from the environment. To achieve this goal, agents use their autonomous, social, reactive, and proactive features. Thanks to their autonomy, agents exercise partial control of their actions and internal states and try to influence outcomes without interference. Agents use their social features to communicate with other agents or units, coordinate actions, and achieve their goals. While their timely response to changes in their environment results from their reactive nature, their target-oriented behavior and taking the initiative to achieve goals emerge as a result of their proactive nature. The system in which many agents come together is called a multi-agent system, and offers control strategy approaches for microgrids [6].

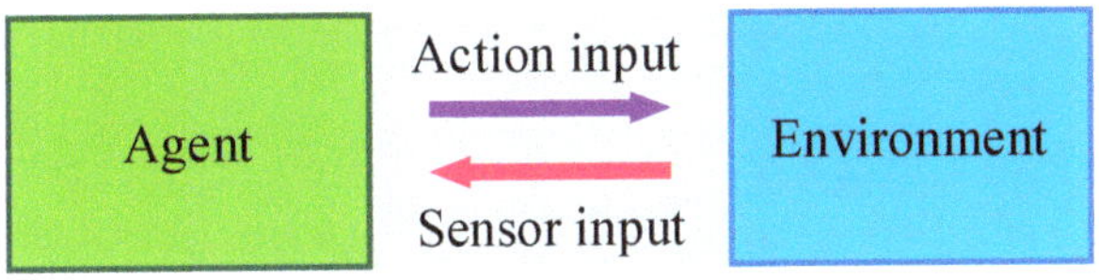

Figure 2. General relationship between an agent and environment.

With the aforementioned features, the multi-agent-based system seems a good solution for the microgrid control problems. Therefore, it has attracted the attention of researchers in recent years. This study provides an overview of agents and multi-agent systems concepts. In addition, application of multi-agent systems for microgrid operation and control are introduced, and recent studies in the literature are reviewed. The advantages of microgrid control systems designed based on multi-agent systems, their superiority over other methods, their limitations, and trends in this direction are discussed. In addition, to highlight multi-agent-based control system performance in DC microgrids, a case study is provided. For this purpose, a DC microgrid is designed in the study. In the designed microgrid, there is the PV system, the wind turbine, the synchronous generator, the battery-based energy storage system, the critical DC load, and the non-critical DC load. Moreover, a multi-agent-based controller is designed to control this DC microgrid. The proposed multi-agent-based controller has a distributed generation agent, a battery agent, a load agent and a grid agent. The DC microgrid and proposed multi-agent-based controller are modelled and simulated with MATLAB/Simulink. The simulation results carried out for different operation conditions validate the performance of the multi-agent-based controller in terms of system stability and power quality at the common DC bus.

The rest of the paper is organized as follows. A general overview of multi-agent systems is given in Section 2. The applications of multi-agent systems for microgrid control are discussed in Section 3. In Section 4, a case study is performed, including the modeling of the DC microgrid and the design of the multi-agent-based controller for the microgrid. The simulation results are provided for different scenarios in Section 5. Finally, the conclusion is drawn in Section 6.

2. Multi-Agent Systems

A multi-agent system is a complex system of autonomous agents with local knowledge and limited capabilities that are able to interact to achieve a global goal. This system, created with artificial intelligence-based techniques as well as traditional control methodologies, offers an additional advantage in creating hybrid controllers in microgrids. Fast communication possibilities such as fiber optic, microwave, and 4G are now becoming integral parts of power systems. This integration makes it easier and more convenient to integrate the multi-agent system into power system applications [7]. The main features that distinguish multi-agent systems from other distributed systems are as follows:

- Any one agent in the system does not have all the information about the solution to the problem.
- None of the agents in the system have all the required capabilities to solve the problem.
- The system control is distributed.
- The data are not kept at a central location; they are distributed.
- The operation is asynchronous.

Software or hardware-based agents and multi-agent systems are designed with different features, such as working with a certain degree of autonomy in a certain environment in order to fulfill their duties, perceiving the dynamic changes in the environment with their sensors and re-evaluating their knowledge and goals according to the perceptions they have obtained, planning in line with their goals and taking actions regarding these plans, and having the ability to communicate with other agents through the language of communication [8–10]. The conceptual design process for building a multi-agent system includes the following four-stage development process:

- Analysis: modeling agent roles and behaviors, identifying the application domain and the problem.
- Design: defining the solution architectures for the problems defined in the analysis step.
- Development: programming agent targets, ontologies, and functionalities.
- Deployment: initialization of the created multi-agent system, runtime agent management, message, and data processing.

Multi-agent systems have emerged as a powerful technology that can overcome the difficulties encountered during the application of information and communication technology in a wide variety of fields. They are considered autonomous software environments and defined as a system that detects the environment with the help of its sensors, and affects the environment they perceive. They are also considered software components that have the ability to act on behalf of the user to perform certain tasks [11]. Agents must be able to deal with changes in the environment appropriately and in a timely manner, address multiple targets, change active targets according to the situation, and perform tasks from a broad perspective. In addition, agents should interact with other agents because goals are achieved by collaborating and competing with other agents. Agents have a planning mechanism that shapes their behavior, and their behavior is formed by the way the planners use the plans in the plan library, alone or in combination with other plans, at the appropriate time and condition.

Systems formed by agents that come together to solve problems that a single agent cannot solve effectively using their own knowledge and individual abilities, in a coordinated manner, by cooperating with each other are called multi-agent systems. In multi-agent systems, each agent has partial knowledge of its environment. The agent in the system can obtain various information about other agents, monitor the actions of other agents or share information with other agents. In this context, one of the most important elements in multi-agent systems is the trust between agents. There are many studies in the literature on how to model the trust and relationship between agents in a multi-agent system [12–14].

The skills of agents are often limited, as is their knowledge. Each agent may need other agents to perform an action related to its own task. Therefore, the characteristic feature of multi-agent systems is their ability to interoperate. An important problem in the cooperation of agents in the system is the need to ensure that the agents work in coordination within a plan. There are various approaches in the field of multi-agent systems to achieve this coordination [15,16]. For example, in a centralized approach, one agent undertakes the task of coordinating other agents [17]. In multi-agent systems, approaches within the framework of virtual organizations are very important to determine this responsible agent to ensure coordination among agents. Multi-agent systems in which multiple agents interact with each other are included in many studies in different disciplines [18,19]. An e-commerce multi-agent system includes buyer and seller agents representing buyers and sellers. Buyer and seller agents try to achieve their own goals by interacting with other agents in various ways in the same environment. Suppliers, shippers, etc. and many other agendas are included in this system [20]. An environment intelligence-based multi-agent system has been proposed to improve assistance and healthcare services for patients who suffer from Alzheimer's. It utilizes various context-aware agent technologies that allow it to automatically and evenly receive information from users and the environment, each focusing on defined concepts such as ubiquity, awareness, intelligence, and mobility [21]. A multi-agent system application is presented for distributed energy resource management in a microgrid consisting of distributed generation units, storage units, and controllable loads. In order to coordinate the distributed energy resources, an agent-based approach based on coordination and networking has been developed, and its performance is demonstrated by software simulation [22].

Communication is another important subject in multi-agent systems. Agents must be able to communicate successfully with each other in order to perform all kinds of actions, such as sharing information, coordinating, and negotiating. The main way to achieve this is to develop communication methods that will support the features of agents, such as autonomy. Communication methods have been studied in the literature of distributed systems for many years [23–25]. The general approach in this field is to provide communication between components via network protocols at various levels [26]. However, this approach is insufficient when the needs of multi-agent systems and the properties of agents are considered. Here, the main problem is the autonomous structure of agents in the multi-agent system. In a communication approach based on network protocols, there are expressions that can be used during communication. The contents of these expressions, and details on how these expressions can be used (in which situations and in what order) are determined by precise rules. However, rule-based communication methods defined in this way are not suitable for multi-agent systems because they constrain the autonomous nature of agents. In applications that a single agent cannot solve or effectively solve using his own knowledge and individual abilities, multi-agent systems, in which many agents come together to solve them in a coordinated manner, are becoming more and more important. Each role given to the agents in the system has responsibilities, abilities, authorizations, and rules depending on the goals of the system.

3. Multi-Agent Systems for Microgrid Control

Multi-agent systems consist of multiple intelligent agents that interact to solve problems that may be beyond the capabilities of the system. For many years, multi-agent designs and architectures have been proposed for applications in power systems and power engineering [27–29]. Distributed energy resources used in microgrid applications are increasing day by day, and making microgrid control more complex. The multi-agent system is well suited for managing this complexity.

There are many advantages of multi-agent systems in microgrid control applications [30]:

Distributed architecture: The structure of distributed generation resources conforms to multi-agent system architectures based on local knowledge and decision making.

Flexibility: In the microgrid system modeled with distributed energy resources and loads, agents can be easily deployed and provide flexibility for future expansion in the system, thanks to its "plug and play" capabilities.

Resilience: The multi-agent system can respond quickly and adapt to faults. In addition, changes in grid topology (disconnection of a load or generator) do not interrupt both local and global system goals (for example, stability and efficiency).

Multi-agent-based control systems for microgrids have some limitations that hinder their widespread adoption but also offer an opportunity for future research [30]:

Emerging behavior: The autonomous and distributed nature of intermediaries can lead to unpredictable consequences. While the intents and targets of agents are programmable, the effect of runtime interactions is not always predetermined. Such immediate behavior may be beneficial in some situations (e.g., market transactions). However, in some applications, this uncertainty can be a disadvantage.

Portability: Hardware implementation of multi-agent system designs and architectures can be difficult. The most recent applications of multi-agent system-based control of microgrids are virtual test software simulations (e.g., MATLAB Simulink). The performance of many multi-agent systems approaches on real microgrid hardware has not been widely tested yet.

Scalability: The higher computational power available today allows researchers to model larger microgrids with many agents coordinating actions on a single platform. However, the ability of multi-agent systems to scale with increases in problem size (with agents across multiple platforms) or diversity (with agents of multiple types) is not well understood.

Security: The massive shift from physical infrastructure to smarter technology increases the risk of security and privacy breaches from malicious outside actors and disruptors.

When examining the active research areas of multi-agent systems in the context of microgrids presented in the literature in order to understand their current involvement in microgrid development, it can be seen that most studies have focused on distributed microgrid control [31–35]. Electrical energy trading, optimization, and power restoration are other popular application areas. The areas wherein multi-agent systems are used in microgrid control, and their properties, are presented in Table 1.

Table 1. Control subjects in the microgrid in which multi-agent system-based control is applied.

Author and Year	Applications of MAS in Microgrid	Aim	Application
Victorio et al., 2021 [31] Chung et al., 2013 [32] Almada et al., 2021 [33] Jabeur et al., 2022 [34] Zheng and Cai. 2010 [35]	Distributed Control	Solving the real and reactive power mismatch arising from distributed generation and maintaining the balance between supply and demand in microgrid.	Multi-agent system-based microgrid control models are created using artificial neural networks and fuzzy systems for tasks such as generation planning and load forecasting for operations planning
Logenthiran et al., 2010 [36] Jin et al., 2021 [37] Khan et al., 2019 [38] Khan and Wang. 2017 [39]	Optimization	Increase efficiency by optimizing the actions of microgrid components.	An artificial immune system-based algorithm is used to optimize the efficiency of renewable energy sources in the system and maximize power generation.
Alhasnawi et al., 2021 [40] Wang et al., 2020 [41] Mohamed et al., 2019 [42]	Power Restoration	Provide power restoration in the event of a large-scale power outage in microgrids.	A hierarchical control strategy is implemented along with a multi-agent immunity algorithm for rapid restoration of strength.

Table 1. *Cont.*

Author and Year	Applications of MAS in Microgrid	Aim	Application
Luo et al., 2018 [43] Gomes et al., 2020 [44] Sesetti et al., 2018 [45]	Electrical Energy Trading	Maximizing the revenue from the microgrid.	Creates a pricing mechanism for the microgrid in the competitive electricity market and algorithms for price determination based on demand and supply strategies.

Many studies have used multi-agent system-based controllers to optimize microgrid operations [36–38]. Awareness of green energy technologies in microgrids has been widely adopted for reducing CO_2 emissions and for a clean environment. Distributed energy resources such as the PV system, diesel engines, gas turbines, small wind turbines, and fuel cell technologies are developing within the power system. The control and maintenance of this power have a great impact on power systems. Multi-agent system technology is adopted for optimum use of electrical power in microgrids. In [39], multi-agent system technology used for microgrid control, optimization, and market distribution [39]. In [42], the multi-agent system with a time-varying microgrid topology is expressed as the best control strategy to address all data restoration problems in microgrids.

Because loads and resources within a microgrid can be diverse and distributed, the real-time response and the distributed generation resource management are critical in preventing local power outages. It is also important to do this efficiently and cost-effectively to achieve an economically viable microgrid. In addition, some studies consider the characteristics of source or load types, and self-regulate themselves with other agents to optimize for cost and efficiency globally. Multi-agent systems have also been used for the power restoration of microgrids [40,41]. The load restoration algorithm consists of agents that make synchronized load restoration decisions based on information learned directly from their neighbors. The global knowledge is discovered based on the mean consensus theorem, although only direct connections are made to neighbors. Multi-agent systems have been applied for electrical energy trading or market model analysis [43–45]. Efforts have been made to establish a power market model for the efficient operation of the microgrid. A multi-agent system electricity trading algorithm is proposed to maximize the revenue from the microgrid [46].

4. The Case Study: Multi-Agent-Based Control of DC Microgrid

4.1. Designed DC Microgrid

As a case study, the multi-agent-based control of a DC microgrid is designed and presented in this study. The designed DC microgrid model is shown in Figure 3. The microgrid includes the wind turbine, the solar PV system, the battery energy storage system (BESS), the synchronous generator, DC loads, and the grid. This system is designed with MATLAB/Simulink.

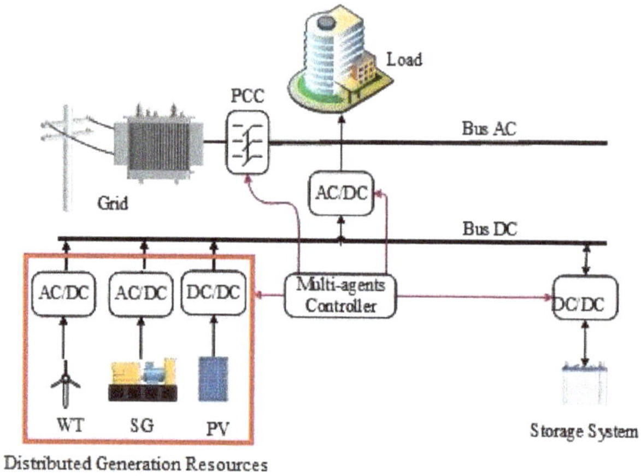

Figure 3. The configuration of the DC microgrid under study.

4.1.1. PV System Model

The PV system model consists of 25 series-connected Suntech STP270S-24 PV modules. The solar irradiation to the system is defined as time-varying. The PV modules produce direct current when the solar irradiation falls on them. However, the voltage and current values depend on natural conditions such as solar irradiation and ambient temperature. Therefore, a maximum power point tracking (MPPT) algorithm is used to ensure maximum power generation for any operation condition. The parameters of the designed PV system in the MATLAB/Simulink model are shown in Table 2.

Table 2. The block parameters of the designed PV system.

The Block Parameters of PV	Value
Open circuit voltage	44.49 (V)
Short-circuit current	8.19 (A)
Voltage at maximum power point	35.00 (V)
Current at maximum power point	7.71 (A)
Temperature coefficient of open circuit voltage	0.1504% (V/°C)

4.1.2. Wind Turbine Model

The wind energy system model, whose parameters are given in Table 3, consists of a three-phase salient pole permanent magnet synchronous generator (PMSG), a wind turbine and a blade angle control system with an output power of 10 kW. PMSG produces three-phase alternating current with the kinetic energy generated by the rotation of the blades with the wind. However, the voltage and frequency values depend on the wind speed. The wind speed is defined to the system in a time-varying manner. These values are set to change at certain time intervals. Since the PMSG generates variable voltage and frequency, a rectifier is used next to the PMSG. Additionally, similar to the PV system, a MPPT algorithm is utilized with a DC-DC converter to get the maximum available power from the wind system and to ensure maximum energy conversion efficiency. The MPPT algorithm generates a current reference for the DC-DC converter by using the torque reference, output DC voltage (the common DC bus voltage) and the wind speed. A hysteresis controller compares this current reference and the actual current value to generate the switching signal for the DC-DC converter.

Table 3. The block parameters of the designed wind turbine.

The Block Parameters of Wind Turbine	Value
Nominal mechanical output power	10 (kW)
Base power of electrical generator	10/0.9 (kVA)
Base wind speed	12 (m/s)
Maximum power at base wind speed	0.8 (pu)
Base rotational speed	1.2 (pu)

4.1.3. Synchronous Generator Model

A synchronous generator is included in the system to take precautions against natural variables and sudden output power changes occurring in the output power of the wind turbine and PV system. Actually, it is used to support the sustainability and stability of the microgrid. The block parameters of the synchronous generator are given in Table 4.

Table 4. The block parameters of the designed synchronous generator.

The Block Parameters of Synchronous Generator	Value
Nominal power	1000 (VA)
Line-to-line voltage	400 (V)
Frequency	50 (Hz)
Stator resistance	0.00285 (pu)

4.1.4. Battery Energy Storage System (BESS)

The BESS is connected to the common DC bus via a bidirectional DC-DC converter to balance the differences between the instant supply power and the demand. Thus, it provides sustainable energy to the loads. Besides, it also mitigates power fluctuations in the wind energy conversion system output power. The block parameters of the 650 V, 20 Ah Li-ion battery used as a BESS are given in Table 5.

Table 5. The block parameters of the Li-ion battery used as a BESS.

The Block Parameters of Li-Ion Battery	Value
Nominal voltage	650 (V)
Rated capacity	20 (Ah)
Initial state-of-charge (SoC)	60 (%)

In the system, the energy management is carried out together with the battery control system, and the balance between the generated power and the demand is ensured. The voltage control action is performed with the common DC bus reference voltage value. If the total output power of the generation units is higher than the load power, the DC bus voltage will increase. Similarly, if the total output power of the generation units is lower than the load demand power, then the DC bus voltage will decrease. This change shows the imbalance between the generation and demand. Based on this variation, a simple voltage controller is used as an energy management system. If the common DC voltage is higher than the specified reference voltage level, the controller generates a negative power reference for the BESS, and the BESS is charged. Conversely, when the common DC bus voltage is lower than the reference voltage level, the controller generates a positive power reference for the BESS, and the BESS is discharged.

4.2. Proposed Multi-Agent-Based Control Strategy

The conventional power systems use a master controller that collects all system information to manage the network and make decisions. With the increase in distributed energy resources in power systems, the power network is getting more complex. In a microgrid, the central control may introduce many drawbacks while it manages and controls many distributed energy resources, loads and storage units [3]. In this study, a distributed control with a multi-agent system is proposed instead of centralized control to overcome the problems caused by diversity in production and load resources. In addition to providing power quality and supply–demand balance by creating an effective management and control mechanism in the microgrid controlled by a multi-agent system, tasks such as battery charge–discharge, battery life improvement, reference bus voltage control, and reduction of voltage fluctuations are also performed by agents. A distributed generation agent, battery agent, load agent and grid agent are designed in the system. Agents communicate with each other to perform their role in the control of the microgrid. The roles of agents in the system can be described as follows:

Distributed Generation Agent: This agent represents distributed generation units such as wind turbines and PV modules in the microgrid. It receives the voltage and current information separately from PV modules, wind turbines and synchronous generators, and uses these data in control. It performs MPPT for the PV systems and wind turbines. It also controls the total power produced in the PV systems, wind turbines and synchronous generators, and share the data with other agents.

Battery Agent: This agent represents the BESS in the microgrid. It controls the battery charge and discharge action by using the battery, the generation and consumption data that it receives from other agents. It also shares the BESS power information (produced/stored) with other agents. It monitors the state of charge (SoC) level and requests power from the distributed generation agent and/or the grid agent when the SoC level is low.

It also controls the charging and discharging of the battery and ensures that the common DC bus voltage is stabilized. P_S refers to the supply power in the microgrid system, and its equation is given in Equation (1). P_S is the sum of the wind turbine output power (P_{wt}), the photovoltaic system output power (P_{pv}), the synchronous generator power (P_{sg}), and the battery power (P_b).

$$P_S = P_{wt} + P_{pv} + P_{sg} + P_b \quad (1)$$

P_d refers to the demand power in the microgrid system, and its equation is given in Equation (2). P_d is the sum of the power of DC loads (P_{dc_load}) and power of AC loads and the grid (P_{ac_load}).

$$P_d = P_{dc\_load} + P_{ac_{load}} \quad (2)$$

In cases in which P_S is greater than P_d, the battery is charged by communicating with other agents thanks to the battery agent, due to the excess power produced in the microgrid system. However, in cases in which P_S is smaller than P_d, the battery is discharged by communicating with other agents due to the lack of enough power generation in the microgrid system.

Load Agent: This agent represents loads in the microgrid. It receives the consumed power information from all loads in the system and transfers it to other agents. It has the ability to monitor, control and negotiate the power level and link status of the controllable load. Particularly when the microgrid is in island mode, it may interrupt, depending on the available power and SoC of the BESS. It has critical load and non-critical load separation capability. In addition, it plays a critical role in the system by sending the information on whether or not the supply–demand balance is provided to other agents.

Grid Agent: This agent represents the grid side within the microgrid. It monitors the grid voltage, phase angle and frequency, and is responsible for notifying other agents of changes in the microgrid state. It provides power and current control exported to or imported from the grid. With the data it receives from other agents, the PCC point is disabled, allowing the system to operate in island mode. It is responsible for monitoring and negotiating power from generation units and importing or exporting power when the microgrid is in on-grid mode.

The cooperation diagram between the agents in the designed multi-agent system is shown in Figure 4. Additionally, Table 6 gives a general idea of how and for what tasks these agents are designed, and also contains the necessary messages for interactions.

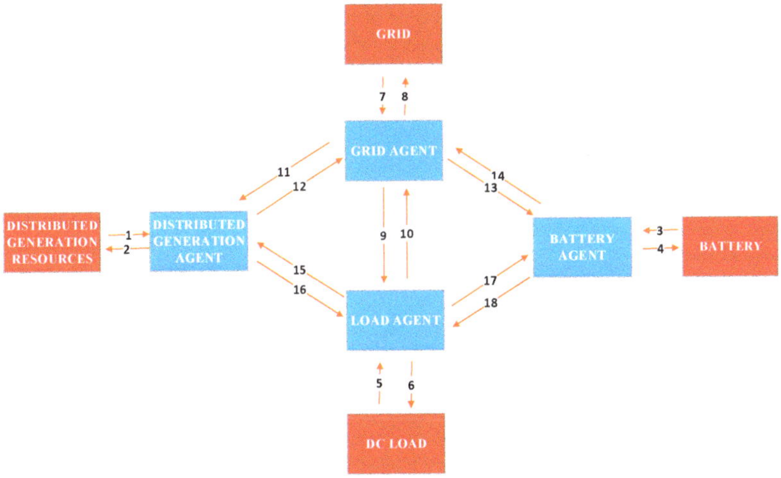

Figure 4. General overview of the proposed multi-agent-based control for the microgrid.

Table 6. Communication and coordination of agents.

Number	Communication and Coordination
1	Distributed generation agent receives power, voltage and current information from distributed generation sources.
2	The distributed generation agent makes MPPT with the information it receives and transfers it to the resources.
3	Battery agent receives voltage, current, power and SoC information from the battery.
4	Battery agent implements the battery control algorithm and transfers it to the battery.
5	Load agent receives power consumed from critical and non-critical loads.
6	Load agent transmits information for the exit of non-critical loads according to the system supply/demand situation.
7	Grid agent receives voltage, current and power information from the grid.
8	Grid agent converts the common DC bus voltage to AC with DQ control and transfers it to the grid.
9	Load agent requests network agent to open/close PCC.
10	Grid agent notifies the installation agent of the mode of the system (on/off grid).
11	Grid agent notifies mode to distributed generation agent (on/off grid).
12	Distributed generation agent transmits the generated power to the grid agent.
13	Grid agent notifies battery agent mode (on/off grid).
14	Battery agent reports common DC bus voltage information to the grid agent.
15	Load agent requests power from the distributed generation agent.
16	Distributed generation agent gives production information to the load agent and requests load shedding in underproduction.
17	Load agent requests power from the battery agent.
18	Battery agent provides production information to load agent and requests charge shedding when SoC is low.

5. Simulation Results

The designed DC microgrid and proposed and multi-agent-based microgrid controller are modelled with MATLAB/Simulink, and simulations representing different cases are carried out. Natural variables such as solar irradiation and wind speed affect the generated power level of the microgrid system. Therefore, the system has been tested on different scenarios depending on the renewable generation unit's situation to test the effectiveness of the designed multi-agent-based control.

5.1. Scenario I: Solar and Wind Power Are Both Available

In order to test the performance of the system in variable solar irradiation conditions, the solar irradiation is changed from 400 to 1000 W/m^2 in the simulation studies, as given in Table 7.

Table 7. Variable solar radiation values.

Time (sn)	Value (W/m^2)
0–2	1000
2–4	400
4–6	800

Similarly, the variable wind speed, defined in Table 8, is used in the simulations to validate the performance of the proposed controller for different operation conditions.

Table 8. Variable wind speed values.

Time (sn)	Value (m/s)
0–0.5	5
0.5–2.5	12
2.5–6	10

The output power of the solar PV system, together with the solar irradiation value that changes with time, is given in Figure 5. Since the available solar power increases and decreases with the solar irradiation level, thanks to the MPPT controller, the output power of the PV system also tracks it and generates maximum power for any operation condition.

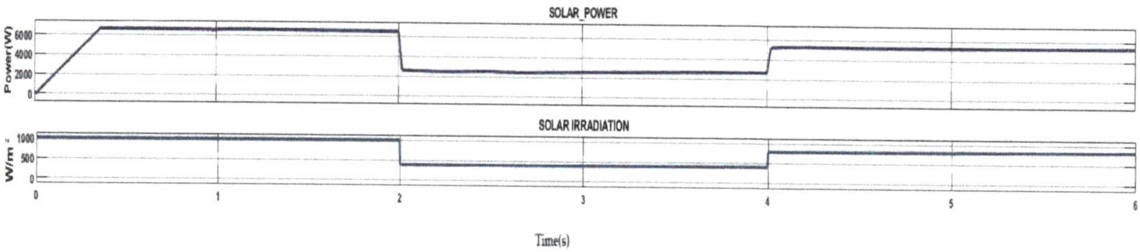

Figure 5. The output power of the PV system.

Simultaneously, the wind speed also varies and the available wind power changes. To achieve maximum use of the renewable energy resources, the microgrid controller should control these units properly. Figure 6 shows the changes of power values of all components in the system. The PV and wind system power levels are obtained with variable values of wind speed and solar irradiation as mentioned before. The controller controls the BESS's charge and discharge condition and power of the batteries to ensure the power sustainability and stable common DC bus voltage. AC and DC loads are the loads in the system. One can observe that for variable generation (because of variable wind speed and solar irradiation) and variable load conditions, the DC bus voltage is kept constant, and the microgrid stability is ensured. It is seen that the system still works efficiently in the time period when the load increases. In the system, the synchronous generator is programmed to run if the battery SoC level is below a certain limit. Since the supply and demand balance can be achieved and the SoC is not below this limit, the synchronous generator is not activated, except at the start. In addition, it is proven in the figure that the agents provide full control and take an active role in the system.

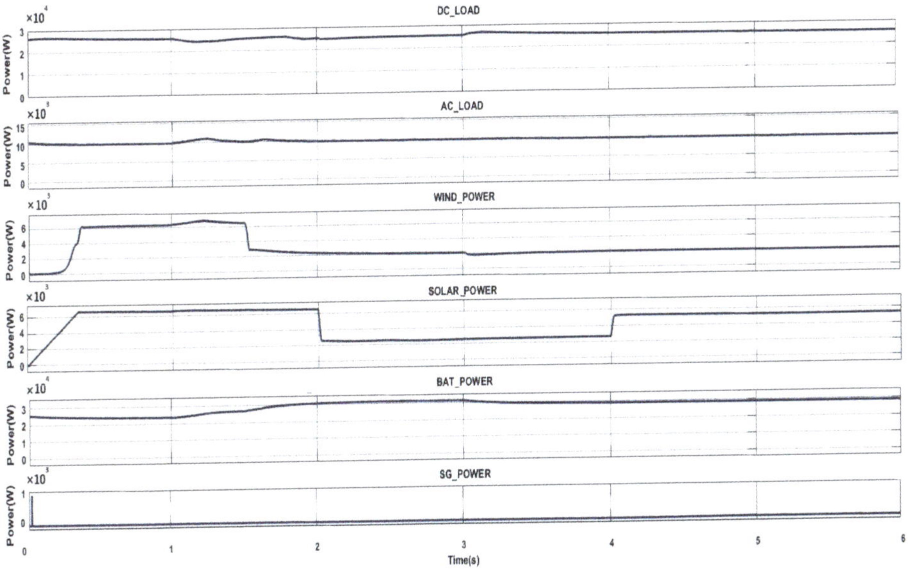

Figure 6. The variation in system components' power level for different conditions.

Figure 7 shows the voltage level of the common DC bus. The reference value for the common DC bus voltage is determined as 800 V, and it is controlled by the battery agent in the system. Thus, it is possible to provide the voltage and current values demanded by the grid agent. As can be seen in the figure, the DC bus voltage is well controlled for any operation condition.

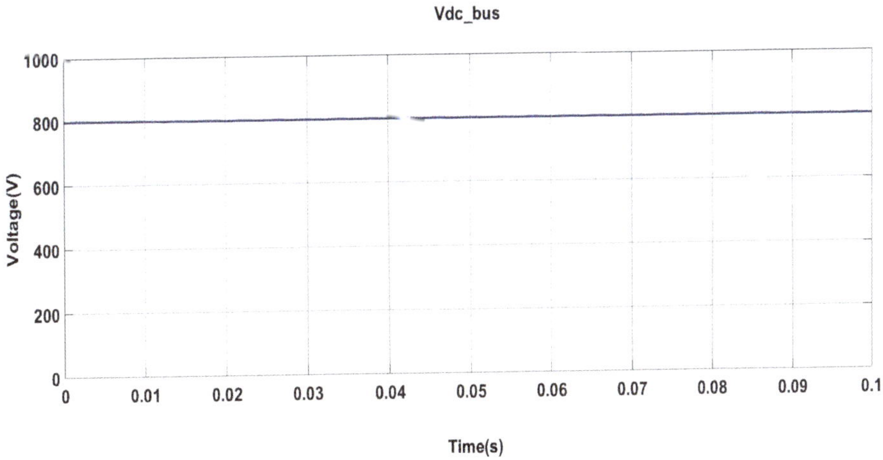

Figure 7. The voltage of the common DC bus.

To explore the operation of the other units, the voltage, current, charge level and power values of the BESS are given in Figure 8. The BESS is controlled by the battery agent. When the generated power cannot meet the total load in the system, the BESS discharges and transfers power to the system, in light of the information transferred by the distributed generation agent and the load agent to the battery agent. The battery current, voltage and power values vary with the generated and load power level.

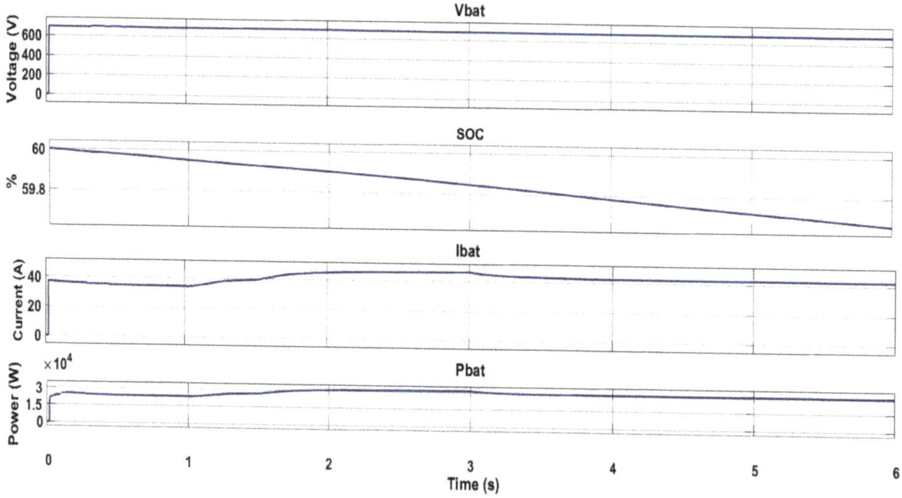

Figure 8. The battery voltage, current, SOC level and charge/discharge power variation.

In Figure 9, the supply, demand and the supply demand difference in the designed microgrid system are given. Accordingly, the generation in the system can always meet demand. Thanks to the monitoring of the difference between supply and demand by the load agent, non-critical DC loads are removed from the system when the power obtained from the distributed generation sources and the battery does not meet the consumption; at the same time, the microgrid is switched to the island mode operation by deactivating the grid connection at the PCC. When the supply power is more than the demand, the system continues to feed the non-critical DC loads and the grid and operation in grid-connected mode.

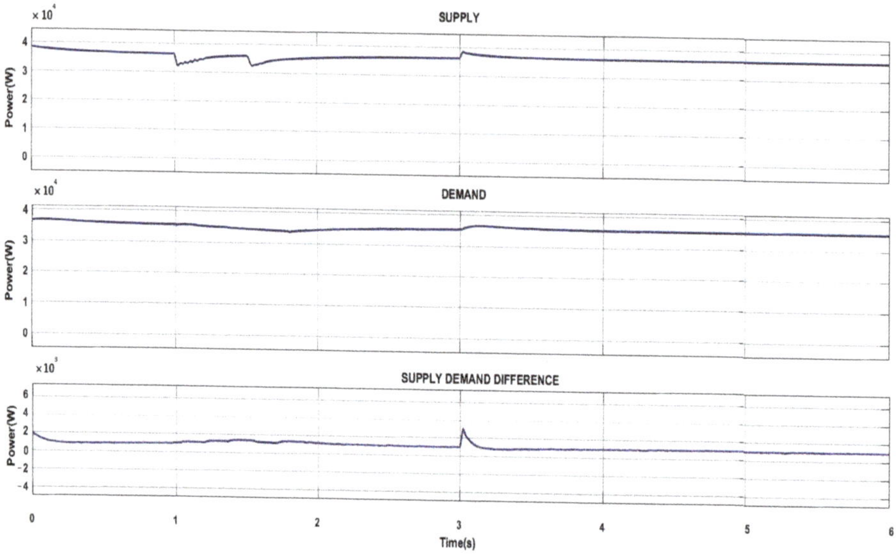

Figure 9. Variation in the generated power and load power in the system.

5.2. Scenario II: Only Solar Power Is Available

In this scenario, while the same solar irradiation value defined as in Table 7 is used for PV system, the wind speed is kept at 0 during the simulation. Therefore, the wind turbine was disabled during the simulation of this scenario. Figure 10 shows the variation in the power values of all components in the system for this condition. While there is no change in the PV system power level, AC and DC load, compared to the first scenario, it is observed that more battery power is required to supply the system. It is determined that the maximum power level provided by the battery (discharge power) to the microgrid system increased from 30 kW to 36 kW, representing a 20% increase compared to the first scenario. As mentioned before, the system is used to provide power if the renewable resources and the battery cannot supply the load. The PV system and BESS supply power to the microgrid system in this scenario, and the synchronous generator does not run again except at the start, because the SOC level of the BESS is bigger than the SOC limit.

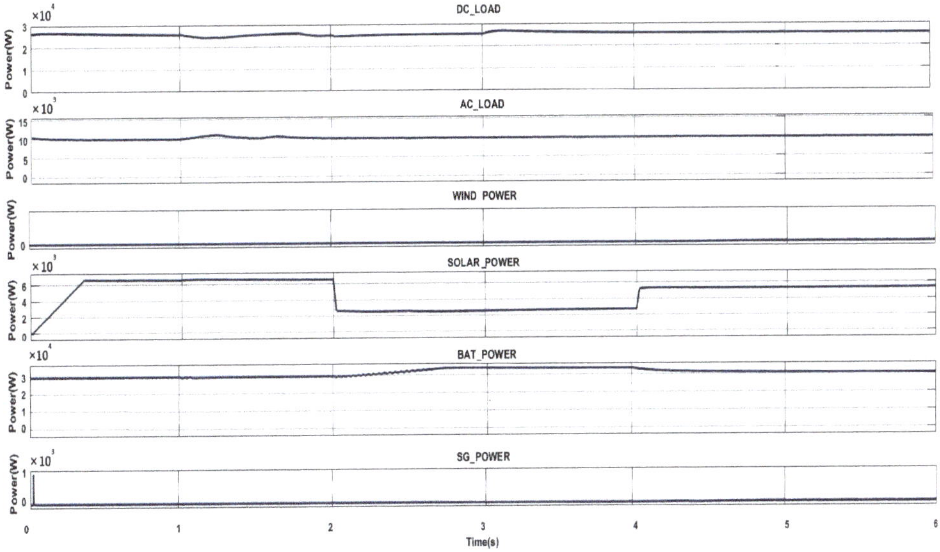

Figure 10. The variation in the power level of each unit in the microgrid for different operation conditions with no wind speed.

Figure 11 shows the voltage level of the common DC bus. A reference value of 800 V is ensured for any solar irradiation and load level thanks to the multi-agent-based microgrid control system. The controller agents define the operation point and determine the required operation mode for any unit. Thus, it is possible to keep the bus voltage constant in order to supply the loads. In addition, it has been proven that the agents provide full control even when the PV power is variable and there is no wind power, and ensure system stability and stable DC bus voltage for any condition.

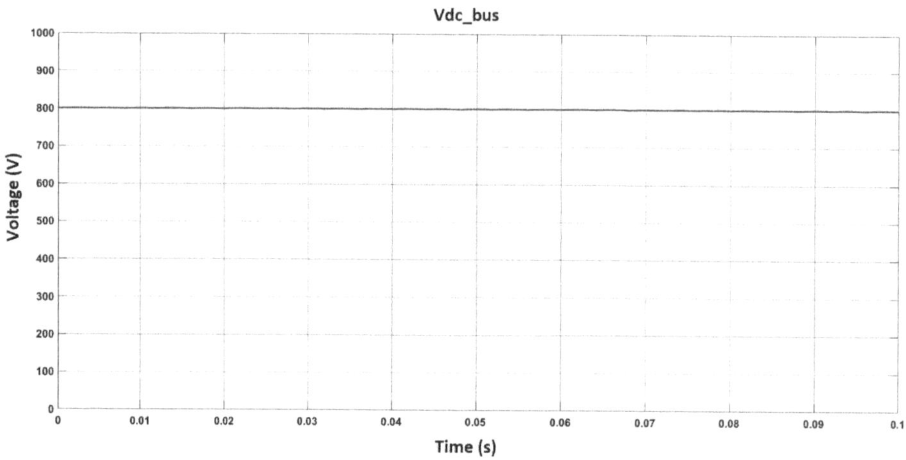

Figure 11. The common DC bus voltage for no wind speed.

6. Conclusions

In this study, an overview of multi-agent-based system applications in microgrids is presented. Agent theory, multi-agent systems and concepts that facilitate microgrid operation and control are introduced, and agent interaction, coordination, and cooperation are discussed in the context of multi-agent system features. The application of multi-agent systems in microgrids for different purposes such as market operations, fault detection, and fault location has been highlighted. In addition, a multi-agent-based controller for the DC microgrid system is proposed and tested as a case study. The presented DC microgrid consists of a solar PV system, wind turbine, synchronous generator, BESS, loads and grid connection. To control this microgrid, agents, their roles and interactions were designed. This system was tested in MATLAB/Simulink, and the performance of the multi-agent-based system was validated. The simulation results show that the multi-agent-based controller improves the stability and power quality of the microgrid.

With the growing interest in distributed generation systems and control of microgrids, multi-agent system-based control development studies are becoming more and more important. Microgrid control and protection applications using multi-agent systems, intelligent distribution control, modeling, and optimization have good research potential. Agent communications continue to evolve, adapting to changing network communication protocols. The improved agent provides a faster response time and adaptability. The standardization of multi-agent system architecture and applications will lead to more system interoperability in the microgrid and a smart grid environment. However, the inherent uncertainty of software complexity, hardware incompatibility, and security risk of malicious external actors limit the widespread adoption of the multi-agent systems to control microgrids.

Author Contributions: Conceptualization, S.E.E., N.A. and A.N.; methodology, N.A. and A.N.; software, S.E.E. and N.A.; validation, S.E.E. and N.A.; formal analysis, A.N.; investigation, S.E.E.; resources, N.A.; data curation, S.E.E.; writing—original draft preparation, S.E.E.; writing—review and editing, N.A. and A.N.; visualization, S.E.E.; supervision, N.A.; project administration, N.A. All authors have read and agreed to the published version of the manuscript.

Funding: This research received no external funding.

Data Availability Statement: Not applicable.

Conflicts of Interest: The authors declare no conflict of interest.

References

1. Altin, N.; Eyimaya, S.E. A combined energy management algorithm for wind turbine/battery hybrid system. *J. Electron. Mater.* **2018**, *47*, 4430–4436. [CrossRef]
2. Hirsch, A.; Parag, Y.; Guerrero, J. Microgrids: A review of technologies, key drivers, and outstanding issues. *Renew. Sustain. Energy Rev.* **2018**, *90*, 402–411. [CrossRef]
3. Altin, N.; Eyimaya, S.E. A Review of Microgrid Control Strategies. In Proceedings of the 2021 10th International Conference on Renewable Energy Research and Application (ICRERA), Istanbul, Turkey, 26–29 September 2021; pp. 412–417.
4. Espina, E.; Llanos, J.; Burgos-Mellado, C.; Cardenas-Dobson, R.; Martinez-Gomez, M.; Sáez, D. Distributed control strategies for microgrids: An overview. *IEEE Access* **2020**, *8*, 193412–193448. [CrossRef]
5. Shabbir, J.; Anwer, T. Artificial intelligence and its role in near future. *arXiv* **2018**, arXiv:1804.01396.
6. Dou, C.X.; An, X.G.; Yue, D. Multi-agent system based energy management strategies for microgrid by using renewable energy source and load forecasting. *Electr. Power Compon. Syst.* **2016**, *44*, 2059–2072. [CrossRef]
7. Khan, M.W.; Wang, J.; Xiong, L.; Huang, S.; Lopez-Ruiz, R. Architecture of a microgrid and optimal energy management system. In *Multi Agent Systems-Strategies and Applications*; Intechopen: London, UK, 2020.
8. Muise, C.; Belle, V.; Felli, P.; McIlraith, S.; Miller, T.; Pearce, A.R.; Sonenberg, L. Efficient multi-agent epistemic planning: Teaching planners about nested belief. *Artif. Intell.* **2022**, *302*, 103605. [CrossRef]
9. Shahinzadeh, H.; Nikolovski, S.; Moradi, J.; Bayindir, R. A Resilience-Oriented Decision-Making Model for the Operation of Smart Microgrids Subject to Techno-Economic and Security Objectives. In Proceedings of the 2021 9th International Conference on Smart Grid (icSmartGrid), Setubal, Portugal, 29 June 2021–1 July 2021; pp. 226–230.
10. Qasem, M.H.; Hudaib, A.; Obeid, N.; Almaiah, M.A.; Almomani, O.; Al-Khasawneh, A. Multi-agent systems for distributed data mining techniques: An overview. In *Big Data Intelligence for Smart Applications*; Springer: Berlin/Heidelberg, Germany, 2022; pp. 57–92.
11. Merabet, G.H.; Essaaidi, M.; Talei, H.; Abid, M.R.; Khalil, N.; Madkour, M.; Benhaddou, D. Applications of multi-agent systems in smart grids: A survey. In Proceedings of the 2014 International Conference on Multimedia Computing and Systems (ICMCS), Marrakech, Morocco, 14–16 April 2014; pp. 1088–1094.
12. Huynh, T.D.; Jennings, N.R.; Shadbolt, N.R. FIRE: An integrated trust and reputation model for open multi-agent systems. In Proceedings of the 16th European Conference on Artificial Intelligence, Valencia, Spain, 22–27 August 2004.
13. Bagherzadeh, L.; Shahinzadeh, H.; Gharehpetian, G.B. Scheduling of distributed energy resources in active distribution networks considering combination of techno-economic and environmental objectives. In Proceedings of the 2019 International Power System Conference (PSC), Tehran, Iran, 9–11 December 2019; pp. 687–695.
14. Khalid, R.; Samuel, O.; Javaid, N.; Aldegheishem, A.; Shafiq, M.; Alrajeh, N. A secure trust method for multi-agent system in smart grids using blockchain. *IEEE Access* **2021**, *9*, 59848–59859. [CrossRef]
15. Consoli, A.; Tweedale, J.; Jain, L. The link between agent coordination and cooperation. In Proceedings of the International Conference on Intelligent Information Processing, Adelaide, Australia, 20–23 September 2006; Springer: Boston, MA, USA, 2006; pp. 11–19.
16. Castelfranchi, C. Modelling social action for AI agents. *Artif. Intell.* **1998**, *103*, 157–182. [CrossRef]
17. Andreadis, G.; Klazoglou, P.; Niotaki, K.; Bouzakis, K.D. Classification and review of multi-agents systems in the manufacturing section. *Procedia Eng.* **2014**, *69*, 282–290. [CrossRef]
18. Bellifemine, F.L.; Caire, G.; Greenwood, D. *Developing Multi-Agent Systems with JADE.*; John Wiley & Sons: Hoboken, NJ, USA, 2007.
19. Ferber, J.; Gutknecht, O. A meta-model for the analysis and design of organizations in multi-agent systems. In Proceedings of the Proceedings International Conference on Multi Agent Systems (Cat. No. 98EX160), Paris, France, 3–7 July 1998; pp. 128–135.
20. Bhavsar, V.C.; Boley, H.; Yang, L. A Weighted-Tree Similarity Algorithm for Multi-Agent Systems in E-Business Environments. *Comput. Intell.* **2004**, *20*, 584–602. [CrossRef]
21. Tapia, D.I.; Corchado, J.M. An ambient intelligence based multi-agent system for alzheimer health care. *Int. J. Ambient. Comput. Intell. (IJACI)* **2009**, *1*, 15–26. [CrossRef]
22. Nematollahi, A.F.; Shahinzadeh, H.; Nafisi, H.; Vahidi, B.; Amirat, Y.; Benbouzid, M. Sizing and Sitting of DERs in Active Distribution Networks Incorporating Load Prevailing Uncertainties Using Probabilistic Approaches. *Appl. Sci.* **2021**, *11*, 4156. [CrossRef]
23. Ding, Z.; Huang, T.; Lu, Z. Learning individually inferred communication for multi-agent cooperation. *Adv. Neural Inf. Process. Syst.* **2020**, *33*, 22069–22079.
24. Hägg, S. A sentinel approach to fault handling in multi-agent systems. In *Australian Workshop on Distributed Artificial Intelligence*; Springer: Berlin/Heidelberg, Germany, 1996; pp. 181–195.
25. Rangwala, M.; Williams, R. Learning multi-agent communication through structured attentive reasoning. *Adv. Neural Inf. Process. Syst.* **2020**, *33*, 10088–10098.
26. Roscia, M.; Longo, M.; Lazaroiu, G.C. Smart City by multi-agent systems. In Proceedings of the 2013 International Conference on Renewable Energy Research and Applications (ICRERA), Madrid, Spain, 20–23 October 2013; pp. 371–376.
27. Albarakati, A.J.; Azeroual, M.; Boujoudar, Y.; ELIysaouy, L.; Aljarbouh, A.; Tassaddiq, A.; ELMarkhi, H. Multi-Agent-Based Fault Location and Cyber-Attack Detection in Distribution System. *Energies* **2023**, *16*, 224. [CrossRef]

28. Stennikov, V.; Barakhtenko, E.; Mayorov, G.; Sokolov, D.; Zhou, B. Coordinated management of centralized and distributed generation in an integrated energy system using a multi-agent approach. *Appl. Energy* **2022**, *309*, 118487. [CrossRef]
29. Domyshev, A.; Sidorov, D. Optimization of the Structure of Power System Multi-Agent Control. *IFAC-Pap* **2022**, *55*, 250–255. [CrossRef]
30. McArthur, S.D.; Davidson, E.M.; Catterson, V.M.; Dimeas, A.L.; Hatziargyriou, N.D.; Ponci, F.; Funabashi, T. Multi-agent systems for power engineering applications—Part I: Concepts, approaches, and technical challenges. *IEEE Trans. Power Syst.* **2007**, *22*, 1743–1752. [CrossRef]
31. Victorio ME, C.; Kazemtabrizi, B.; Shahbazi, M. Price Forecast Methodologies Comparison for Microgrid Control with Multi-Agent Systems. In Proceedings of the 2021 IEEE Madrid PowerTech, Madrid, Spain, 28 June 2021–2 July 2021; pp. 1–6.
32. Chung, I.Y.; Cheol-HeeYoo SJ, O. Distributed intelligent microgrid control using multi-agent systems. *Engineering* **2013**, *5*, 1–6. [CrossRef]
33. Almada, J.B.; Leao, R.P.; Almeida, R.G.; Sampaio, R.F. Microgrid distributed secondary control and energy management using multi-agent system. *Int. Trans. Electr. Energy Syst.* **2021**, *31*, e12886. [CrossRef]
34. Jabeur, R.; Boujoudar, Y.; Azeroual, M.; Aljarbouh, A.; Ouaaline, N. Microgrid energy management system for smart home using multi-agent system. *Int. J. Electr. Comput. Eng.* **2022**, *12*, 2088–8708. [CrossRef]
35. Zheng, W.D.; Cai, J.D. A multi-agent system for distributed energy resources control in microgrid. In Proceedings of the 2010 5th International Conference on Critical Infrastructure (CRIS), Beijing, China, 20–22 September 2010; pp. 1–5.
36. Logenthiran, T.; Srinivasan, D.; Khambadkone, A.M.; Aung, H.N. Multi-Agent System (MAS) for short-term generation scheduling of a microgrid. In Proceedings of the 2010 IEEE international Conference on Sustainable Energy Technologies (ICSET), Kandy, Sri Lanka, 6–9 December 2010; pp. 1–6.
37. Jin, S.; Wang, S.; Fang, F. Game theoretical analysis on capacity configuration for microgrid based on multi-agent system. *Int. J. Electr. Power Energy Syst.* **2021**, *125*, 106485. [CrossRef]
38. Khan, M.W.; Wang, J.; Ma, M.; Xiong, L.; Li, P.; Wu, F. Optimal energy management and control aspects of distributed microgrid using multi-agent systems. *Sustain. Cities Soc.* **2019**, *44*, 855–870. [CrossRef]
39. Khan, M.W.; Wang, J. The research on multi-agent system for microgrid control and optimization. *Renew. Sustain. Energy Rev.* **2017**, *80*, 1399–1411. [CrossRef]
40. Alhasnawi, B.N.; Jasim, B.H.; Sedhom, B.E. Distributed secondary consensus fault tolerant control method for voltage and frequency restoration and power sharing control in multi-agent microgrid. *Int. J. Electr. Power Energy Syst.* **2021**, *133*, 107251. [CrossRef]
41. Wang, Y.; Nguyen, T.L.; Xu, Y.; Tran, Q.T.; Caire, R. Peer-to-peer control for networked microgrids: Multi-layer and multi-agent architecture design. *IEEE Trans. Smart Grid* **2020**, *11*, 4688–4699. [CrossRef]
42. Mohamed, A.; Lamhamdi, T.; El Moussaoui, H.; El Markhi, H. A multi-agent system for fault location and service restoration in power distribution systems. *Multiagent Grid Syst.* **2019**, *15*, 343–358. [CrossRef]
43. Luo, F.; Dong, Z.Y.; Liang, G.; Murata, J.; Xu, Z. A distributed electricity trading system in active distribution networks based on multi-agent coalition and blockchain. *IEEE Trans. Power Syst.* **2018**, *34*, 4097–4108. [CrossRef]
44. Gomes, L.; Vale, Z.A.; Corchado, J.M. Multi-agent microgrid management system for single-board computers: A case study on peer-to-peer energy trading. *IEEE Access* **2020**, *8*, 64169–64183. [CrossRef]
45. Sesetti, A.; Nunna, H.K.; Doolla, S.; Rathore, A.K. Multi-agent based energy trading platform for energy storage systems in distribution systems with inter-connected microgrids. In Proceedings of the 2018 IEEE Industry Applications Society Annual Meeting (IAS), Portland, OR, USA, 23–27 September 2018; IEEE: Piscataway, NJ, USA, 2018; pp. 1–8.
46. Dong, X.; Li, X.; Cheng, S. Energy management optimization of microgrid cluster based on multi-agent-system and hierarchical stackelberg game theory. *IEEE Access* **2020**, *8*, 206183–206197. [CrossRef]

energies

Article

Sizing of Hybrid Supercapacitors and Lithium-Ion Batteries for Green Hydrogen Production from PV in the Australian Climate

Tarek Ibrahim [1], Tamas Kerekes [1,*], Dezso Sera [2], Shahrzad S. Mohammadshahi [2] and Daniel-Ioan Stroe [1]

1 Department of Energy, Aalborg University, 9220 Aalborg, Denmark
2 School of Electrical Engineering & Robotics, Queensland University of Technology, Brisbane 4000, Australia
* Correspondence: tak@energy.aau.dk

Abstract: Instead of storing the energy produced by photovoltaic panels in batteries for later use to power electric loads, green hydrogen can also be produced and used in transportation, heating, and as a natural gas alternative. Green hydrogen is produced in a process called electrolysis. Generally, the electrolyser can generate hydrogen from a fluctuating power supply such as renewables. However, due to the startup time of the electrolyser and electrolyser degradation accelerated by multiple shutdowns, an idle mode is required. When in idle mode, the electrolyser uses 10% of the rated electrolyser load. An energy management system (EMS) shall be applied, where a storage technology such as a lithium-ion capacitor or lithium-ion battery is used. This paper uses a state-machine EMS of PV microgrid for green hydrogen production and energy storage to manage the hydrogen production during the morning from solar power and in the night using the stored energy in the energy storage, which is sized for different scenarios using a lithium-ion capacitor and lithium-ion battery. The mission profile and life expectancy of the lithium-ion capacitor and lithium-ion battery are evaluated considering the system's local irradiance and temperature conditions in the Australian climate. A tradeoff between storage size and cutoffs of hydrogen production as variables of the cost function is evaluated for different scenarios. The lithium-ion capacitor and lithium-ion battery are compared for each tested scenario for an optimum lifetime. It was found that a lithium-ion battery on average is 140% oversized compared to a lithium-ion capacitor, but a lithium-ion capacitor has a smaller remaining capacity of 80.2% after ten years of operation due to its higher calendar aging, while LiB has 86%. It was also noticed that LiB is more affected by cycling aging while LiC is affected by calendar aging. However, the average internal resistance after 10 years for the lithium-ion capacitor is 264% of the initial internal resistance, while for lithium-ion battery is 346%, making lithium-ion capacitor a better candidate for energy storage if it is used for grid regulation, as it requires maintaining a lower internal resistance over the lifetime of the storage.

Keywords: green hydrogen; PV; supercapacitor; DC microgrids; lithium-ion batteries; sizing

Citation: Ibrahim, T.; Kerekes, T.; Sera, D.; Mohammadshahi, S.S.; Stroe, D.-I. Sizing of Hybrid Supercapacitors and Lithium-Ion Batteries for Green Hydrogen Production from PV in the Australian Climate. *Energies* **2023**, *16*, 2122. https://doi.org/10.3390/en16052122

Academic Editors: Alexander Micallef and Zhaoxia Xiao

Received: 25 January 2023
Revised: 19 February 2023
Accepted: 20 February 2023
Published: 22 February 2023

1. Introduction

Energy can be stored in different forms, and one of these forms is fuel. Fuel was used since the industrial revolution in the 18th century to run the machines and used for thousands of years earlier in heating, lighting and cooking, which makes it a more conventional means of energy that humans use in their life. The main challenge of using fuel is its chemical irreversibility and burning releases compounds such as CO_2, which contributes to the greenhouse gas effect. One method to overcome the side effect of fuel is to use fuels that do not release CO_2 when consumed, such as hydrogen. The problem with using hydrogen as a fuel is that its industrial production comes mostly from fossil and nonrenewable sources. From that, the importance of green hydrogen becomes clear, and researchers are working on more efficient ways of producing hydrogen from renewable sources.

The fact that green hydrogen production only requires water and a renewable energy source, promoted the development of microgrids for green hydrogen production. Renewable microgrids have been widely used for providing power to homes in rural areas where the power grid is not available [1].

Pairing hydrogen production/storage with solar or wind microgrid batteries powers residential areas in a zero-emission manner.

One of the challenges of using solar power for hydrogen production is its intermittent and unpredictable power output. The hydrogen production cell is sensitive to power fluctuations where multiple shutdowns and restarts accelerate the degradation of the cell. Extending the life of the hydrogen cell happens by putting the cell in idle mode, where it becomes 10% of the production load.

Overcoming the power intermittency and securing the idle mode required power from PV is achieved by including energy storage in the microgrid. Currently, the most used storage technology integrated with solar and wind power systems is based on Lithium-ion (Li-ion) batteries [2]; nevertheless, due to the high cost, the need for stable temperature, and the limited lifetime of Li-ion batteries [3], there is a need for alternative electrical energy storage solutions to keep up with the development and integration of renewable plants. One such technology is represented by Li-ion capacitors (LiCs). LiCs are a promising solution for energy storage [4], which can overcome some of the disadvantages of Li-ion batteries, which were mentioned before.

Conventional systems for green hydrogen production are based on grid-connected facilities supplied by renewable sources and use the excess energy for hydrogen production as a method of storing the energy. The current focus in the field of green hydrogen production is on utilizing distributed microgrids or retrofitting existing renewable energy plants to optimize production [5]. Energy storage (ES) based on LiC is compared with LiB ES, with a focus on size, remaining capacity, and internal resistance, to determine the best fit for green hydrogen production. The energy management system (EMS) is being developed to ensure the efficient operation of the electrolyser and the ES production [5].

The ES sizing optimization problems are divided into two parts: system modeling with input parameters and real data from the system location, and output part for evaluating the outputs such as SoC [6]. A sizing algorithm is then used to iterate the simulation for different input parameters and search for the optimal solution according to the system criteria that shall be met. Searching methods can be used for sizing such systems, such as directed search [7] or the sparrow search algorithm, if required optimization is multi-dimensional [8].

There is increasing attention on establishing a DC microgrid to eliminate AC power conversions and further increase efficiency [9]. The goal is to maximize the production of green hydrogen and supply commercial needs through the main production of the microgrid or through excess energy instead of storing it in batteries.

The sizing of the storage needs to be optimized for outdoor climatic conditions at the grid's location (ES has no temperature-controlled environment). The capacity of Li-ion batteries and Li-ion capacitors is dependent on the temperature at the site location. The degradation rate of both storage technologies is also dependent on the temperature and number of full equivalent cycles (FEC) of the cell. Increasing the size will reduce the FEC of the storage and will increase its lifetime but will also increase its cost. Thus, an energy management system (EMS) shall be used to maximize the hydrogen production and the expected life of storage cells and secure the power of the load in idle mode. This work aims to propose and provide an analysis of two storage technologies and compare them to suggest the best technology for green hydrogen production applications.

This paper is divided into the following sections: Section 2, Green hydrogen production system; Section 3, Li-ion capacitor and Li-ion battery degradation behavior; Section 4, Sizing results for the Li-ion capacitor and Li-ion battery; Section 5, Sensitivity analysis of Li-ion capacitor and Li-ion battery; Section 6, Discussion and future work Section 7, and Conclusion.

2. Green Hydrogen Production System

In this paper, the main focus is to optimize the ES size that assists hydrogen production from PV power. This section gives an insight into the components and operation of the system.

The green hydrogen production from a PV facility consists of PV arrays, an energy storage system, and a hydrogen generator through electrolysis (Electrolyser).

There are different water electrolysis technologies such as Alkaline Electrolysis Cells, Proton Exchange Membrane (PEM), and Solid Oxide Electrolysis Cells. PEM electrolysis technology is widely established as an efficient and suitable option for hydrogen production from renewable-powered systems due to its high efficiency, and ability to produce high-purity hydrogen [10,11]. PEM has some challenges, such as the use of expensive catalyst materials and shorter-life membrane materials. To maintain a long lifetime of the electrolyser, an EMS is required.

Replicating the system model of the solar hydrogen pilot plant in greater Brisbane [5], the electrolyser power is 66% of the PV nominal power. The aim of this work is to demonstrate the DC off-grid hydrogen production as shown in Figure 1. The figure presents the DC Microgrid for green hydrogen production used [9,12], as DC microgrids produce hydrogen more efficiently by eliminating unnecessary AC power conversions from the system. In [9] as in Figure 1, the power conversion between a 380V DC bus supplied by PV powers the electrolyser, while a 48V bus is tied to the ES, and the 380 V–48 V is powered through a bidirectional DC-DC converter. The power electronics of the system are not considered in this work, but the storage and EMS components of the system are the focus. The modeled system in this work consists of a nominal 1500 W PV and 1000 W PEM electrolyser, an ES that uses LiC or LiB cells, and an EMS that governs the operation of the electrolyser and the power flow from/to the ES.

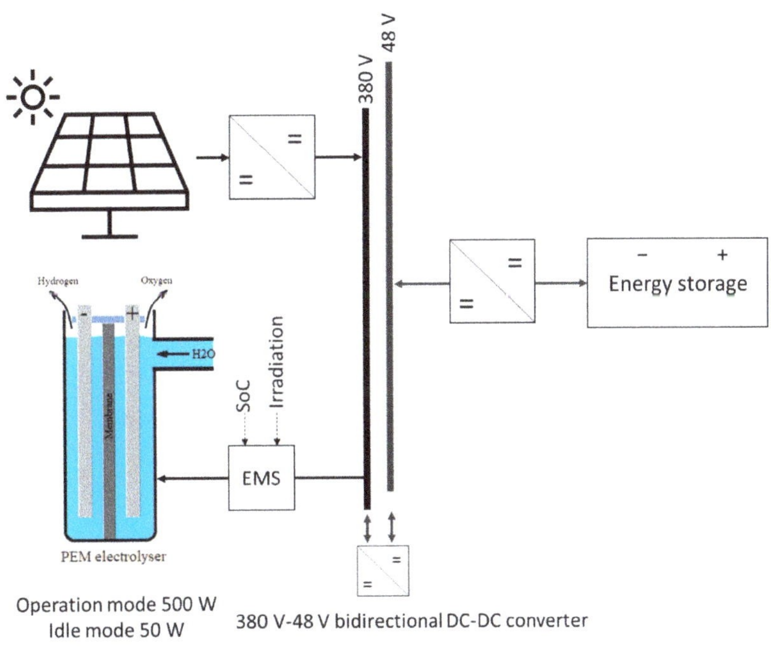

Figure 1. The considered microgrid for green hydrogen production.

In Figure 2, The PEM electrolyser has three stats of operations managed by the EMS as follows [13]:

- H2 Production mode: During this state, the electrolyser is producing hydrogen at its rated production capacity powered by the PV panels generation; if the PV generation drops during the day, the ES will substitute.
- Standby/Idle mode: During this state, the electrolyser consumes 10% of its rated production capacity power to sustain the required temperature and pressure of the electrolyser. The electrolyser is ready to return to the production state in a short mode transitioning time.
- Off/Failure mode: During this state, the electrolyser is in a total shutdown, where it losses the pressure and the temperature required by the electrolyser.

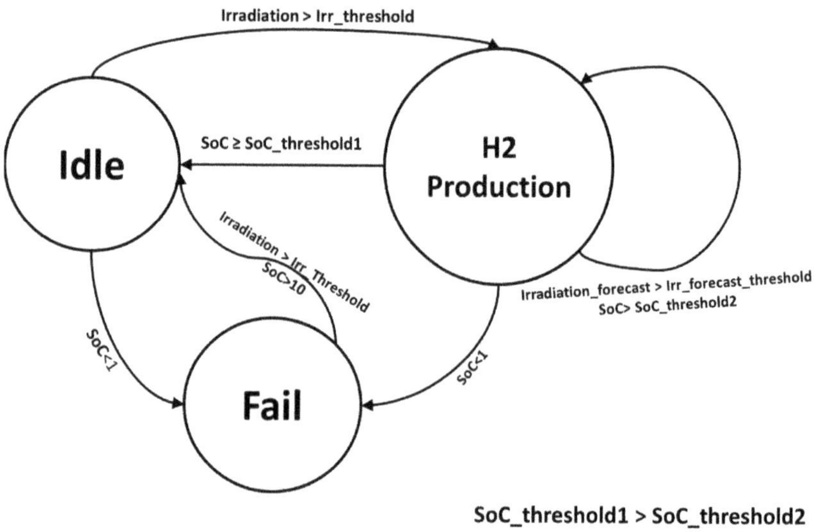

Figure 2. EMS state machine for the electrolyser.

There are data inputs that govern the electrolyser load transitions from one state to the other such as the state of charge (SoC) of the ES, solar irradiation, and solar irradiation forecast. The irradiation is the current solar irradiation, and the irradiation forecast is the irradiation predicted for the next day. Transitions from each state to the other are as follows:

- The transition from idle to production mode takes place when the solar irradiation is higher than a specific threshold (Irr_threshold). This is considered a hot start of the electrolyser because it sustains its required startup pressure and temperature.
- The transition from production to idle mode takes place when the SoC of the ES drops to a certain threshold where the remaining stored energy in the ES will power the electrolyser with 10% of its rated power to avoid the total shutdown situation. The electrolyser remains in the production mode even after crossing the ES SoC threshold if the Irradiation forcast the next day is enough to generate energy that will sustains the system for the next 24 h.
- A transition to the fail mode where a total shutdown and depressurization of the electrolyser occurs when the ES is fully discharged and SoC is almost zero. Transitioning from the fail mode to the idle mode is considered a cold start where the temperature increases and pressure builds up to be in the idle state conditions, which takes more time than the hot start.

Irradiation forecast signal selects whether the transition from production to idle mode happens at a high SoC (SoC_threshold1) and the night production is reduced or at a low SoC (SoC_threshold2) where the night production is increased.

3. Li-ion Capacitor and Li-ion Battery Degradation Behavior

3.1. Li-ion Capacitor and Li-ion Battery under Consideration

Hybrid supercapacitors, also known as lithium-ion capacitors (LiC), are an emerging technology that combines the advantages of an electric double-layer capacitor (EDLC) with those of a lithium-ion battery. LiC has a high-power density and can withstand temperature fluctuations, making it suitable for outdoor applications. It also has a longer cycle lifetime and higher power density than lithium-ion batteries (LiB). LiB, on the other hand, is more susceptible to degradation due to temperature fluctuations and has a lower power density. Additionally, LiB is more sensitive to fluctuations in power supply and requires a higher oversizing to prevent degradation.

In outdoor applications where ES is not placed in a temperature-controlled environment, LiC offers several advantages over LiB as an energy storage solution. Its ability to withstand temperature fluctuations and its longer cycle lifetime make it a more reliable option. Additionally, having an optimized size for the ES reduces the overall cost of the system. A comparison of the degradation behavior between LiB and LiC showed that LiC had better performance in terms of cycle life compared to LiB but the bottleneck is in the calendar life of the LiC [14]. In Table 1 the specifications of the used LiB and LiC in this study are presented.

Table 1. Datasheet parameters of LiB and LiC.

Property	LiC	LiB
Nominal capacity	4 Ah	13 Ah
Nominal voltage	3.2 V	2.26 V
Maximum voltage	4 V	2.9 V
Minimum voltage	2.5 V	1.5 V
Max. charge/discharge current	30 A	130 A
Calendar life	5 years	25 years
Internal resistance	6 mΩ	1.5 mΩ
Specific energy	48.8 Wh/kg	74 Wh/kg
Energy density	77.7 Wh/l	146 Wh/l
Operation and storage temperature	−25 °C to 65 °C	−40 °C to 50 °C

ES is needed for PV systems that are used for hydrogen production because of the intermittent and fluctuating nature of solar power and the need to keep the electrolyser in production or idle mode and avoid a total shutdown. In Australia, PV systems are exposed to various weather conditions, including extreme heat in the summer and cool temperatures in winter, which affects the performance of the PV panels and the efficiency of the hydrogen production process.

The use of ES in this system allows for the storage of excess solar energy that can be used during times of low solar irradiance or high demand for hydrogen. This helps to ensure a consistent and reliable supply of hydrogen, which is essential for industrial and transportation applications [15].

3.2. Li-ion Capacitor and Li-ion Battery Degradation

Laboratory testing on LiC cells was performed to study their degradation behavior in different operating and environmental conditions. The aging tests were conducted by continuously charging and discharging the LiCs with a 100% cycle depth, and various C-rates and temperatures, as presented in [12]. The results of the tests were used to develop an aging model for the LiC cells, which was then used to predict the capacity fade of the LiC storage over a period of time and in variable temperature and load conditions [16]. From [16] the cycling aging was measured, where the capacity fade of the LiC ($Cf_{Cycling\ LiC}$) was approximated using Equations (2) and (3). In (2) A is the degradation factor of the LiC as a function of current (I) and temperature (T). $Cf_{Cycling\ LiC}$ is a function of the cycle count during simulation (nc), I and T. nc as a function of time is counted during the simulation using (1), where $Q_{present}$ is the cell storage capacity. In Figure 3, the expected number of full equivalent cycles (FECs) of the LiC before reaching the 20% capacity fade at various temperatures and currents is presented.

$$\text{nc(t)} = \frac{\int I}{2 \cdot Q_{present}} dt \quad (1)$$

$$A(I,T) = 0.0027 - 0.00036 \cdot I + 7.79 \cdot 10^{-6} \cdot I^2 + 4.3 \cdot 10^{-7} \cdot I \cdot T + 8.85 \cdot 10^{-7} \cdot T \quad (2)$$

$$Cf_{Cycling\ LiC}(t,I,T) = 100 - A(I,T) \cdot nc + A(I,T) \cdot nc^{0.95} \quad (3)$$

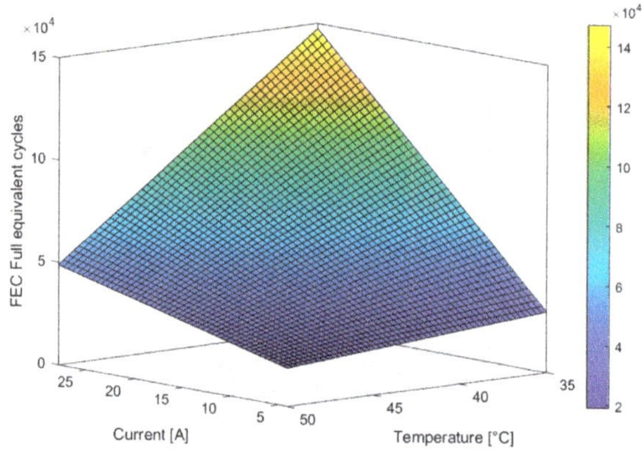

Figure 3. Full equivalent cycles (FEC) of LiC to drop to 80% of beginning-of-life capacity.

The data for the LiB degradation behavior analysis was obtained from the literature [17–19], and the cycling capacity fade is estimated using the model formulated in (4) and (5). $End\_of\_life\_FEC_{LiB}$ is the expected number of FECs that the LiB can withstand at various temperatures and currents before reaching the 20% capacity fade, which is presented in Figure 4.

$$End\_of\_life\_FEC_{LiB}(I,T) = 2.4 \cdot 10^4 + 573.1 \cdot I - 904 \cdot T - 9.6 \cdot I \cdot T + 9.1 \cdot T^2 \quad (4)$$

$$Cf_{Cycling\ LiB}(t,I,T) = \frac{20\% \ \cdot nc}{End\_of\_life\_FEC_{LiB}} \quad (5)$$

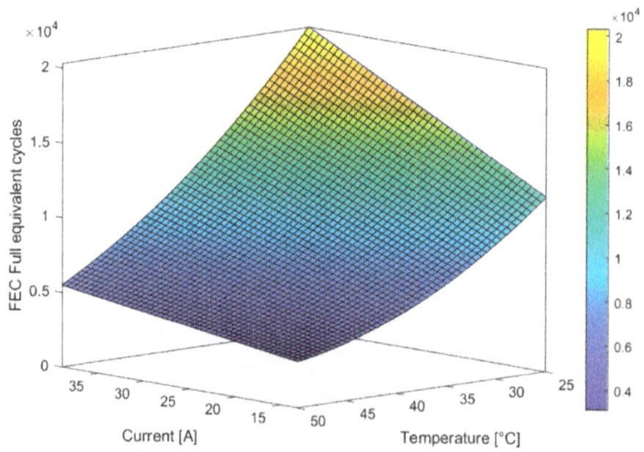

Figure 4. Full equivalent cycles (FEC) of LiB to drop to 80% of beginning-of-life capacity.

Overall, the laboratory testing and degradation behavior analysis demonstrated the superiority of LiC over LiB as an energy storage solution in applications that makes high FEC and are placed outdoors; however, by considering the calendar aging of LiC and LiB it is found that the calendar aging effect on the LiB is very small (0.8% capacity loss per year) [18] compared to the calendar aging of the LiC cells which is dependent on the idle time conditions of the LiC, such as SoC and the temperature, as shown in Figure 5. The capacity fade caused by calendar aging of the LiC $Cf_{calendar\_LiC}$ as a function of SoC and T is formulated in (6) and derived from the calendar aging test performed in [20] and aligned with the shelf life from the LiC datasheet (5 years shelf life at room temperature and 0% SoC). $Cf_{calendar\_LiC}$ is the percentage of the faded capacity over the time (t) before the end of life (EOL) which is a 20% loss of the starting capacity.

$$Cf_{calendar\_LiC}(SoC, T) = \frac{20\% \cdot \left(101.7 + 1.018 \cdot SoC - 0.57 \cdot T + 0.0053 \cdot SoC \cdot T - 0.01 \cdot T^2\right)}{t} \quad (6)$$

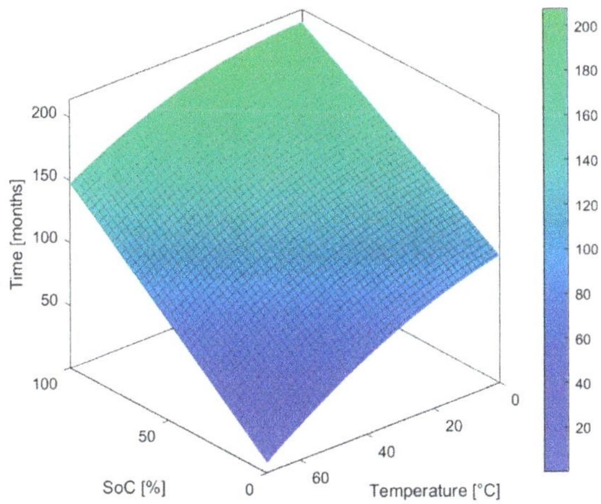

Figure 5. The expected calendar life, expressed in months of the LiC at various SOCs and temperatures, for an EOL criterion of a 20% capacity fade.

3.3. *Li-ion Capacitor and Li-ion Battery Internal Resistance Development*

The internal resistance (IR) development during the lifetime of the ES is an important parameter that governs the response behavior of the ES in regulating the grid. IR increase due to the aging of LiC and LiB is studied and modeled.

The [20] IR increase due to calendar aging is modeled in Figure 6 which presents the required time in months to have a 200% increase in the LiC IR. The LiC IR increase due to calendar aging was modeled as a function of temperature and SoC.

In [16] the LiC IR increase due to cycling aging ($IR_{CYCLING\_LiC}$) was approximated using Equations (7) and (8). In (7), B is the IR increasing factor of the LiC as a function of current (I) and temperature (T) and $ESR_{CYCLING\_LiC}$ is a function of the cycle count during simulation (nc), I and T.

$$B(I, T) = -0.0076 - 0.0013 \cdot I + 3.3 \cdot e^{-5} \cdot I^2 - 3.7 \cdot e^{-6} \cdot I \cdot T + 0.0008 \cdot T - 1.8 \cdot e^{-7} \cdot \mathrm{T}^2 \quad (7)$$

$$IR_{CYCLING\_LiC}(t, C - rate, T) = 100 - B_{C-rate,T} \cdot nc + B_{C-rate,T} \cdot nc^{0.95} \quad (8)$$

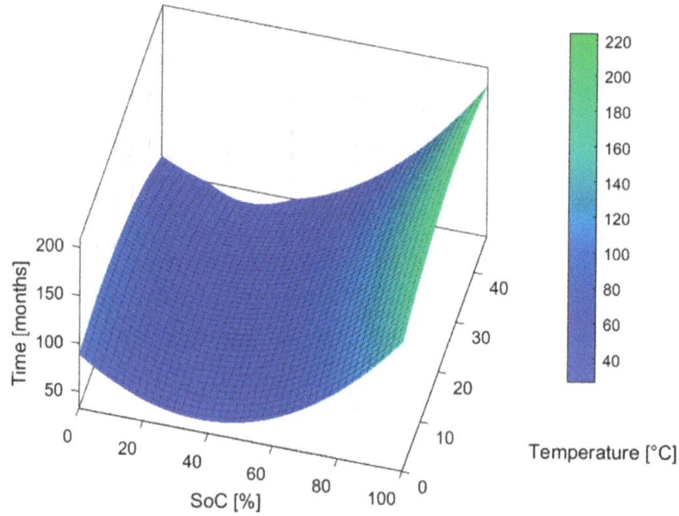

Figure 6. The expected time, expressed in months, for the LiC to increase the IR to 200% due to calendar aging.

From [18], the $IR_{CYCLING\_LiB}$ is approximated to increase by 10% for each 3000 FEC. The time required for the IR of the LiB to increase by 200% was modeled as a function of idling SOC and temperature and the results are presented in Figure 7.

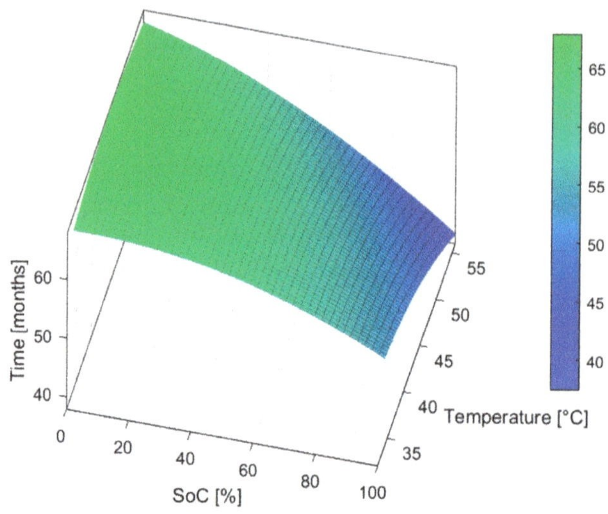

Figure 7. The expected time, expressed in months, for the LiB to increase the IR to 200% due to calendar aging.

It can be noticed that for the case of the IR, idling at lower SoCs for the LiBs has the opposite effect than for LiCs and elongates the calendar lifetime.

4. Storage System Sizing for the Australian Climate

In Figure 8, the simulation model for the considered system is presented showing the PV source, electrolyser load, EMS and ES. The system model is used for iterative simulations for 10 years period (2012 to 2022) to be used for optimizing the system size. The components of the system are described as follows:

- PV source is a 1500 W nominal capacity. PV capacity size is fixed for all test cases.
- Electrolyser for hydrogen generation is 1000 W in generation mode and 100 W in idle mode (Two electrolyser units are used).
- EMS as described in Figure 2. The state machine governs the load operation and the states where the power from the PV is delivered to the load or the energy storage.
- Energy storage unit (ES) represents the LiC or the LiB models including the aging models of each energy storage technology.
- System model inputs are the temperature and irradiation for 10 years at Brisban, Australia. Output signals such as the SoC and ES capacity fade over 10 years [21]. Irradiation forecast is also considered as an input to the EMS state machine.

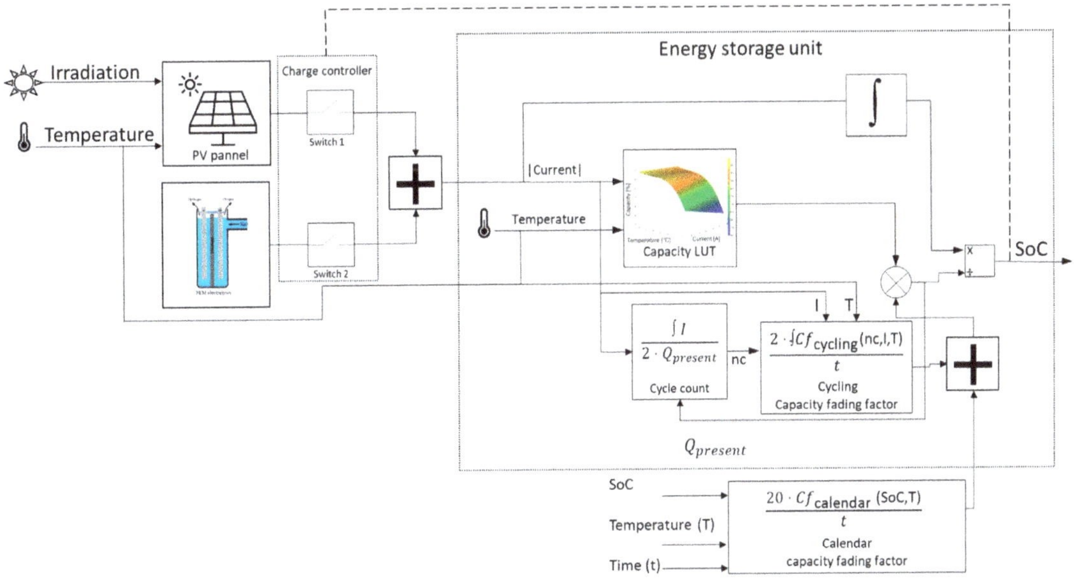

Figure 8. Green hydrogen production from the PV system model for simulation.

After setting the system model the sizing process starts.

In the literature, several optimization algorithms are used depending on the complexity and the number of optimized parameters. While the sparrow search algorithm can be used for a multi-dimensional problem such as day-ahead active power scheduling [8], hill climbing sizing problems only need a directed search algorithm that changes one parameter, simulate, evaluate and iterate [7]. The goal of the algorithm, which is presented in Figure 9, is to find the optimal ES size with a similar method to perturb and observe what increases or decreases the size in such a manner to speed up the sizing process.

The objective of the sizing is to have the optimal storage size that sustains the operation of the system during the night. The used constraint for sizing was fail mode avoidance where fail mode happens if the ES could not meet the demand of the electrolyser. The sizing process has two parts: the simulating modeled system in Figure 8 and the evaluation of the outputs through the algorithm in Figure 9.

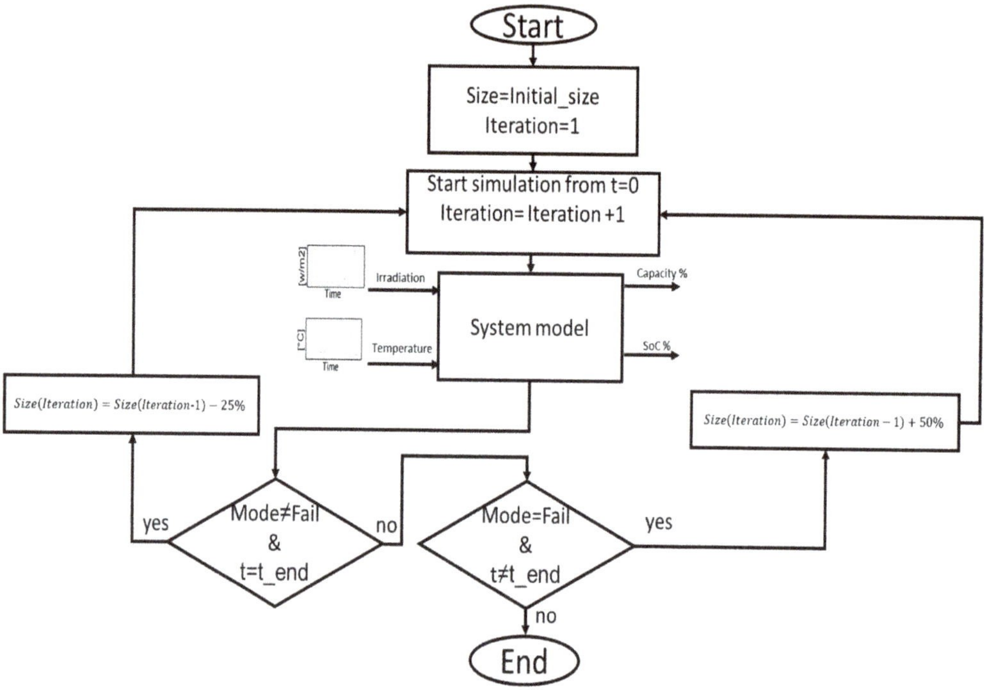

Figure 9. ES optimization algorithm using system simulation of green hydrogen production from PV.

The process starts by setting the initial size of the simulation and the iteration number and then running the simulation. If the output load signal does not show a fail state (Load = 0) and the simulation has reached the end time (t = t_end), the size of the simulation is reduced by 25% and the iteration number is incremented. The process then runs the simulation again from t = 0 with the new size. If the mode of the system is "Fail" and the simulation has not reached the end time (t ≠ t_end), the size of the simulation is increased by 50% and the iteration number is incremented and the simulation starts over. The sizing process ends when the system is not in the fail mode and the time reached t_end in the simulation.

In the next section, the system is sized for different cases and the results are analyzed.

5. Sizing Results for the Li-ion Capacitor and Li-ion Battery

In Table 2, six different sizing cases for both the LiC and LiB as ES for the green hydrogen system are presented. Each case has a different EMS state machine setting named idle,1, 2, and 3. Each EMS state machine number represents a different SoC threshold for the transition from production mode to idle mode as shown in Figure 2 to match the production time during the night with the ES size. Depending on the Irradiation forecast SoC threshold is selected, and whether the transition from production to idle mode happens at a high SoC (SoC_threshold1= 60 to 30%) and the night production is reduced or at low SoC (SoC_threshold2 = 25% in all cases) where the night production is increased. While the idle state machine secures the required power for the electrolyser to remain in idle mode during the night. The size is presented as the number of cells and Ah size. The size of the LiC is considered 100% for comparison purposes between LiC and LiB and illustrates the oversizing. The capacity fade is measured at the end of 10 years of operation where the fade is due to both cycling and calendar aging. For each case, the energy delivered to the electrolyser is represented in the parameter electrolyser load per year.

Table 2. Comparison between LiC and LiB ES for the green hydrogen production from PV (*refers to the remaining capacity after 10 years of operation).

	LiC					LiB						
Cases	**LiC Size [Cells]**	**Size [Ah]**	**LiC Size [%]**	**Remaining Capacity * [%]**	**LiC IR [%]**	**LiB Size [cells]**	**Size [Ah]**	**LiB Size [%]**	**Remaining Capacity * [%]**	**LiB IR [%]**	**State Machine for ES (SoC_threshold1)**	**Electrolyser Load per Year [MWh]**
1	1610	6440	100	80.8	260	620	8680	135	89.4	342	3	5.23
2	910	3640	100	80.60	262.6	368	5152	142	87.87	344.7	(SoC $\leq$ 30%)	4.83
3	710	2840	100	80.3	263.6	287	4018	141	87	346	2 (SoC $\leq$ 40%)	4.58
4	610	2440	100	80.2	264	247	3458	142	86.45	346.7	1	4.39
5	528	2112	100	79.8	265.6	214	2996	142	85.8	348	(SoC $\leq$ 50%)	4.24
6	280	1120	100	80.3	270	117	1638	146	83.3	353	Idle (SoC $\leq$ 60%)	3.2

As it can be seen from Table 2 and Figure 10, increasing the size of the ES is reflected in the delivered energy to the electrolyser which means higher production of hydrogen. However, the relation is exponential as shown in (9) and (10).

$$\text{Electrolyser\_load\_per\_year}_{LiC} = 4.8 \cdot e^{(1.326 \cdot 10^{-5} \cdot \text{size})} - 3.4 \cdot e^{(-0.00072 \cdot \text{size})} \quad (9)$$

$$\text{Electrolyser\_load\_per\_year}_{Lib} = 4.67 \cdot e^{(1.347 \cdot 10^{-5} \cdot \text{size})} - 3.6 \cdot e^{(-0.00057 \cdot \text{size})} \quad (10)$$

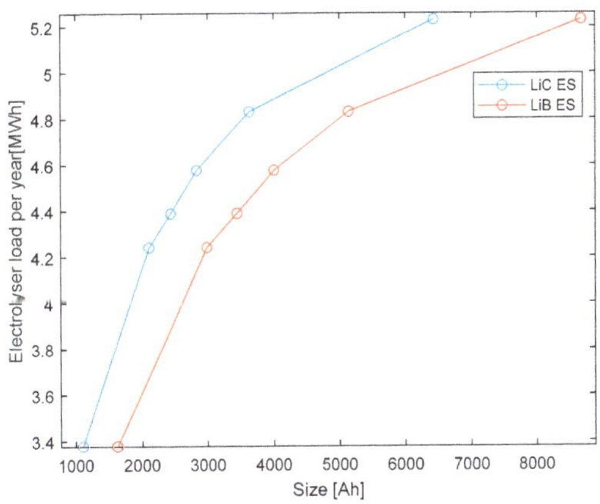

Figure 10. Sizing of LiC and LiB ES and the energy delivered to the electrolyser per year.

From Figure 10, it was found that LiB is oversized in all cases compared to LiC. In Figure 11, the green hydrogen production from the PV system was simulated for 14 days for cases 1, 3 and 6. Case 6 (blue) represents the smallest ES size where the storage only secures the power to the electrolyser to remain in the idle mode and avoid total shutdown, while Case 1 (red) represents the biggest ES size which powers the electrolyser load during the night to produce hydrogen. Case 1 also shows that the ES is not charged to 100% SoC during the day because it takes a longer time to be fully charged (PV capacity is fixed for all cases while the ES size increases) and this causes a higher effect of calendar aging in the case of LiC ES. The LOAD (H2 Electrolyser) shows that the electrolyser works with full capacity during the day in Case 6 because the ES is small in this case to deliver 10% of the load power during the night while in Case 1 the ES is large enough to keep the electroryser during the night. The results in Figure 11 are from LiB ES simulation but it applies for both LiC and LiB in terms of the delivered power to the load as they are sized in each case to provide the same power.

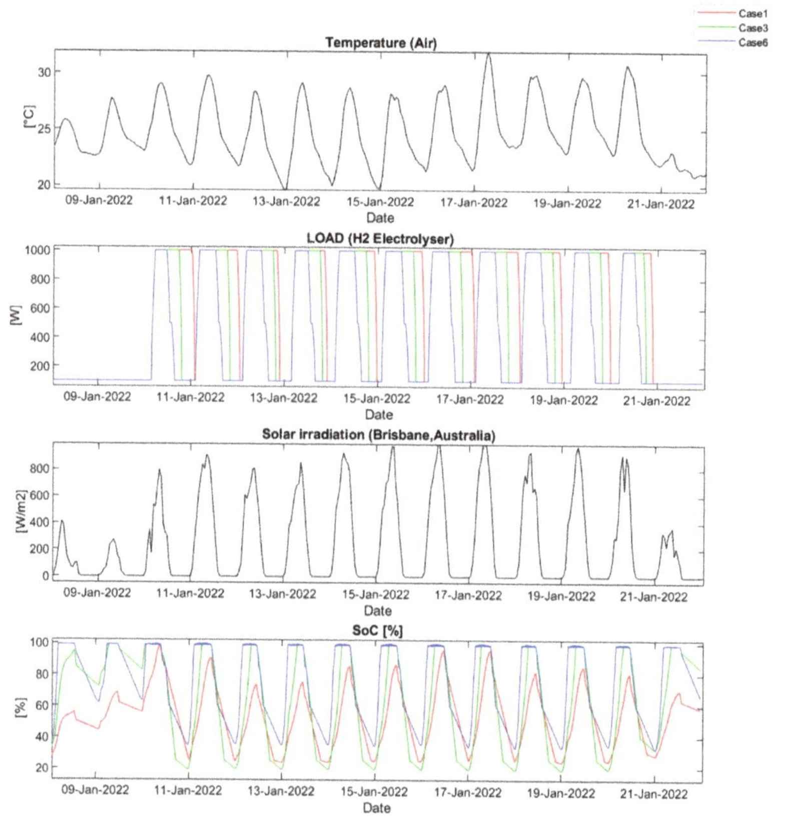

Figure 11. Simulation of 14 days of the system showing the temperature, irradiation, electrolyser load and SoC for cases 1,3 and 6. These results are from LiB ES simulation.

In Figure 12, the histogram plots show that the ES spends more time in low SoC in big ES such as Case 1 and spends more time in higher SoC in small ES such as Case 6, and from Figure 5, it is known that lower SoC for longer time shortens the calendar life of the ES; however, smaller ES sizes have a higher cycling aging compared to the bigger ES scenarios. Figure 13 represents the relation between the electrolyser energy in MWh and the capacity fading percentage. Increasing the energy storage size leads to increased electrolyzer load energy, and decreased aging due to cycling, but it also increases the aging due to staying in a low state of charge (SoC) for longer periods. This impact is more visible in the LiC energy storage system. LiB has longer retention of capacity but comes with an increased IR to around 350% after 10 years of operation, while LiC has a less IR increase to around 265% in the same period as shown in Figure 14.

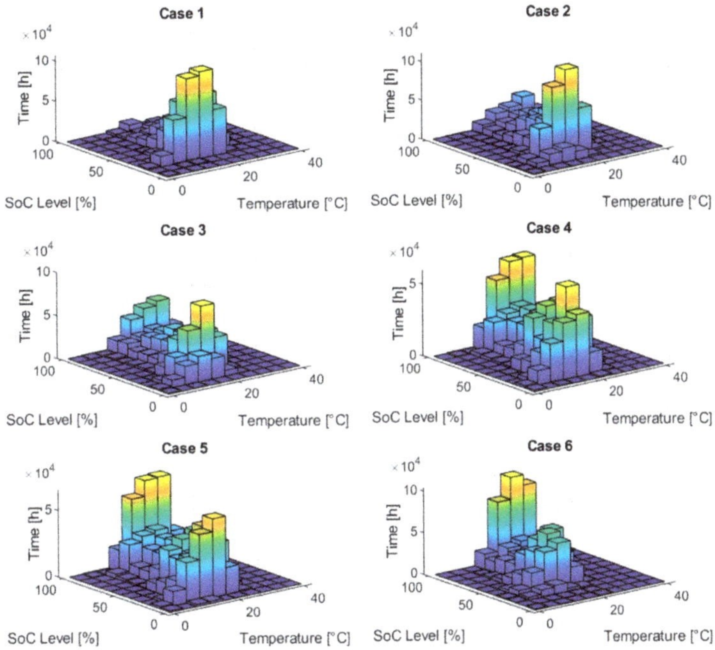

Figure 12. 3D histogram of the simulated 6 cases showing the time ES spent on each temperature and SoC state.

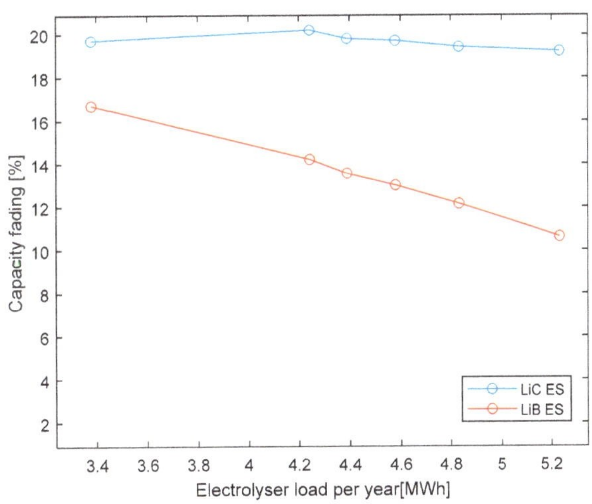

Figure 13. Capacity fade of the LiC and LiB ES as a function of the energy delivered to the electrolyser load.

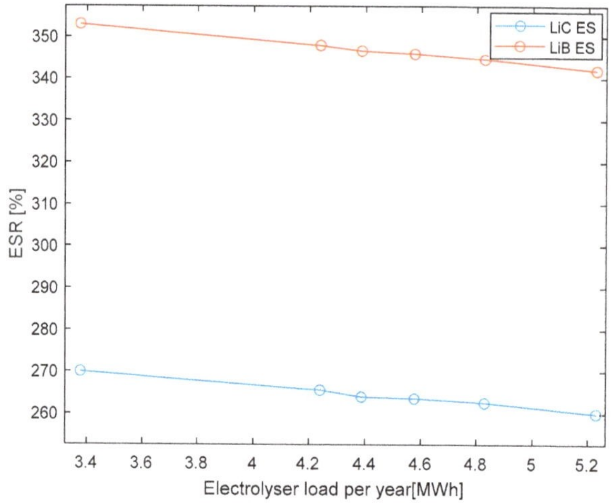

Figure 14. IR of the LiC and LiB ES after 10 years of operation as a function of the energy delivered to the electrolyser load.

Maintaining low IR is important when using ES not only for energy storage but also for grid regulation, where the ES is used to provide a rapid response by either absorbing or injecting power into the grid. High IR is also a sign of lower efficiency of the ES and increasing power losses [14].

In Table 2, while each case was sized using the algorithm in Figure 9 to achieve the same Energy Delivered to Electrolyser for LiC and LiB, it was found that LiB is oversized in all cases compared to LiC. However, LiC has a smaller remaining capacity compared to LiB after 10 years of operation which makes the LiB a better ES for the application, but considering the IR and the need to use the ES for grid regulation, LiC is the better option because lower IR means better response to the grid which suffers from the power fluctuations from PV. From Figure 12 it is found that case 4 has an even distribution of high and low SoC stay time, where the effect of calendar aging and cycling aging is minor in the case of LiC. LiB performs better in cases 5 and 6 because LiB is not affected much by the calendar capacity fade, and upsizing the ES in the case of LiB means a higher energy production for green hydrogen and a lower capacity fade by 6%. LiC has a high potential in the green hydrogen production from PV if the low IR is required in the grid design for better response and lower power losses due to ES degradation.

6. Discussion and Future Work

The motivation behind this work was driven by the increasing trend of green transitioning initiatives undertaken by governments worldwide. One such project is the Redlands Research Facility located east of Brisbane, Australia, which aims to establish a green hydrogen plant using locally produced solar energy. The green hydrogen pilot plant consists of a microgrid that uses concentrated-PV (CPV) and PV arrays, backed with commercially available batteries, to enhance its energy efficiency and a 40 kW PEM electrolyser [5] supplied with 5.5 L per hour of water. The project aimed to provide experimental validation and an integrated modeling of a hybrid renewable energy process that utilizes solar power, energy storage, and non-potable water to produce, store, and use hydrogen.

Another approach that arised within the project was to develop a fully DC microgrid that eliminates the unnecessary AC conversion components from the hydrogen production microgrid [12]. This led to the development of a downscaled fully DC hydrogen generation microgrid and a 500 W electrolyser and energy storage were required [9].

Using LiC was proposed for this application as an alternative energy storage to LiB, because of LiC's promising characteristics and the advantages provided by the hybridization of supercapacitors and Li-ion batteries [4].

Commercially available LiC cells were tested for characterization under different conditions and went through an extensive aging test, where each cell went under different aging conditions to model its performance and aging behavior and predict its performance in different climates using system simulation [16].

The aging models for both LiC and LiB were derived from lab tests and the literature. An EMS was modeled to imitate the operation of the green hydrogen microgrid, and the system was modeled and used to make a size optimization and estimation of the degradation happening to the energy storage after an operation of 10 years. The results show that LiC has a lower optimal size in terms of Ah compared to LiB because of its temperature tolerance and it degrades much slower in terms of FEC; however, LiC loses much of its capacity due to calendar aging which is not very effective on the LiB. Though, calendar aging increases the internal resistance of the LiB more than the LiC.

Future work will consider the usage of LiC for grid regulation and power smoothing as well as using it for storage. Also, it will consider the power generation data from the CPV and PV at the Redlands facility and make a comparison between the performance of the LiC regulating the grid at the beginning and after years of operation where the performance is dependent on the remaining capacity and the internal resistance.

7. Conclusions

The study provides an analysis of Li-ion capacitors and Li-ion batteries as energy storage systems for green hydrogen production from PV in the Australian climate. The study compares the performance of LiC and LiB in six different cases using four modes of EMS state machines and a sizing algorithm that optimize the energy storage to secure a safe operation of the electrolyser and increase its productivity. The EMS state machine has an idle mode where the size is minimum and the ES is required only to deliver 10% of the electrolyzer load power to avoid a total shutdown. The other EMS modes are operating the system for different sizes. The EMS uses inputs such as irradiation and irradiation forecast and temperature from the plant location. The system model is used for iterative simulations for 10 years period to be used for optimizing the system size. A directed search algorithm was used to reduce the number of search iterations in the sizing process. The results show that LiB is oversized in all cases compared to LiC, but LiC has a smaller remaining capacity after 10 years of operation due to the higher calendar aging. However, considering the IR and the need to use energy storage for grid regulation, LiC can be a better option because a lower IR means a better response to the grid. Additionally, IR after 10 years of operation for LiC is 264% of its initial IR, while for LiB it is 346%. Furthermore, the average remaining capacity after 10 years of operation for LiC is 80.2%, while for LiB it is 86%. These results indicate that while LiB may have a higher remaining capacity, it also has a higher IR. The study shows that Li-ion battery performs better in cases not affected much by the calendar capacity fade and upsizing the energy storage in the case of Li-ion battery means a higher energy production for green hydrogen and a lower capacity fade by 6%. Overall, LiC has a high potential in green hydrogen production from PV if the low IR is required in the grid design for better response and lower power losses due to energy storage degradation.

Author Contributions: Methodology, T.I. and S.S.M.; Investigation, T.I. and D.S.; Resources, T.K., D.S. and D.-I.S.; Writing – original draft, T.I.; Writing – review & editing, T.I., T.K. and D.-I.S.; Supervision, T.K., D.S. and D.-I.S. All authors have read and agreed to the published version of the manuscript.

Funding: This research received no external funding.

Data Availability Statement: The data supporting the reported results can be provided upon request from the corresponding authors.

Conflicts of Interest: The authors declare no conflict of interest.

Nomenclature

Notation	Definition	Unit
LiC	Li-ion Capacitor	-
LiB	Li-ion Battery	-
Cf	Capacity Fade	-
nc(t)	Cycle count during simulation	-
IR	Internal Resistance	Ω
SoC	State of Charge	%
ES	Energy Storage	-
PV	Photovoltaic	-
CPV	Concentrated Photovoltaic	
EMS	Energy Management System	-
MWh	Megawatt-hours	-
EOL	End of Life	-
T	Temperature	°C
I	Current	A
t	Time	s
t_end	End time of simulation	-
Q_present	Cell storage capacity	Ah
SoC_threshold1	State of Charge threshold for transition from Production to Idle mode	%
SoC_threshold2	State of Charge threshold to remain in production mode	%
fail	Fail state of the electrolyser (total shut down)	-
idle	idle state of the electrolyser (10% of the rated power)	-

References

1. Huang, F.; Liu, J.; Wang, Z.; Shuai, C.; Li, W. Of jobs, skills, and values: Exploring rural household energy use and solar photovoltaics in poverty alleviation areas in China. *Energy Res. Soc. Sci.* **2020**, *67*, 101517. [CrossRef]
2. Jung, W.; Jeong, J.; Kim, J.; Chang, D. Optimization of hybrid off-grid system consisting of renewables and Li-ion batteries. *J. Power Sources* **2020**, *451*, 227754. [CrossRef]
3. Thampan, T.; Ding, Y.; Toomey, L.; Hundich, A.; Babu, V. Accelerated Degradation of Li-Ion Batteries for High Rate Discharge Applications. In Proceedings of the Ground Vehicle Systems Engineering and Technology Symposium, Novi, Michigan, 11–13 August 2020; pp. 13–15.
4. Ibrahim, T.; Stroe, D.; Kerekes, T.; Sera, D.; Spataru, S. An Overview of Supercapacitors for Integrated PV—Energy Storage Panels. In Proceedings of the 2021 IEEE 19th International Power Electronics and Motion Control Conference (PEMC), Gliwice, Poland, 25–29 April 2021; pp. 828–835.
5. Mohammadshahi, S.S.; Boulaire, F.A.; Love, J.; Gorji, S.A.; Mackinnon, I.D. A flexible analytical model for operational investigation of solar hydrogen plants. *Int. J. Hydrog. Energy* **2022**, *47*, 782–808. [CrossRef]
6. Hernández, J.C.; Sanchez-Sutil, F.; Muñoz-Rodríguez, F.J. Design criteria for the optimal sizing of a hybrid energy storage system in PV household-prosumers to maximize self-consumption and self-sufficiency. *Energy* **2019**, *186*, 115827. [CrossRef]
7. Hernández, J.C.; Gomez-Gonzalez, M.; Sanchez-Sutil, F.; Jurado, F. Optimization of battery/supercapacitor-based photovoltaic household-prosumers providing self-consumption and frequency containment reserve as influenced by temporal data granularity. *J. Energy Storage* **2021**, *36*, 102366. [CrossRef]
8. Wenzhi, S.; Zhang, H.; Tseng, M.L.; Weipeng, Z.; Xinyang, L. Hierarchical energy optimization management of active distribution network with multi-microgrid system. *J. Ind. Prod. Eng.* **2022**, *39*, 210–229. [CrossRef]
9. Ganjavi, A.; Gorji, S.; Hakemi, A.; Moradi, A.; Sera, D. Design and Implementation of an SiC-Based 48 V-380 V Dual Active Bridge DC-DC Converter for Batteries Employed in Green Hydrogen Microgrids. In Proceedings of the 7th IEEE Southern Power Electronics Conference (SPEC 2022), Denarau Island, Fiji, 5–8 December 2022.
10. Schmidt, O.; Gambhir, A.; Staffell, I.; Hawkes, A.; Nelson, J.; Few, S. Future cost and performance of water electrolysis: An expert elicitation study. *Int. J. Hydrog. Energy* **2017**, *42*, 30470–30492. [CrossRef]
11. Carmo, M.; Fritz, D.L.; Mergel, J.; Stolten, D. A comprehensive review on PEM water electrolysis. *Int. J. Hydrog. Energy* **2013**, *38*, 4901–4934. [CrossRef]
12. Gorji, S.; Sera, D. Projects: Fully DC Microgrid for Green Hydrogen Production. QUT Centre for Clean Energy Technologies and Practices. Available online: research.qut.edu.au/ccetp/projects/fully-dc-microgrid-for-green-hydrogen-production (accessed on 20 October 2022).
13. Matute, G.; Yusta, J.M.; Beyza, J.; Correas, L.C. Multi-state techno-economic model for optimal dispatch of grid connected hydrogen electrolysis systems operating under dynamic conditions. *Int. J. Hydrog. Energy* **2021**, *46*, 1449–1460. [CrossRef]

14. Soltani, M.; Ibrahim, T.; Stroe, A.I.; Stroe, D.I. Comparison of High-Power Energy Storage Devices for Frequency Regulation Application (Performance, Cost, Size, and Lifetime). In Proceedings of the IECON 2022–48th Annual Conference of the IEEE Industrial Electronics Society, Brussels, Belgium, 17–20 October 2022; pp. 1–6.
15. Cerniauskas, S.; Grube, T.; Praktiknjo, A.; Stolten, D.; Robinius, M. Future hydrogen markets for transportation and industry: The impact of CO_2 taxes. *Energies* **2019**, *12*, 4707. [CrossRef]
16. Ibrahim, T.; Kerekes, T.; Sera, D.; Stroe, D.I. Degradation Behavior Analysis of High Energy Hybrid LITHIUM-Ion Capacitors in Stand-Alone PV Applications. In Proceedings of the IECON 2022–48th Annual Conference of the IEEE Industrial Electronics Society, Brussels, Belgium, 17–20 October 2022; pp. 1–6.
17. Stroe, A.I.; Knap, V.; Stroe, D.I. Comparison of lithium-ion battery performance at beginning-of-life and end-of-life. *Microelectron. Reliab.* **2018**, *88*, 1251–1255. [CrossRef]
18. Stroe, A.I. *Analysis of Performance and Degradation for Lithium Titanate Oxide Batteries*; Aalborg University Press: Aalborg, Denmark, 2018.
19. Stroe, A.I.; Stroe, D.L.; Knap, V.; Swierczynski, M.; Teodorescu, R. Accelerated Lifetime Testing of High Power Lithium Titanate Oxide Batteries. In Proceedings of the 2018 IEEE Energy Conversion Congress and Exposition (ECCE), Portland, OR, USA, 23–27 September 2018; pp. 3857–3863.
20. Soltani, M.; Ronsmans, J.; Van Mierlo, J. Cycle life and calendar life model for lithium-ion capacitor technology in a wide temperature range. *J. Energy Storage* **2020**, *31*, 101659. [CrossRef]
21. Solcast. Solar Irradiance and temperature Data, 2012–2022 Brisbane, Australia. Available online: https://solcast.com (accessed on 21 September 2022).

energies

Article

Handling Computation Hardness and Time Complexity Issue of Battery Energy Storage Scheduling in Microgrids by Deep Reinforcement Learning

Zeyue Sun, Mohsen Eskandari *, Chaoran Zheng and Ming Li

School of Electrical Engineering and Telecommunications, University of New South Wales, Sydney, NSW 2052, Australia
* Correspondence: m.eskandari@unsw.edu.au

Abstract: With the development of microgrids (MGs), an energy management system (EMS) is required to ensure the stable and economically efficient operation of the MG system. In this paper, an intelligent EMS is proposed by exploiting the deep reinforcement learning (DRL) technique. DRL is employed as the effective method for handling the computation hardness of optimal scheduling of the charge/discharge of battery energy storage in the MG EMS. Since the optimal decision for charge/discharge of the battery depends on its state of charge given from the consecutive time steps, it demands a full-time horizon scheduling to obtain the optimum solution. This, however, increases the time complexity of the EMS and turns it into an NP-hard problem. By considering the energy storage system's charging/discharging power as the control variable, the DRL agent is trained to investigate the best energy storage control method for both deterministic and stochastic weather scenarios. The efficiency of the strategy suggested in this study in minimizing the cost of purchasing energy is also shown from a quantitative perspective through programming verification and comparison with the results of mixed integer programming and the heuristic genetic algorithm (GA).

Keywords: battery energy storage systems; deep reinforcement learning; energy management system; microgrid; optimization; renewable energy resources

Citation: Sun, Z.; Eskandari, M.; Zheng, C.; Li, M. Handling Computation Hardness and Time Complexity Issue of Battery Energy Storage Scheduling in Microgrids by Deep Reinforcement Learning. *Energies* **2023**, *16*, 90. https://doi.org/10.3390/en16010090

Academic Editors: Alexander Micallef, Zhaoxia Xiao and Mohamed Benbouzid

Received: 3 November 2022
Revised: 14 December 2022
Accepted: 19 December 2022
Published: 21 December 2022

1. Introduction

A microgrid (MG) is a compact grid, including distributed energy resources (DERs) and local loads, and gained great attention to address the issues of integrating renewable energy resources (RESs) into the grid [1,2]. Because of this, a typical MG often consists of a variety of renewable energy power production devices, energy storage systems (ESSs), loads, as well as ancillary equipment, including energy converters and controllers [3]. The study on MGs covers a wide range of topics, including research on MG architecture, power electronics control [4], investment and operating costs [5], dynamic and transient stability [6–8], protection [9], safety, and maintenance. The ESSs control approach drew the most attention among them as the study area for MG energy dispatching [10].

The problem with developing the optimal strategy to control the dispatchable DERs and ESSs is significant, as the stability and efficiency of MG are suffering from the intermittent and stochastic characteristics of RESs [11]. The energy management system (EMS) is responsible for maintaining the MG operating in a low-cost and stable way. Particularly when MG is operating in grid-connected mode, EMS also works on the management of electricity trading between MG and the utility grid [12]. Therefore, implementing proper optimization algorithms to organize the EMS determinants of the performance of MG economic operation [2,5,10]. Exploiting the battery energy storage system (BESS) is essential for preserving the MG's power balance and minimizing the effect of intermittent and uncontrollable renewable energy.

The requirement for a proper continuous joint economy–dynamics model of EMS operation that appropriately incorporates accurate battery cycle age, degradation cost, and price is stressed in [13]. It was thoroughly studied in [13] that energy management and the BESS optimal scheduling can be challenging when the amount of data is large, and the operational strategy is defined by nonlinear/nonconvex mathematical models. Deep learning algorithms recently offered fresh approaches for tackling challenging MG control and energy management issues as a result of the growth of artificial intelligence [14]. An effective technique for the realization of artificial intelligence without historically labeled data is DRL [15,16]. First, developments in computer power, particularly highly parallelized graphical processing unit (GPU) technology, enabled deep neural networks to be trained with thousands of weight parameters. Second, DRL took advantage of a sizable deep convolutional neural network (CNN) to improve representation learning. Third, experience replay was employed by DRL to solve the correlated control issues.

The MG energy management problem fits inside the deep reinforcement learning solution framework as a real-time control problem, and there was some excellent research in this area [17–19]. Reference [20] applied a novel model-free control to determine an optimal control strategy for a multi-zone residential HVAC system to minimize the cost of generating energy consumption while maintaining user comfort. To analyze the influence of different scenario combination models on the MG energy storage disposition strategy, a problem environment model of the energy storage disposition was created using the example of the MG system for private users [21]. Reference [22] proposed an EMS for the real-time operation of a dynamic and stochastic pilot MG on a university campus in Malta, consisting of a diesel generator, photovoltaic modules, and batteries. Reference [23] performed reinforcement learning training for the unpredictability of the solar output of the MG to lower the MG's power cost using the data anticipated by the neural network. Reference [24] proposes a model-based approximate dynamic programming algorithm and thoroughly considers load, photovoltaic, real-time electricity price fluctuation, and power flow calculation. It then uses a deep recurrent neural network to approximate the value function. From the standpoint of ensuring the security of power grid operation, reference [25] suggested a deep reinforcement learning-based control technique for power grid shutdown. Retail pricing strategies are provided by [26] using Monte Carlo reinforcement learning algorithms from the viewpoint of distribution system operators, with the objectives of lowering the demand-side peak ratio and safeguarding user privacy. The advantages of applying deep reinforcement learning for online progress optimization of building energy management systems in a smart grid setting are explored in [27], and a sizable Pecan Street Inc. database is used to confirm the method's efficacy. Additionally, a model-free DRL was used to improve the reliability and resiliency of (distribution) grids in the context of Internet of Things (IoT) [28] and by forming islanded [29] and multi-MG systems [30].

However, a critical problem with the optimization of the BESS scheduling by the EMS in MG is time complexity. The optimal operation (charge/discharge) of the BESS in each time step (e.g., each hour) depends on its operation point (and consequence state of charge (SoC)) in the previous time step and also affects the optimal point in the future time step. To address this issue, the common method is finite horizon predictive EMS [31]. Nevertheless, a full-day (24 h) time horizon is needed to achieve the optimum solution for designing the BESS charging profile. Notably, the time complexity of the optimization problem of the BESS in a full-day period is $\mathcal{N}^{24}$, where $\mathcal{N}$ is the discretized number of possible charge/discharge levels based on the power (current) and energy ratings of the BESS. To handle this problem, dynamic programming, given by Bellman's optimality principle, was used to tackle the scalability and time complexity of the EMS optimization problems using deep Q-learning methods [32]. Yet, the curse of dimensionality is the problem with discrete Q-learning, which turns it into an NP-hard problem. Furthermore, including the dispatchable DGs in the EMS as well as energy trading with the grid (or other MGs [33]) along with the BESS, increases the dimension space for the EMS and makes the problem computationally intractable (particularly for real-time applications [34,35] and market clearing).

To tackle the computational hardness of the BESS scheduling in the MG, we propose using the model-free DRL. The DRL agent randomly selects the charge/discharge rate of the BESS and dispatchable unit, such as the diesel generator (DG), for each timestep, based on which the EMS schedules electricity and trading for MG and the grid. The selected values are sent to the reward function, obtaining a reward or penalty towards the state–action pair. DRL learns from the reward and updates the scheduling policy to avoid penalized actions and practice highly rewarded actions. This process is repeated for a large number of randomized episodes to guarantee the optimality of the solution. Using deep neural networks with an appropriate training algorithm that trains the DRL to maximize future expected rewards helps to realize the problem objectives. The trained DRL can observe the intermittently generated power from RESs, organize the dispatchable resources and grid power to keep the power balance of MG, sell electricity to the grid and earn some profit at the time of high electricity prices, and purchase electricity from the grid at the time of low electricity price.

The contributions of the paper are summarized as follows:

- In this paper, the DRL technique is utilized to handle the time complexity and large dimension space associated with the NP-hardness of optimal charge/discharge scheduling of the BESS. For this purpose, the DRL structure and the state–action–reward tuple are appropriately designed. The continuous deep deterministic policy gradient (DDPG) is used as the training algorithm to avoid the curse of dimensionality issue.
- The DRL can also handle time complexity associated with the nonconvexity of optimization problems for the BESS scheduling and nonlinear power flow. Complementarity constraints should be imposed to avoid simultaneous charge/discharge of the BESS that makes the problem non-convex. Alternatively, using slack integer variables increases the computational burden of the optimization problem.
- Therefore, the trained DRL is practicable for real-time BESS scheduling in MGs for different applications, such as frequency (dynamics) support and ancillary services that are needed to cover intermittent RESs.
- The searching space algorithm is proven to be environment-free and adaptable for EMS in various MG architectures with different scales.
- In order to comprehensively reveal the advantages of this method, the optimization results are compared with the results of the mixed integer nonlinear programming (MINLP) and genetic algorithm (GA).

2. Microgrid Architecture and Modeling

In this paper, we consider the MG including the BESS, a DG under technical constraints, a wind turbine (WT), and a photovoltaic generation (PV) with both deterministic and stochastic power generation, a load with uncertain demand. In order to minimize the 24-h accumulated electricity purchase fee, electricity purchased from the utility grid and hourly electricity prices are also taken into account. The configuration of the MG model is shown in Figure 1.

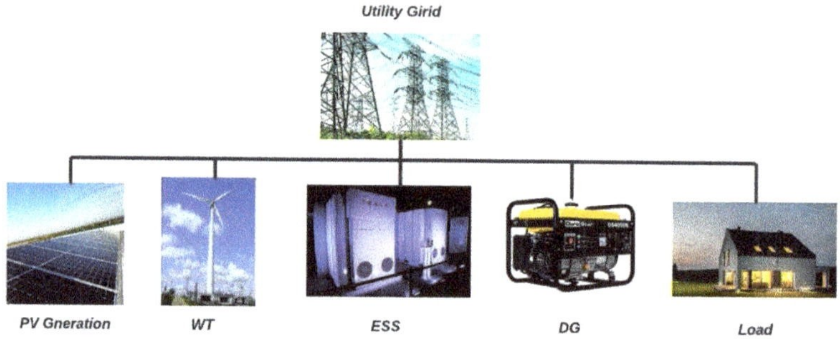

Figure 1. Microgrid architecture, which consists of DERs including RESs (PV and WT), DG, and the battery ESS, and loads. The MG can be disconnected from the utility grid and work in autonomous mode, where DERs supply the local load and the BESS holds the consumption–production balance.

2.1. PV Generation

In community MGs, photovoltaic power generation equipment is a key renewable energy resource that can transform light energy directly into electrical energy using photovoltaic power production panels. It has a great deal of economic potential and is renowned for its cleanliness and economy. The photovoltaic power generating system, which includes photovoltaic elements including solar panels, AC and DC inverters, solar charge and discharge controllers, and loads, is often positioned on the roof of the building. Grid-connected type, off-grid type, and multi-energy complementing type are the three categories into which solar power-generating devices fall according to the system network structure of the MG. Off-grid photovoltaic power generating systems are appropriate for isolated islands, remote mountainous areas, and other locations, since they cannot interchange electricity with external power networks. A grid-connected photovoltaic power generating system is employed to generate electricity for the community MG under study in this research, which may interchange electrical energy with the main grid or other microgrids.

$$P_{grid}(t) + P_{pv}(t) + P_{wt}(t) = P_{load}(t) \quad (1)$$

The output of the photovoltaic power generation system in the MG primarily depends on its power generation, which is inextricably linked to weather factors, such as the highest and lowest temperatures, the average temperature, and the intensity of the day's light, all of which have an impact on the solar panels' ability to produce electricity. This paper makes the assumption that the sun shines from 7 am to 8 pm, that its output power varies nonlinearly with light intensity and ambient temperature, and that its light intensity swings from weak to strong and then to weak. The probability density function and mathematical model of a solar power-generating panel are as follows, and the output power of a typical photovoltaic power generation device roughly obeys the distribution:

$$f\left(\frac{I_T}{I_{max}}\right) = \frac{\gamma(\alpha+\beta)}{\gamma(\alpha)+\gamma(\beta)}\left(\frac{I_T}{I_{max}}\right)^{\alpha-1}\left(1-\frac{I_T}{I_{max}}\right)^{\beta-1} \quad (2)$$

$$\begin{cases} \alpha = \mu(\ \mu(1-\mu)/\sigma\hat{}2 - 1) \\ \beta = (1-\mu)(\ \mu(1-\mu)/\sigma\hat{}2 - 1) \end{cases} \quad (3)$$

$$P_{PV} = \varphi_{PV} S_{PV} \frac{I_T}{I_S}[1 + \alpha_{PV}(T_r - T_s)] \quad (4)$$

where γ is the gamma function, and α and β are the relevant parameters of the β distribution. In Formulas (2)–(4), P_{PV} is the rated output power of the photovoltaic power generation system; φ_{PV} is the power frequency reduction coefficient of the photovoltaic power generation system; *Tr* and *Ts* represent the actual temperature and standard temperature of the photovoltaic module, respectively; IT is the actual light intensity of the solar panel at 25 °C ambient temperature; and α_{PV} is the power temperature coefficient of the PV panel. The value of α_{PV} for different materials of the PV panel is not the same, and the general value is $\alpha_{PV} = -0.45$.

2.2. Wind Turbine

Due to its high construction cost, wind power generation equipment, another significant renewable energy resource in MGs, is not as commonly employed as solar power generation equipment. Additionally, as the production of wind energy is very volatile, unpredictable, and is significantly influenced by external variables such as the weather, it is sometimes challenging to make precise short-term projections. Currently, probability distributions are frequently employed to suit the wind speed distribution of wind turbines, with Weibull, Normal, Rayleigh, and other distributions among the most popular choices. The Weibull distribution is one of them and has the following probability density function expression:

$$f(v) = \frac{k}{a}\left(\frac{v}{a}\right)^{k-1} exp\left(-\frac{v^k}{a^k}\right) \tag{5}$$

$$\begin{cases} k = \frac{\sigma}{\mu} \\ a = \frac{\mu}{\gamma\left(1+\frac{1}{k}\right)} \end{cases} \tag{6}$$

where the wind turbine's wind speed v, its mean value, and its standard deviation are represented by the letters σ, and μ, respectively. The probability density function of the Welsh distribution, which is currently in widespread usage, can be used to calculate the wind speed distribution of the wind turbine. It is therefore possible to determine the wind turbine's output power by substituting the wind speed into the mathematical model of the wind turbine. The specific mathematical equation is as follows:

$$P_{WT} = \begin{cases} \frac{P_e}{v_e - v_{in}} v - \frac{P_e}{v_e - v_{in}} v_{in} & v_{in} \le v \le v_e \\ P_e & v_e \le v \le v_{out} \\ 0 & other \end{cases} \tag{7}$$

where p_{wt} is the output power of the wind turbine, Pe is the rated power, v is the current wind speed, v_e, v_{in}, and v_{out} are the rated wind speed, cut-in wind speed, and cut-out wind speed, respectively.

2.3. Battery Energy Storage System (BESS) Modeling

To maintain the stable operation of the MG and balance the system power, an energy storage system must be implemented due to the intermittent and unstable properties of distributed power generating modes, such as solar and wind power. Batteries are frequently employed as high-efficiency energy storage devices, and their internal energy state conforms to the following equation:

$$E_{ESS}(t) = E_{ESS}(t-1) + \eta_{ch} P_{ESS,ch}(t) * \Delta t - \frac{1}{\eta_{disch}} P_{ESS,disch}(t) * \Delta t \tag{8}$$

where η_{ch} and η_{disch} denote charge and discharge efficiency. The battery's capacity to charge and discharge at time t is represented by $E_{ESS}(t)$, while the time t between two charging and discharging operations is represented by Δt. The charging and discharging of power $P_{ESS}(t)$ of the energy storage system in the MG system is typically employed as an essential control variable to take part in the MG energy scheduling, which is the subject of this study. To avoid the charge and discharge happening at the same time, while solving EMS as an optimization problem, we define binary variables.

2.4. Diesel Generator (DG)

The diesel generator (DG) is used to boost its flexibility and independence. When the power production of renewable energy resources becomes stochastic and unreliable, MG relies on electricity stored in the BESS and acquired from the main grid. If the BESS runs out of energy and the main grid has a high electricity price, generating power from DG is the most cost-effective solution to keep loads supplied by MG. The power output and cost factors of DG were modelled as the function of fuel cost:

$$F(P_{DG}(t)) = aP_{DG}^2(t) + bP_{DG}(t) + c \tag{9}$$

where $P_{DG}(t)$ represents the power output of DG; a, b, and c represent the cost factor of DG.

2.5. Loads and Utility Grid (UG)

In a MG system, the component that uses the most electricity is referred to as the load. The load demand for a fixed MG system is often not customizable since it depends on the MG's characteristics and the surrounding climate. The load curve is employed as a set quantity of input to the MG system in this paper's energy scheduling problem. The load is expressed as $P_{load}(t)$ for time step t.

For the modelling of the utility grid, the power purchased from the grid to MG and the power sold from MG to the grid are mainly considered, representing as P_{grid}^{+} and P_{grid}^{-}, respectively. In addition, further descriptions of load and grid will be discussed with the objective function and power balance equation in the next section.

3. DRL-Based MG Energy Management System

3.1. Objective Function

Depending on the requirements of the microgrid system, various control goals can be established, such as minimizing pollutant emissions, cutting back on fuel costs for power generation, reducing voltage offset, cutting back on network active power loss, or increasing voltage stability. Typically, only one control objective or a mix of control objectives can be chosen. This paper seeks to reduce the sum of the cost of power from the external grid and the cost of fuel costs. The form is given in (10):

$$\begin{gathered} \mathrm{F}\,(\boldsymbol{s_t}, \boldsymbol{a_t}, \mathrm{t}) = \min\,[C_{Total}(t)] \\ C_{Total}(t) = \int_{t=0}^{T} C_{Grid}(t) * P_{Grid}(t)dt + \int_{t=0}^{T} F(P_{DG}\,(t))dt \end{gathered} \tag{10}$$

where C_{Grid}, P_{Gird} and $F(P_{DG}\,(t))$ represent the real-time electricity price, the power generated by the external grid and the fuel cost of DG in time t, respectively, and P_{grid}, P_{DG} satisfies the power balance constraint in (11). C_{total} represents the total cost of the entire MG to obtain electrical energy from the main grid and DG.

$$P_{PV}(t) + P_{WT}(t) + P_{ess}(t) + P_{DG}(t) + P_{Grid}(t) = P_{Load}(t) \tag{11}$$

Additionally, in this paper, time $t \in D$, $D = \{1, 2 \ldots 24\}$ and the unit of t is one hour.

3.2. State Space

The reinforcement learning algorithm DDPG transforms the mentioned control target into a solution form. The energy storage charging and discharging power controller is the agent in the reinforcement learning issue, and the MG mathematical model created in Section 2 serves as the environment. Utilizing the agent and the environment's ongoing interaction to get the optimal control strategy is the aim of reinforcement learning. As a result, it is necessary to identify the unique representation of the Markov state sequence quadruple (S, A, R, π) for this issue.

For the MG model, the information provided by the environment to the agent is generally renewable energy output, time-of-use electricity price, load, and state of charge of electric energy storage. Therefore, the state space of the microgrid model is defined as:

$$s_t = [P_{PV}(t),\ P_{WT}(t), P_{Grid}(t), P_{Load}(t), SOC(t), C_{Grid}(t)] \tag{12}$$

In the state space: P_{PV}, P_{WT} are the power output of the renewable energy in the t period, kW; $P_{Load}(t)$ is the load demand of the MG in the *t* period, kW; C_{Grid} is the time-of-use electricity price (purchased from external grid by microgrid) in the t period (time of use price, TOU), AUD/(kWh); and $SOC(t)$ is the state of charge of the energy storage system in time t.

3.3. Action Space

After the agent observes the state information of the environment, it selects an action from the action space A according to its own policy set π. Based on controllable devices in MG, the output of DG and the charge/discharge power of the BESS were introduced to the action space. Therefore, the action space of the MG considered in this paper is expressed by:

$$a_t = [P_{ESS}(t), P_{DG}(t)] \tag{13}$$

3.4. System Constraints

To ensure the simulation result is realistic, the energy storage system is set to operate under the constraints of charging and discharging power with a practicable range of the state of charge, and its state of charge also has both limitations to ensure the BESS will not have overheating problems caused by over-voltage or over-current issues. Considering the size and minimum power output of DG, a power constraint is applied by (10):

$$P_{ESS}^{min} \le P_{ESS}(t) \le P_{ESS}^{max} \tag{14}$$

$$SOC^{min} \le SOC(t) \le SOC^{max} \tag{15}$$

$$P_{DG}^{min} \le P_{DG}(t) \le P_{DG}^{max} \tag{16}$$

3.5. Reward Function

The objective function and the constraints must be combined when creating the reward function *R*. The following is the definition of the reward function from state *t* to state $t+1$:

$$R(s_t, a_t, t) = -\left[\int_t^{t+1} C_{grid}(t) * P_{grid}(t)dt + \beta(t)\right] \tag{17}$$

The first component represents the amount of power used, during this time, $\beta(t) = 0$, and the second component is the penalty item provided to define the limitations. When the constraints (9)–(11) are not satisfied, $\beta(t)$ is given a constant with a very big value. When the state–action pair does not exceed the system limitations, the value of $\beta(t)$ is zero, ensuring the estimation of the agent only depends on the price and power of the utility grid. However, if the state–action pair exceeds the limit, $\beta(t)$ is assigned a large penalty value to penalize the taken action by the agent. Reward function maximization is the goal of reinforcement learning; hence a minus sign must be added before these two elements.

4. Deep Reinforcement Learning Algorithm

4.1. DRL Structure

Based on the Markov decision process, reinforcement learning (RL) helps intelligent agents choose actions that will result in the greatest overall reward during their interactions with the environment. An environment and an agent are typically present in RL models, see Figure 2. The agent learns how to react to the environment depending on its current state, while the environment rewards the agent in return. RLs can be categorized as either model-based or model-free, depending on whether explicit environment modeling is necessary. Some typical RL algorithms are as follows: (1) Q-learning, which generates the action for the following step using the quality values Q(s, a) stored in the Q-table, and updates the quality value, where a stands for the learning rate, g for the deduction factor, R for the reward, a and s for the action and state in the current step, and a′ and s′ for the action and state in the following step; (2) the deep Q-network (DQN), which employs deep learning algorithms (such as DNN, CNN, and DT) to produce a continuous Q-quality value in order to get around Q-exponentially learning's rising computational cost; (3) The policy gradient algorithm, which generates the next-step action based on the policy function (which quantifies the state and action values at the current step) rather than a quality value such as Q; (4) The actor–critic algorithm, which uses the actor to generate the next-step action based on the current-step state and then adjusts its policy based on the score from the critic, whereas the critic uses the critic function to score the actor at the current step [21].

$$\begin{gathered} Q(s,a) \leftarrow Q(s,a) + \alpha[R + \gamma max_{a'} Q(s',a') - Q(s,a)] \\ s \leftarrow s' \end{gathered} \tag{18}$$

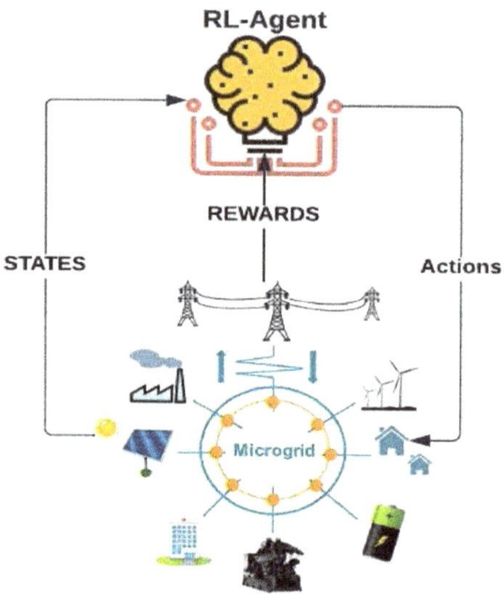

Figure 2. Schematic diagram of RL.

Three types of algorithms may be used to solve reinforcement learning optimization problems: value function-based algorithms, policy gradient-based algorithms, and search-based and supervised algorithms [24]. The solution techniques based on the value function are the main topic of this work. Examples include the dynamic programming algorithm, the Monte Carlo algorithm, the time series difference algorithm, etc. Among these, the dynamic programming approach is useful for addressing problems when there is a model, and the state space has a small dimension. The downside of the Monte Carlo approach is that it requires entire state sequence information, which is challenging to get in many aperiodic systems. It is a straightforward algorithm that is not model-based. A complete state sequence is not necessary for the time series difference approach to estimate the value function. The online difference algorithm SARSA and the offline difference algorithm Q-learning algorithm are two examples of the traditional time series difference approach. Both approaches retain a Q-table to tackle minor issues. The problem with reinforcement learning is when the state and action space is continuous or discrete on a very large scale, it is necessary to keep an exceptionally big Q-table, which poses storage challenges. However, this issue was resolved by the advancement of deep neural networks. A deep reinforcement learning technique that is more suited for difficult issues may be produced by using deep neural networks rather than Q-tables. Deep Q-learning (deep Q-network, DQN) is a common algorithm.

$$Q(s, a) \leftarrow Q(s, a) + \alpha\delta \tag{19}$$

The action value function is represented by $Q(s, a)$, while the learning rate is represented by δ. The ideal reinforcement learning control method can be discovered when the update formula converges. The DQN replaces the Q-function $Q(s, a/Z)$ in Q-learning by using a deep neural network. Following the calculation of the current target Q-value using Formula (3), the neural network's parameter is adjusted based on the mean square error between the current target Q-value and the Q-value provided by the Q-network.

4.2. Deep Deterministic Policy Gradient (DDPG)

Based on the DRL algorithm that was introduced above, DDPG is adopted as the optimization algorithm to reduce the MG's cost. DDPG is an RL algorithm that learns the policy and Q-function simultaneously. This approach processes off-policy data through the Bellman equation to achieve the learning goals towards the Q-function, and DDPG then learns the policy from the Q-function. According to the Q-function: $\mathcal{Q}^*(s, a)$ and current state space, the optimized action $a^*(s)$ can be obtained from this equation:

$$a^*(s) = \mathbf{argmax}_{a} \mathcal{Q}^*(s, a) \tag{20}$$

The loss function of the Q-network can be described as the learning process of the Q-network under the guidance of the reward function. By using the temporal difference principle, the loss function can be defined as (16):

$$\mathcal{L}\left(w^C\right) = \left[\mathcal{Q}\left(s, a \middle| w^Q\right) - \left(r + \gamma\left(\mathcal{Q}\left(s\_, a\_ \middle| w^Q\right)\right)\right)\right]^2 \tag{21}$$

In this formula, $\mathcal{Q}(s, a|w^Q)$ represents the Q-function in the current states and the accumulated future reward of the agent when the action of this state was executed. For the next states, s_, the Q-function is defined as the same as the last states. In $\mathcal{Q}(s\_, a\_|w^Q)$, s_ and a_ represent the state–action pair in the next states and w^Q is the weight of the Q-network. The actions in DDPG are determined by the policy; r represents the reward corresponding to the excused action from the current states s to the next states s_.

The objective function for the Q network is defined as:

$$F\left(w^C\right) = \mathbf{min}\left[\mathcal{L}\left(w^C\right)\right] \tag{22}$$

By introducing the learning rate θ, the update date mode can be described as:

$$w^C \longleftarrow w^C + \theta\nabla_{w^C}\mathcal{L}\left(w^C\right) \tag{23}$$

The DDPG is presented in Algorithm 1.

Algorithm 1 Deep deterministic policy gradient

Input: Initialize Q-function and policy, clean out replay buffer $\mathscr{D}$;
Define objective parameters for Q-function and policy in θ and ϕ;
Loop;
    Based on current observation, generate state-action pair $(\boldsymbol{s}, \boldsymbol{a})$;
    Practice action in the environment;
    Obtain the reward $\boldsymbol{r}$ and move to next state $\boldsymbol{s}'$, check the ending signal $\boldsymbol{e}$;
    Append $(\boldsymbol{s}, \boldsymbol{a}, \boldsymbol{r}, \boldsymbol{s}', \boldsymbol{e})$ to relay buffer $\mathscr{D}$;
    if $\boldsymbol{s}'$ is the last state and ending signal $\boldsymbol{e}$ is true;
        for training episodes **do**;
            Import transients $\mathscr{T}(\boldsymbol{s}, \boldsymbol{a}, \boldsymbol{r}, \boldsymbol{s}', \boldsymbol{e})$ from $\mathscr{D}$;
            Solve the objectives by transients
            $O(\boldsymbol{r}, \boldsymbol{s}', \boldsymbol{d}) = r + \gamma(1 - e)Q_{obj}\left[s', P_{obj}(s')\right]$;
            Obtain updated Q-function
            $\nabla \frac{1}{|\mathscr{T}|} \sum_{(\boldsymbol{s}, \boldsymbol{a}, \boldsymbol{r}, \boldsymbol{s}', \boldsymbol{e}) \in \mathscr{T}} (Q(\boldsymbol{s}, \boldsymbol{a}) - O(\boldsymbol{r}, \boldsymbol{s}', \boldsymbol{d}))^2$;
            Obtain updated policy
            $\nabla \frac{1}{|\mathscr{T}|} \sum_{\boldsymbol{s} \in \mathscr{T}} Q_{obj}\left(s', P_{obj}(s')\right)$;
            Update objective networks' weight
            $\boldsymbol{w}^C \longleftarrow \boldsymbol{w}^C + \theta \nabla_{\boldsymbol{w}^C} \mathscr{L}(\boldsymbol{w}^C)$;
            end for;
        end if;
    until reward convergence.

5. Case Study

5.1. Simulation Settings

Table 1 demonstrates the trading price between the utility grid and MG. Table 2 and Figure 3 present system parameters, including the power generation of PV and WT, the load demand, and the electricity price from the utility grid side. Each column has 24 rows, and each row indicates the corresponding parameter in one hour, which means the sampling period time is one hour and the whole period of simulation is 24 h. The parameters in Table 2 are highly determined by realistic weather and load demand. This case study considers a deterministic scenario on a sunny day. PV output reaches a peak in the afternoon and WT randomly generates electricity fluctuating around its average value. Load demand has two peaks during both daytime and night. Market price increases when load demand is high. Otherwise, electricity sells in a cheap way when a demand valley is shown. Through this simplified MATLAB simulation model, the effectiveness of the DDPG algorithm is demonstrated. The program of the model made use of the Reinforcement Learning Toolbox in MATLAB to train the Q-agent and policy. In addition, to make comparisons of simulation results, YALMIP/GUROBI, which is a commercial optimization solver, is practiced for processing the nonlinear programming model. Here, we further explain the profits and optimality of the proposed algorithms.

Table 1. Purchases and sell price.

Electricity Purchase (AUD/kWh)	Electricity Sells (AUD/kWh)
1.1	0.85

Table 2. Input parameters.

PV (kW)	WT (kW)	Load (kWh)	Price (AUD/kWh)
0	50	50	0.434137
0	50	60	0.42391
0	50	60	0.42
0	50	54	0.426393
0	40	40	0.448192
10	30	40	0.503548
25	30	65	0.598517
30	40	79	0.63099
50	45	100	0.650717
60	45	120	0.654483
65	50	110	0.661257
75	45	77	0.645917
75	45	70	0.626667
70	55	68	0.633886
70	55	75	0.639901
60	60	90	0.647722
40	65	117	0.667376
27	60	125	0.683024
10	55	130	0.7
0	55	125	0.673981
0	60	130	0.623744
0	60	90	0.558902
0	55	80	0.493186
0	55	75	0.456695

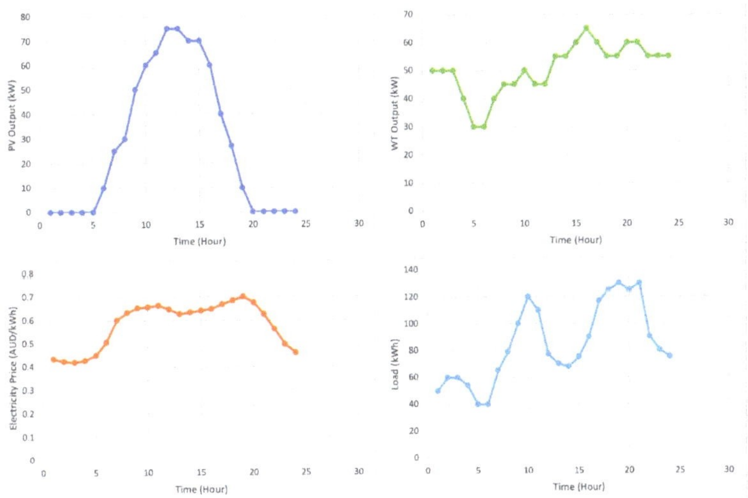

Figure 3. Profiles of input parameters.

5.2. Simulation Results and Discussion

From the beginning, Figure 4 demonstrates the curve of the reward function. The graph shows that after around 550 episodes of training, the agent's reward infinitely approaches zero. As a consequence, the agents are effectively trained, and the simulation results are effective. Part of the code can be shown in Algorithm 1.

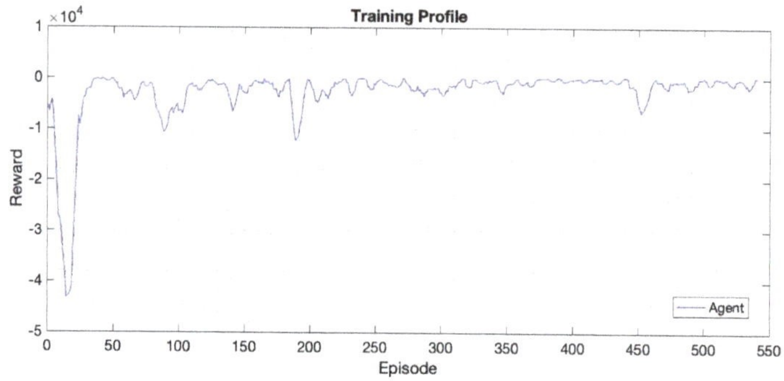

Figure 4. Training profile.

Having illustrated the realism and validity of the simulation, Figure 5 shows the power output profiles of ESS, DG, load, and utility grid. The load profile shows that, during the daytime, electricity demand keeps rising from 6 to 10 am, resulting in the first peak value at 125 kW. Then, demand descends to 75 kW at 1 pm, before it quickly lifts to the second peak value, which is 130 kW at 7 pm. The peak value remains for 2 h and demand decreases to 70 kW at the end of the day. Regarding the power exchange between the MG and the utility grid, most of the time, MG sells electricity to the utility grid and purchases electricity from the utility grid when the load demand increases to the second peak value during the evening. DG power output rises from midnight for 1 h to 40 kW and slowly drops to 30 kW in 17 h. DG profile then directly drops to zero in one hour and returns to 30 kW at 11 pm.

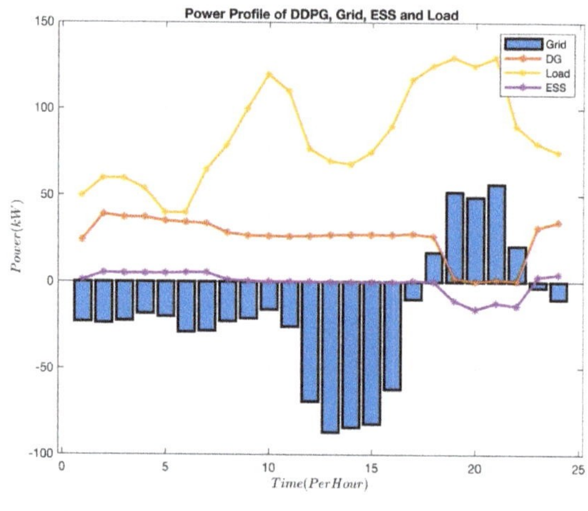

Figure 5. Power profile of DDPG, grid, BESS, and load.

Figure 6 demonstrates the SOC profile of ESS. Considering the SOC profile with the BESS charge and discharge profile, the state of charge starts from 40% and charges to approximately 60% in 0–6 am. Then the BESS module turns to float charging mode, keeping the SOC at 60% until the second load peak comes. From about 18 pm, a dramatic decreasing trend is found. The BESS discharges and supplies the load with the utility grid and RES from 18 pm to 23 pm. Finally, the BESS is slightly charged from RES and grid for 1 h.

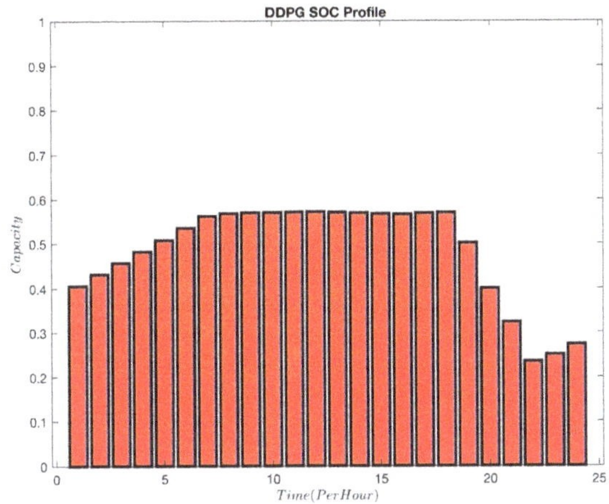

Figure 6. BESS SOC state of charge.

In order to describe the effectiveness of the DDPG algorithm, the charge/discharge profile of ESS, state of charge, and electricity price are plotted on one graph. As shown in Figure 7, the BESS charges when the electricity price is relatively low, resulting in a rise in SOC. On the other hand, the BESS discharges when the price is relatively high, resulting in a decrease in SOC. Such a kind of energy management schedule strategy produced by DDPG lowers the cost of power procurement while increasing the profitability of power sales. Figure 8 demonstrates the cost profile in MG. The system only has 5 h at a high cost. It remains at a low cost for 14 h and has a negative cost, i.e., profit for 5 h. Therefore, the MG with DDPG scheduled EMS can effectively reduce the operating cost.

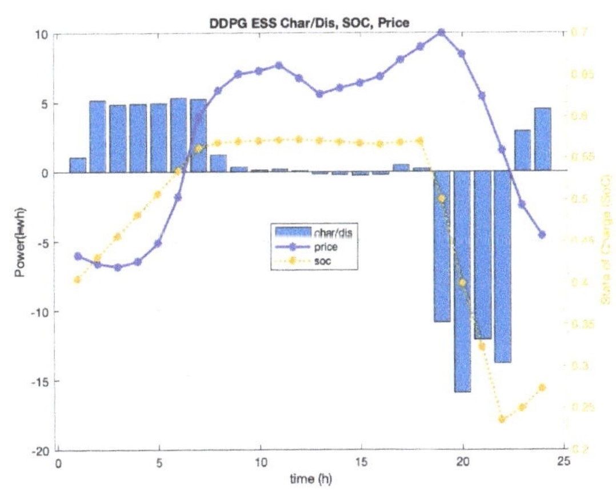

Figure 7. BESS power, SOC, and price profile.

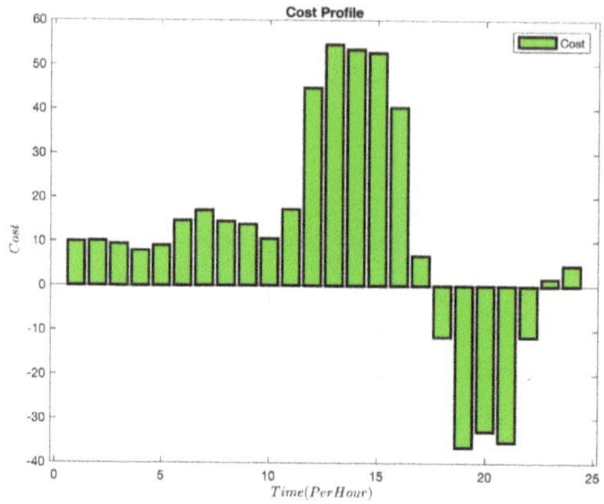

Figure 8. MG cost profile.

5.3. Results Comparision and Analysis

To demonstrate the efficacy of DDPG further, we compared its performance with the GA and the MINLP method, with identical simulation data. MINLP is an adaptable modeling tool for EMS because it can address nonlinear problems with continuous and integer variables. Figure 9 shows the convergence process of the genetic algorithm, illustrating its effectiveness.

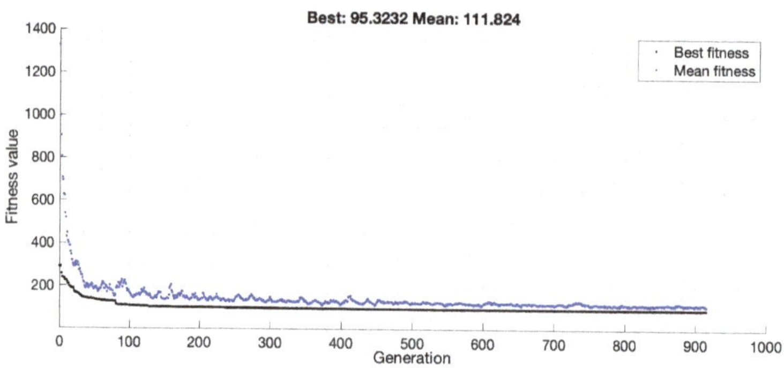

Figure 9. GA convergence process.

Table 3 shows the cost parameters of DG, corresponding to Formula (9).

Table 3. DG cost parameters.

a	b	c
0.001	0.15	77.44

The following three figures demonstrate the difference between DDPG, MINLP, and GA in electricity trading between MG and the utility grid. According to Figure 10a, from 0 am to 17 pm, DDPG has an outstanding performance that sells electricity from MG to the utility grid the whole time, and for the rest of the period, DDPG purchases electricity at an average of 40 kW per hour. Figure 10b shows that MINLP spends half of the whole time period selling electricity from MG to the grid, and the total amount of sold electricity is around 300 kWh, which is the half amount of DDPG. Considering the purchased electricity, MINLP purchased almost the same amount of electricity from the grid as the amount it sells. Figure 10c reveals that from 0 am to 17 pm, GA has two hours for purchase with a lower average power output value than DDPG. Hence, we can conclude that DDPG has the extraordinary capability of selling energy from MG to the utility grid.

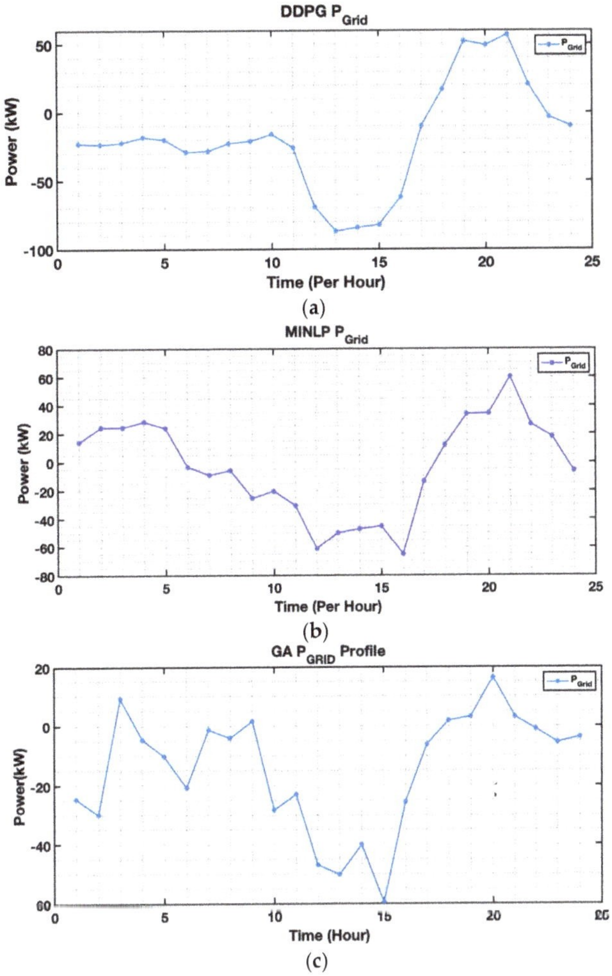

Figure 10. Performance comparison for grid power profile: (**a**) DDPG; (**b**) MINLP; and (**c**) GA grid.

In addition, focusing on the energy storage system, the following six figures demonstrate the BESS performance including charge and discharge power and state of charge, obtained by three introduced algorithms. Tables 2 and 4 present the system constraints, such as the charge and discharge power constraints of the energy storage system, the power output constraints of DG, and the constraints of ESS's state of charge.

Table 4. System Constraints.

P_{ESS}(kW)	P_{DG}(kW)	SOC
[−40, 40]	[0, 60]	[0.2, 0.85]

Based on Figure 11, it is obvious that DDPG stores energy in the BESS when the utility power price is lower in a day, and DDPG discharges ESSs from 6 pm to 10 pm, selling electricity to the grid at the peak value of the price, earning the maximum profit for MG during this period. It is worth noting that the cost of DG was set to 0.5 AUD/kWh, and Figure 3 shows that the electricity price is higher than 0.5 in the period of 6 am to 23 pm. From that perspective, theoretically, DG should remain at its maximum output when its cost is lower than the electricity price, as the results are shown in Figure 11, for the linear programming DG output profile. On the other hand, the BESS should be charged as fully as possible (under the system constraints), which is the most desirable energy management schedule. However, DDPG just slightly charges the BESS for 20% and stops the charging/discharging movement until the second price peak is reached. Additionally, according to Figure 12, we notice that GA has a better performance on both the BESS and DG power profiles. The curve is greatly shaking due to the prediction error, but the charging and discharging depth of the BESS and power output of DG is greater than the corresponding value of DDPG.

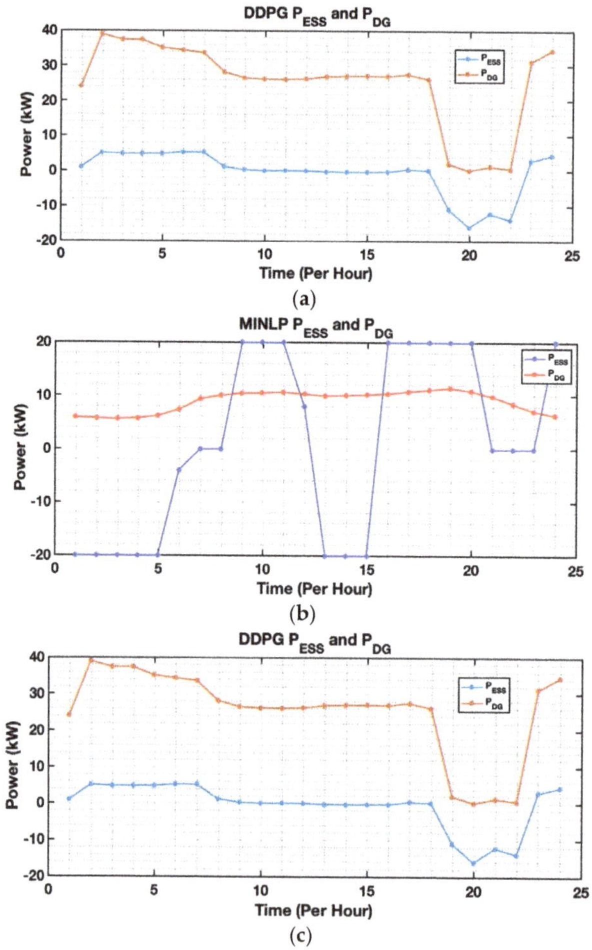

Figure 11. Performance comparison for the BESS and DG output profile: (**a**) DPPG; (**b**) MINLP; and (**c**) GA.

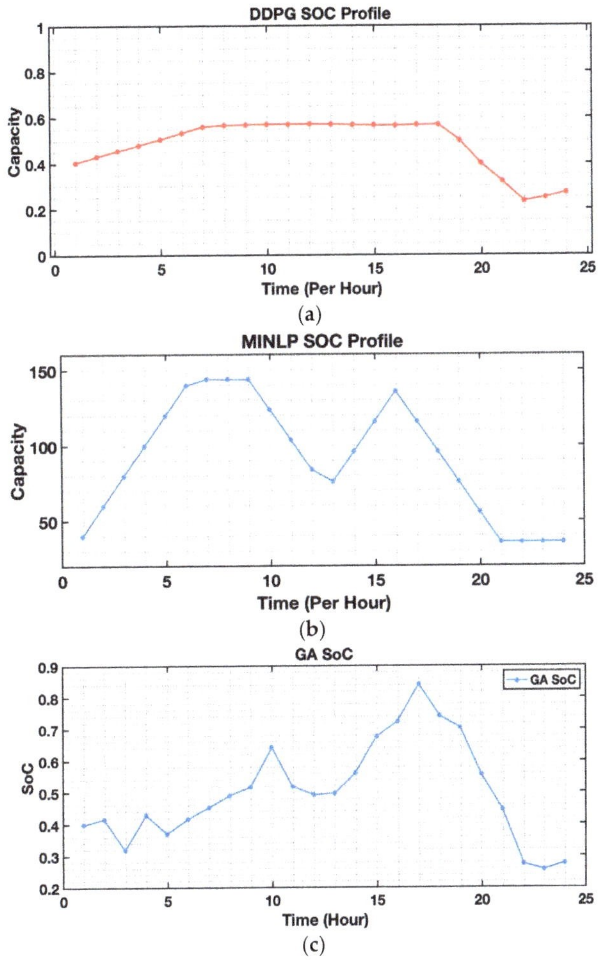

Figure 12. Performance comparison for SOC Profile: (**a**) DDPG; (**b**) MINLP; and (**c**) GA.

However, with the unsatisfied profile produced by DDPG, it can be a kind of advantage that the BESS and DG do not reach their constraints, which results in less degradation cost for long-term operation in MG and reduces the possibility for the happening of safety issues.

Table 5 compares the simulation results of three different techniques. In terms of the overall cost created by EMS, MINLP has the lowest purchasing cost, with a cost of 45.772 AUD per day. GA comes in second, followed by DDPG. However, in terms of computation time, DRL can be processed faster than GA and MINLP.

Table 5. Result comparisons.

	DDPG-Based DRL	MINLP	GA
Purchasing Cost (AUD)	87.333	45.772	88.406
Computation Time (s)	1.202061	2.745373	23.328163

Compared with MINLP and GA, in this small scale and accurately modeled MG, DDPG's exploring strength is not as strong in finding a policy with high accuracy as

MINLP, because during the space searching process, agent actions are effect by system noise. However, if we consider the DDPG in a more realistic scenario, DDPG does have its unique advantages. For example, MINLP can only be implemented on a small-scale MG system, as the large-scale MG with hierarchy architecture will make the mathematical modeling process extremely complex and computationally time-consuming, which is not practical in the real world. In contrast, as DRL is a set of model-free algorithms, DDPG can adapt to any unknown MG structure. Additionally, compared with the GA, DDPG can be quickly used for a new dataset that only spends a very short time for secondary training for its deep neural network. The other two algorithms take a long time for modeling and convergence, respectively. Lastly, DDPG has the capability to handle the task of a high degree of stochastic data, which is very common in RES-based MG. The other two are not compatible due to the same reason. These three unique characteristics of DDPG determine that it is valuable and realistic to reduce the purchasing cost of MG.

6. Conclusions

This paper developed a DRL-based MG energy management system and obtained an energy schedule policy for one day with a sampling time of one hour. EMS policy aimed to reduce the MG electricity cost. The structure of the DRL agent was designed by defining appropriate functions related to the state–action–reward tuple. The DDPG algorithm was adopted to train the DRL agent using a simulated MG. The performance of the DRL agent was compared with the results of the MINLP and GA. This paper revealed both the advantages and disadvantages of the DDPG-based DRL.

The DDPG-based DRL agent for EMS in MGs has the following strengths and drawbacks:

- The DRL agent tries a large number of trial-and-error episodes during training, by which all possible combinations of the BESS charge/discharge schedule, with various initial SoC, are tested. The DDPG optimizes the DRL network to maximize the rewards and thus minimizes the purchasing cost. This training process costs computational costs.
- In an MG with a simple structure and determinant load/weather, the DRL agent would cost more computation time for training compared with GA and MINLP. In the simulated scenario, the DRL achieved higher purchasing costs, but smaller computation time for real-time action.
- The training process of DRL would increase in a large-scale MG system, but after training, the DRL agent reveals superior computation for real-time performance.
- The DRL agent is able to deal with uncertainties that happened in MG, such as stochastic power generation produced by RESs in MG.
- When the EMS is adapted to a new location, e.g., replacing a new database with EMS, the DRL agent can quickly be adapted by training its deep neural network.

In future development, with the aid using the transfer learning principle, GA and MINLP can be used for training the DRL for enhancing its rewards to approach linear programming results.

Author Contributions: Conceptualization, M.E.; Methodology, Z.S., M.E. and M.L.; Software, Z.S., C.Z. and M.L.; Validation, Z.S.; Formal analysis, Z.S., M.E. and C.Z.; Investigation, Z.S.; Data curation, Z.S. and C.Z.; Writing—original draft, Z.S., C.Z. and M.L.; Writing—review & editing, M.E.; Visualization, Z.S.; Supervision, M.E.; Project administration, M.E.; Funding acquisition, M.E. All authors have read and agreed to the published version of the manuscript.

Funding: This research received no external funding.

Conflicts of Interest: The authors declare no conflict of interest.

Abbreviations

BESSs	Battery energy storage system	ESS	Energy storage system
CNN	Conventional neural networks	GA	Genetic algorithm
DDPG	Deep deterministic policy gradient	GPU	Graphical process units
DER	Distributed energy resources	MGs	Microgrids
DG	Diesel generator	MINLP	Mixed integer nonlinear programming
DNN	Deep neural network	PV	Photovoltaic
DRL	Deep reinforcement learning	RESs	Renewable energy resources
DT	Decision tree	SoC	State of charge
DQN	Deep Q-network	TL	Transfer learning
EMS	Energy management system	WT	Wind turbine

References

1. Olivares, D.E.; Mehrizi-Sani, A.; Etemadi, A.H.; Cañizares, C.A.; Iravani, R.; Kazerani, M.; Hajimiragha, A.H.; Gomis-Bellmunt, O.; Saeedifard, M.; Palma-Behnke, R.; et al. Trends in microgrid control. *IEEE Trans. Smart Grid* **2014**, *5*, 1905–1919. [CrossRef]
2. Moradi, M.H.; Eskandari, M.; Hosseinian, S. Operational Strategy Optimization in an Optimal Sized Smart Microgrid. *IEEE Trans. Smart Grid* **2015**, *6*, 1087–1095. [CrossRef]
3. Tabar, V.S.; Jirdehi, M.A.; Hemmati, R. Energy management in microgrid based on the multi objective stochastic programming incorporating portable renewable energy resource as demand response option. *Energy* **2017**, *118*, 827–839. [CrossRef]
4. Asadi, Y.; Eskandari, M.; Mansouri, M.; Savkin, A.V.; Pathan, E. Frequency and Voltage Control Techniques through Inverter-Interfaced Distributed Energy Resources in Microgrids: A Review. *Energies* **2022**, *15*, 8580. [CrossRef]
5. Shi, W.; Li, N.; Chu, C.C.; Gadh, R. Real-time energy management in microgrids. *IEEE Trans. Smart Grid* **2015**, *8*, 228–238. [CrossRef]
6. Eskandari, M.; Li, L.; Moradi, M.H.; Siano, P.; Blaabjerg, F. Optimal voltage regulator for inverter interfaced distributed generation units part I: Control system. *IEEE Trans. Sustain. Energy* **2020**, *11*, 2813–2824. [CrossRef]
7. Eskandari, M.; Savkin, A.V. A Critical Aspect of Dynamic Stability in Autonomous Microgrids: Interaction of Droop Controllers through the Power Network. *IEEE Trans. Ind. Inform.* **2021**, *18*, 3159–3170. [CrossRef]
8. Eskandari, M.; Savkin, A.V. On the Impact of Fault Ride-Through on Transient Stability of Autonomous Microgrids: Nonlinear Analysis and Solution. *IEEE Transactions on Smart Grid* **2021**, *12*, 999–1010. [CrossRef]
9. Uzair, M.; Eskandari, M.; Li, L.; Zhu, J. Machine Learning Based Protection Scheme for Low Voltage AC Microgrids. *Energies* **2022**, *15*, 9397. [CrossRef]
10. Zhou, N.; Liu, N.; Zhang, J.; Lei, J. Multi-objective optimal sizing for battery storage of PV-based microgrid with demand response. *Energies* **2016**, *9*, 591. [CrossRef]
11. Mansouri, M.; Eskandari, M.; Asadi, Y.; Siano, P.; Alhelou, H.H. Pre-Perturbation Operational Strategy Scheduling in Microgrids by Two-Stage Adjustable Robust Optimization. *IEEE Access* **2022**, *10*, 74655–74670. [CrossRef]
12. Rezaeimozafar, M.; Eskandari, M.; Amini, M.H.; Moradi, M.H.; Siano, P. A Bi-Layer Multi-Objective Techno-Economical Optimization Model for Optimal Integration of Distributed Energy Resources into Smart/Micro Grids. *Energies* **2020**, *13*, 1706. [CrossRef]
13. Eskandari, M.; Rajabi, A.; Savkin, A.V.; Moradi, M.H.; Dong, Z.Y. Battery energy storage systems (BESSs) and the economy-dynamics of microgrids: Review, analysis, and classification for standardization of BESSs applications. *J. Energy Storage* **2022**, *55*, 105627. [CrossRef]
14. Zheng, C.; Eskandari, M.; Li, M.; Sun, Z. GA-Reinforced Deep Neural Network for Net Electric Load Forecasting in Microgrids with Renewable Energy Resources for Scheduling Battery Energy Storage Systems. *Algorithms* **2022**, *15*, 338. [CrossRef]
15. Nguyen, T.T.; Nguyen, N.D.; Nahavandi, S. Deep reinforcement learning for multiagent systems: A review of challenges, solutions, and applications. *IEEE Trans. Cybern.* **2020**, *50*, 3826–3839. [CrossRef] [PubMed]
16. Mousavi, S.S.; Schukat, M.; Howley, E. Deep reinforcement learning: An overview. In Proceedings of the SAI Intelligent Systems Conference, London, UK, 7–8 September 2017; pp. 426–440.
17. Hua, H.; Qin, Y.; Hao, C.; Cao, J. Optimal energy management strategies for energy Internet via deep reinforcement learning approach. *Appl. Energy* **2019**, *239*, 598–609. [CrossRef]
18. Khooban, M.H.; Gheisarnejad, M. A novel deep reinforcement learning controller-based type-II fuzzy system: Frequency regulation in microgrids. *IEEE Trans. Emerg. Top. Comput. Intell.* **2020**, *5*, 689–699. [CrossRef]
19. François-Lavet, V.; Taralla, D.; Ernst, D.; Fonteneau, R. Deep reinforcement learning solutions for energy microgrids management. In Proceedings of the European Workshop on Reinforcement Learning (EWRL 2016), Barcelona, Spain, 3–4 December 2016.
20. Du, Y.; Zandi, H.; Kotevska, O.; Kurte, K.; Munk, J.; Amasyali, K.; Mckee, E.; Li, F. Intelligent multi-zone residential HVAC control strategy based on deep reinforcement learning. *Appl. Energy* **2021**, *281*, 116117. [CrossRef]
21. Jin, X.; Lin, F.; Wang, Y. Research on Energy Management of Microgrid in Power Supply System Using Deep Reinforcement Learning. In *IOP Conference Series: Earth and Environmental Science*; IOP Publishing: Bristol, UK, July 2021; Volume 804, Number 3.

22. Yoldas, Y.; Goren, S.; Onen, A. Optimal Control of Microgrids with Multi-stage Mixed-integer Nonlinear Programming Guided Q-learning Algorithm. *J. Mod. Power Syst. Clean Energy* **2020**, *8*, 1151–1159. [CrossRef]
23. Phan, B.C.; Lai, Y.C. Control strategy of a hybrid renewable energy system based on reinforcement learning approach for an isolated microgrid. *Appl. Sci.* **2019**, *9*, 4001. [CrossRef]
24. Zeng, P.; Li, H.; He, H.; Li, S. Dynamic energy management of a microgrid using approximate dynamic programming and deep recurrent neural network learning. *IEEE Trans. Smart Grid* **2018**, *10*, 4435–4445. [CrossRef]
25. Li, J.; Chen, S.; Wang, X.; Pu, T. Research on load shedding control strategy in power grid emergency state based on deep reinforcement learning. *CSEE J. Power Energy Syst.* **2022**, *8*, 1175.
26. Antonopoulos, I.; Robu, V.; Couraud, B.; Kirli, D.; Norbu, S.; Kiprakis, A.; Flynn, D.; Elizondo-Gonzalez, S.; Wattam, S. Artificial intelligence and machine learning approaches to energy demand-side response: A systematic review. *Renew. Sustain. Energy Rev.* **2020**, *30*, 109899. [CrossRef]
27. Kumari, A.; Tanwar, S. A reinforcement-learning-based secure demand response scheme for smart grid system. *IEEE Internet Things J.* **2021**, *9*, 2180–2191. [CrossRef]
28. Lei, L.; Tan, Y.; Dahlenburg, G.; Xiang, W.; Zheng, K. Dynamic energy dispatch based on deep reinforcement learning in IoT-driven smart isolated microgrids. *IEEE Internet Things J.* **2020**, *8*, 7938–7953. [CrossRef]
29. Huang, Y.; Li, G.; Chen, C.; Bian, Y.; Qian, T.; Bie, Z. Resilient distribution networks by microgrid formation using deep reinforcement learning. *IEEE Trans. Smart Grid* **2022**, *13*, 4918–4930. [CrossRef]
30. Zhao, J.; Li, F.; Mukherjee, S.; Sticht, C. Deep Reinforcement Learning based Model-free On-line Dynamic Multi-Microgrid Formation to Enhance Resilience. *IEEE Trans. Smart Grid* **2022**. [CrossRef]
31. Garcia-Torres, F.; Bordons, C.; Ridao, M.A. Optimal Economic Schedule for a Network of Microgrids With Hybrid Energy Storage System Using Distributed Model Predictive Control. *IEEE Trans. Ind. Electron* **2019**, *66*, 1919–1929. [CrossRef]
32. Wei, Q.; Liu, D.; Shi, G. A novel dual iterative Q-learning method for optimal battery management in smart residential environments. *IEEE Trans. Ind. Electron* **2015**, *62*, 2509–2518. [CrossRef]
33. Du, Y.; Li, F. Intelligent multi-microgrid energy management based on deep neural network and model-free reinforcement learning. *IEEE Trans. Smart Grid* **2019**, *11*, 1066–1076. [CrossRef]
34. Rana, M.J.; Zaman, F.; Ray, T.; Sarker, R. Real-time scheduling of community microgrid. *J. Clean. Prod.* **2021**, *286*, 125419. [CrossRef]
35. Mocanu, E.; Mocanu, D.C.; Nguyen, P.H.; Liotta, A.; Webber, M.E.; Gibescu, M.; Slootweg, J.G. On-line building energy optimization using deep reinforcement learning. *IEEE Trans. Smart Grid* **2018**, *10*, 3698–3708. [CrossRef]

energies

MDPI

Article

Renewable Energy Communities in Islands: A Maltese Case Study

Alexander Micallef *, Cyril Spiteri Staines and John Licari

Department of Industrial Electrical Power Conversion, University of Malta, MSD 2080 Msida, Malta
* Correspondence: alexander.micallef@um.edu.mt

Abstract: Renewable energy communities are considered as key elements for transforming the present fossil fuel-based energy systems of islands into renewable-based energy systems. This study shows how renewable energy communities can be deployed in the Maltese context to achieve higher penetration of residential-scale photovoltaic systems. Case studies for five renewable energy communities in the Maltese LV distribution network have been analyzed in detail. A novel community battery energy storage sizing strategy was proposed to determine the optimal storage capacity at each energy community. The main objective of the community battery storage in each REC is to minimize the reverse power injection in the grid (minimize the total reverse energy and reverse peak power values), as well as to reduce the peak evening electricity demand. The optimal sizes for communal BESSs were determined to be of 57 kWh (EC 1), 55 kWh (EC 2), 31 kWh (EC 3), 37 kWh (EC 4) and 10 kWh (EC 5), respectively. The community storage systems were observed to reduce the overall impact of all five energy communities on the grid infrastructure. Power system simulations were performed for a typical spring day to evaluate the impact of communal BESS placement on the node voltages for all five energy communities. The results showed that the community storage was more effective at reducing the node rms voltage magnitudes when deployed at the end of the respective energy communities, rather than at the beginning of the community. During peak generation hours, reductions of up to 0.48% in the node rms voltage magnitudes were observed. This contrasts with reductions of only 0.19% when the community storage was deployed at the beginning of the energy communities.

Keywords: battery sizing; community storage; peak shaving; renewable energy community

Citation: Micallef, A.; Spiteri Staines, C.; Licari, J. Renewable Energy Communities in Islands: A Maltese Case Study. *Energies* **2022**, *15*, 9518. https://doi.org/10.3390/en15249518

Academic Editor: David Macii

Received: 29 November 2022
Accepted: 12 December 2022
Published: 15 December 2022

1. Introduction

The Renewable Energy Directive (2018/2001/EU) of the European Union (EU) [1] aims to make renewables more accessible to citizens, giving them the possibility to engage in joint renewable energy projects. The directive defines citizen-driven renewable energy communities (RECs) as legal entities that can produce, consume, store, and sell renewable energy or provide flexibility services to the grid through demand-response and storage.

Citizens are becoming increasingly environmentally and socially conscious towards energy issues. As a result, there is an increase in demand for more democratic processes in energy policies. Figure 1 summarizes the key elements of these RECs. RECs enable citizens to actively participate in the energy market through the deployment of community-scale storage and data obtained from second-generation smart meters [2]. The aggregation of participants in RECs is also advantageous to the distribution system operators (DSOs) as multiple flexible assets can be accessed through a single reference point. Recent literature on RECs has addressed issues related to the regulatory barriers [3–7], maximization of self-sufficiency [8–11], local energy sharing strategies [12–17] and the interaction of RECs with the power system [3,18].

1.1. Regulatory Challenges of RECs

Di Silvestre et al. [3] provide an in-depth analysis of the European REC experience, focusing on possible regulatory changes, together with technical and financial hurdles.

The authors analyze, in detail, the existing legislation and incentives set up in the EU member states in the wake of RED-II. An overview of the strengths of the RECs as a novel approach to energy management, as well as any critical issues related to their adoption, was also given. Ceglia et al. [4] define the social, economic and technical aspects of RECs, highlighting their advantages with respect to previous EU (and Italian) energy sharing directives. A case study for a REC in Benevento (Italy) was also described through data obtained from the HOMER database. The authors showed that RECs are more advantageous than conventional systems and systems of efficient users (SEU) on social and economic factors. Katsaprakakis et al. [5] provide a model for energy communities in Greek islands that aimed to achieve 100% energy self-sufficiency based on the lessons learned from existing Greek RECs in Sifnos, Crete, and Chalki. The case studies considered various elements, including general information about the existing Greek RECs, local power production (conventional and renewable), energy cooperatives and the Clean Energy transition agenda. The study by O'Neill-Carrillo et al. [6] identified possible practices that could overcome the technical and social challenges leading to the formation of RECs. The governance of the community was deemed as a fundamental pillar for the success of these communities. The three critical aspects that were identified are: sense of community, teamwork, and empowerment. Sense of community implies that any differences between the needs of the individual and the collective should not lead to conflicts within the community. Effective teamwork and coordination are critical to the success of the RECs. These aspects can only be reached if there is a set of rules that people agree on beforehand to collectively manage local resources. Empowerment enables citizens to have the proper tools to participate in the RECs, while ensuring that they are accountable for their decision-making and actions. Chamorro et al. [7] categorize RECs as being either Urban, Rural or Universities. Urban RECs are formed in cities, providing opportunities for sector coupling by combining energy, water, gas, and transportation services. Rural RECs provide inexpensive and reliable electricity supply to households and local businesses, which can also be isolated from the main grid. University RECs are typically living laboratories for green technologies as these emulate small cities. These simulated environments can provide ideal test sites for innovative technologies and management strategies.

1.2. *Energy Optimisation Strategies for RECs*

The maximization of self-sufficiency and self-consumption for prosumers is one of the most researched topics in literature. This is also the case when considering the integration of renewable energy sources and energy storage systems in RECs. Self-consumption (and self-consumption index KPI) defines how much of the local PV production is used locally, while the self-sufficiency (and self-sufficiency index KPI) defines the percentage of energy needs that the local generation can cater for. The authors of [8] developed a multi-variable optimization strategy that minimizes the energy provision costs for a representative REC in Flanders, Belgium. The mixed-integer linear model considers the electricity tariffs, the ratio of electrification of heating and transportation sectors, and the capital expenditure (CAPEX) of renewable energy sources and storage systems. The REC outperformed the other scenarios in the both the financial and environmental KPIs; however, the financial advantages over individual prosumers were not significant. This implies that for a wide adoption of RECs, citizens will need to be suitably incentivized. Similarly, Barone et al. [9] developed a multi-variable dynamic REC simulation model in TRNSYS and applied it to a case study in the island of El Hierro (Canary Islands, Spain). The simulation tool is aimed at achieving energy independence for remote islands, rather than focusing on its applicability to generalized objectives. A limitation of this study is that the solar PV generation was based on weather data from a nearby location, while the consumption data were obtained from real world data. Doroudchi, Khajeh and Laaksonen [10] maximize the self-sufficiency of RECs by introducing community and distributed storage to minimize the energy exchange with the utility grid. Excess generation from the community-owned solar PV generation is stored as thermal energy in both community scale and distributed electric heat energy storage. The study showed that the payback period for the community scale storage is less than the distributed electric heat energy storage, even though the payback period depends on the usage. However, one should note that the sizing of the communal and distributed thermal storage was not a result of the problem formulation but was determined arbitrarily. Cielo et al. [11] proposed an optimization procedure to size community solar photovoltaic systems with integrated energy storage. The optimization procedure is based on the maximization of both self-consumption and self-sufficiency of the RECs. The sizing strategy is, however, heavily skewed towards economic REC models and frameworks. When applied to an Italian context, internal return rates of approx. 11% were obtained, highlighting the attractiveness of exploiting the REC concept.

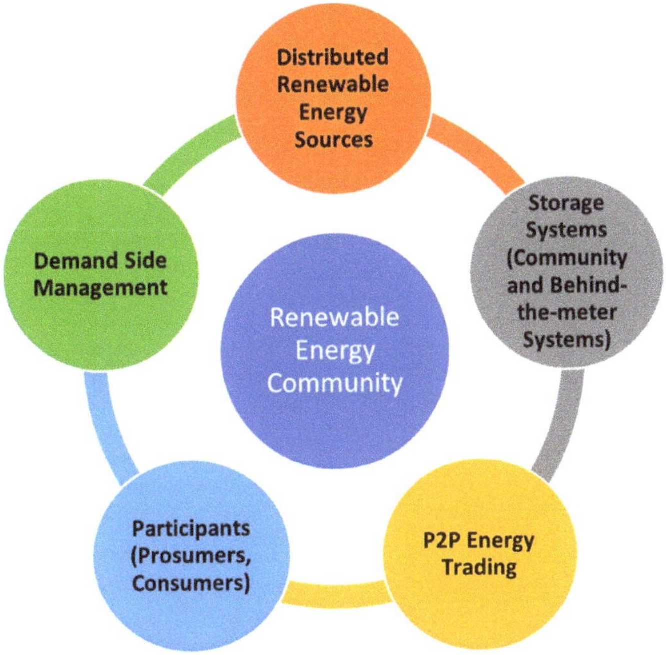

Figure 1. Block Diagram of the key elements in a Renewable Energy Community.

1.3. Citizen Driven Energy Markets

An important feature for RECs is the possibility of energy exchange between all the participants (utility, prosumers, and consumers). The authors of [12] proposed a local energy sharing strategy with a focus on price-forming methods. The proposed methods can be integrated into a net-billing system and adopted for different regulatory set-ups. This strategy can be used to improve local supply-demand balancing, reduce voltage deviations, and improve social welfare. However, the interactions and implications for the utility power system are not clear. Dolatabadi, Siano and Soroudi [13] proposed a real time optimization algorithm that preserves the privacy of prosumers within an REC. Similarly, Di Silvestre et al. [3] show how Blockchain technology can contribute to the development of RECs by enabling P2P energy trading with secure transactions while protecting personal data. The algorithm assesses the impact of energy exchanges within the community for the provision of ancillary services to the utility. The energy exchanges within the REC were considered as virtual self-consumption by the prosumers, together with the provision of ancillary services. However, this strategy would limit the market participation by consumers that cannot own local renewables and/or storage systems. The authors of [14] proposed a peer-to-peer (P2P) energy trading system for RECs that would enable each prosumer to manage their energy consumption, storage scheduling and energy trading. The proposed community-based P2P energy trading system combines an online energy control with a double auction trading algorithm. The authors of [15,16] consider the scheduling of community energy storage to maximize the usage by all consumers while accounting for operational constraints. The scheduling heuristic makes use of a day-ahead auction approach to share the resources within the REC by making use of a time-of-use (TOU) tariffs. Aziz, Dagdougui and Elhallaoui [17] applied mean field game theory to determine Nash equilibrium strategies for an REC that includes 100 prosumers (with solar photovoltaic systems), located in Montreal city. The REC is formed through a virtual power bank that contains the distributed BESSs of the prosumers. The solutions for the community microgrid optimization minimize household individual cost functions and, in turn, reduces the aggregate cost by at least 40%.

1.4. Interaction of RECs with the Main Grid

While there are several studies analyzing the interaction of integrating battery energy storage at utility level (high-voltage and medium-voltage) with the utility grid, the topic of how RECs affect the electricity network has not yet received due attention in the literature [3]. The integration of storage at utility level (high-voltage and medium-voltage) is beyond the scope of this study, as RECs in the Maltese context are being envisaged at the LV network. The authors of [3] discuss how the formation of RECs will impact present interactions of prosumers with the electric power system by focusing on the Italian scenario. The authors also raised a number of questions, that remain to be addressed, for the effective integration of RECs into power systems, including the number of points of common coupling (PCC) with the main grid, the assessment of the impact on the main grid, the capability to provide ancillary services, and the capability to support the widespread adoption of demand response, amongst others. Sudhoff et al. [18] consider the reduction in the peak power exchange between the community and the electric grid through the prosumer PV and battery systems, together with flexible loads. A case study comparing rural, urban, and suburban RECs showed that the establishment of RECs resulted in less energy being required from outside the community. The study also shows that there is no minimum REC size to offer ancillary services to the grid using the local available assets.

1.5. Renewables in the Maltese Islands

The penetration of renewable energy sources in Malta has been increasing over the years, despite the demographic and spatial limitations of the Maltese islands. During 2021, 9.9% of the electricity supply in Malta was obtained from renewable sources (mainly photovoltaics). The domestic sector accounted for 93.6% of the total stock of solar photovoltaic (PV) installations and 46.1% of the total energy production [19]. The scenario depicted in Figure 2 compares the present Maltese RES share to that of other EU countries. However, the challenges that must be overcome to achieve higher penetrations of RES in the Maltese distribution network should not be taken lightly. The RECs provide an attractive solution to overcome some of the existing challenges.

The rest of the paper is organized as follows. Section 2 describes the methodology used for the study reported in this paper. Section 3 contains a description of the five considered renewable energy communities in the Maltese LV distribution network. Section 4 describes the proposed battery sizing strategy that determines the minimum BESS based on the defined battery utilization factor (BUF). Section 5 describes a case scenario for the five RECs on a typical spring day, showing the impact of the placement of the communal BESS on the voltage profiles of the energy communities. A summary of the obtained simulation results is given in Section 6 on the implications of this study.

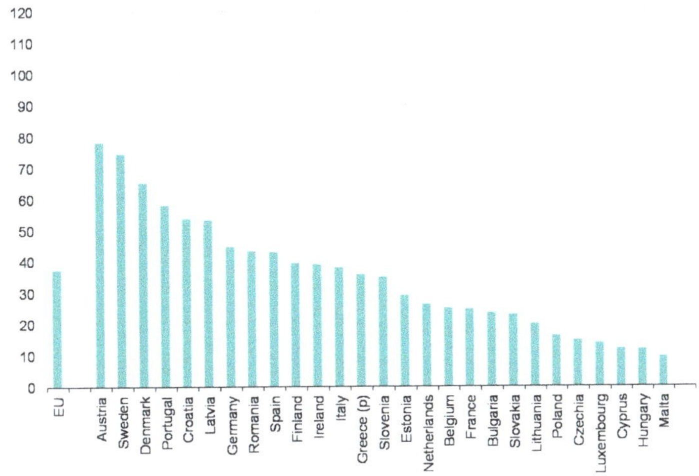

Figure 2. Share of Energy from Renewable Sources in the EU in Gross Electricity Consumption. Source: Eurostat (online data code: nrg_ind_ren).

2. Methodology

This study shows how RECs could be deployed in the Maltese context to improve the penetration of RES. The centralized community storage in the RECs was aimed at increasing the self-sufficiency of the RECs. In addition, the reduction in the peak electricity demand and minimization of the reverse power flows into the utility grid was also addressed. This study also proposes a community-scale battery energy storage system (BESS) sizing strategy, based on the battery utilization factor. This strategy was previously evaluated by the authors of [20] for utility scale storage and adapted to community scale applications. The methodology used in this paper is summarized in Figure 3. The workflow is subdivided into three main parts as follows:

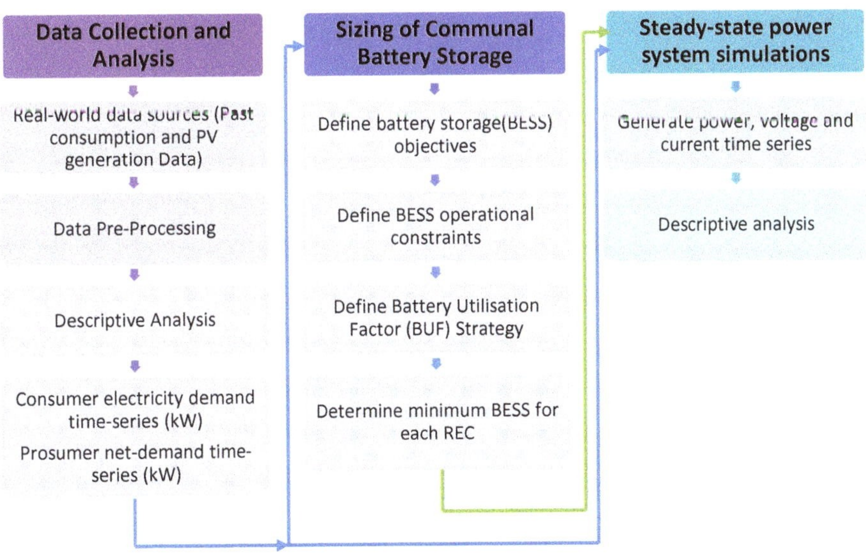

Figure 3. Methodology applied in this study to analyze the potential of RECs in reducing the reverse power flows and reducing the evening peak demand.

(1) collection and analysis of consumption and PV generation data (Section 3)

During the initial stages, the topologies for renewable energy communities in the Maltese context were defined. Data collection and the analysis of the real-world consumption and PV generation data were performed. The aggregated results grouped according to each respective REC are summarized in the next section.

(2) sizing of communal battery storage systems (BESSs) (Section 4)

The historical electricity consumption and PV generation data were used as an input to the battery utilization factor (BUF) sizing strategy. The BUF was used to determine the minimum BESS for each respective REC for the functionality and operational constraints described in this study.

(3) power system simulations to show RECs can improve the local voltage profiles (Section 5)

Steady-state power system simulations for a typical Maltese spring day were then used to produce the voltage profiles of each respective REC. The consumer and prosumer electricity net-demand profiles specify the net active power across each node in the REC, except at the substation transformer. A power-flow solution was implemented in MATLAB/Simulink to obtain the node voltages of each REC, over a 24 h period and with a 15 min resolution.

Problem Formulation and Constraints

As has been defined in the previous sections, the main objective of the community BESS in each REC is to minimize the reverse power injection in the grid (minimize the total reverse energy and reverse peak power values), as well as to reduce the peak evening electricity demand. Therefore, the objective of the communal BESSs aims to minimize the power losses at the PCC of each REC. The objective function can be defined by:

$$P_{loss} = \min\left\{\sum_{t=1}^{t_{total}} \sum_{i,j=1,i\neq j}^{n_t} g_{i,j}\left(V_i^2 + V_j^2 - 2\,V_iV_j\cos(\theta_i - \theta_j)\right)\right\} \tag{1}$$

where t is the time interval; t_{total} = 96 is the total time period (with 15 min resolution); $g_{i,j}$ is the conductance between buses i and j; n_t is the total number of branches in the REC; V_i and V_j are the voltage magnitudes of the buses i and j; and θ_i and θ_j are the voltage angles of buses i and j.

The BESS systems were sized according to the methodology defined in Section 4. The following assumptions and constraints were considered for the BESS sizing and the analysis carried out in the next sections:

- The power balance equation at the PCC of each REC is defined by:

$$\mathrm{P}_t^{\mathrm{grid}} = \mathrm{P}_t^{\mathrm{netload}} + \mathrm{P}_t^{\mathrm{charge,BESS}} - \mathrm{P}_t^{\mathrm{discharge,BESS}} \tag{2}$$

where $\mathrm{P}_t^{\mathrm{grid}}$ is the power from/to the grid; $\mathrm{P}_t^{\mathrm{netload}} = \mathrm{P}_t^{\mathrm{distrbutedPV}} - \mathrm{P}_t^{\mathrm{distirbutedLoad}}$ is the difference between the distributed PV generation and the load consumption; $\mathrm{P}_t^{\mathrm{charge,BESS}}$, $\mathrm{P}_t^{\mathrm{discharge,BESS}}$ are the power charged and discharged by the BESS.
- The BESS round-trip energy efficiency consisting of the battery charging and discharging efficiencies (includes that of the power electronic converter) is assumed constant at 85%. The efficiency of the power electronic converter is assumed to remain constant for any output power from the BESS. Therefore, the charging efficiency, $\eta^{\mathrm{charge,bess}}$, is of 92.2% and the discharging efficiency, $\eta^{\mathrm{discharge,bess}}$, is of 92.2%.
- The BESS SoC is constrained to within the range between an $\mathrm{SoC}_t^{\mathrm{min,bess}}$ of 20% and an $\mathrm{SoC}_t^{\mathrm{max,bess}}$ of 80% to prolong the lifetime, i.e.:

$$\mathrm{SoC}_t^{\mathrm{min,bess}} \leq \mathrm{SoC}_t^{\mathrm{actual,bess}} \leq \mathrm{SoC}_t^{\mathrm{max,bess}} \tag{3}$$

where $\mathrm{SoC}_t^{\mathrm{actual,bess}}$ is the actual battery SoC at any time of the day.

- The BESS useable energy capacity is constrained to use the energy available within the range as defined by the SoC constraints, i.e.:

$$E_t^{\text{min,bess}} \leq E_t^{\text{actual,bess}} \leq E_t^{\text{max,bess}} \tag{4}$$

where $E_t^{\text{actual,bess}}$ is the available battery energy at any time of the day.

- The maximum discharge power $P_t^{\text{discharge,bess}}$ was limited to a maximum value of 1C ($P_{MAX}^{\text{discharge,bess}}$):

$$P_t^{\text{discharge,bess}} \leq P_{MAX}^{\text{discharge,bess}} \tag{5}$$

- The maximum charging power $P_t^{\text{charge,b}}$ was limited to a maximum value of $P_{MAX}^{\text{discharge,bess}}$:

$$P_t^{\text{charge,bess}} \leq P_{MAX}^{\text{charge,bess}} \tag{6}$$

- The actual energy stored in the BESS, $E_t^{\text{Actual,bess}}$, is determined by adding the net-energy of the BESS in the previous time step, $E_{t-1}^{\text{Actual,bess}}$, to the energy charged in the BESS, $E_t^{\text{charge,bess}}$, and subtracting the discharged energy, $E_t^{\text{discharge,bess}}$. The energy conservation equation of the BESS is defined by:

$$E_t^{\text{Actual,bess}} = E_{t-1}^{\text{Actual,bess}} + E_t^{\text{charge,bess}} - E_t^{\text{discharge,bess}} \tag{7}$$

- The peak shaving operation is only functional if the evening maximum demand $P_t^{\text{demand,REC}}$ at the grid operators PCC exceeds the pre-defined maximum limit of the respective REC. In these cases, the BESS discharges to reduce the peak demand according to the available energy in the BESS:

$$P_{MAX}^{\text{demand,REC}} \leq P_t^{\text{demand,REC}} \tag{8}$$

- The BESS is assumed to be discharged at the start of the analysis (initial SoC of 20%).
- The self-discharge rate was considered negligible.
- Other battery-specific characteristics were not considered.

3. Energy Communities in a Maltese Context

Figure 4 shows the simplified line diagram of a typical secondary substation in the Maltese LV distribution network. The substation has 182 nodes, with 192 single-phase consumers and 14 three-phase consumers. The consumers are divided across five feeders connected to the secondary (LV) side of a 250 kVA, 11,000/400 V, 50 Hz, Dyn11 substation transformer with off-load tap changer. The off-load tap setting is set to −5%, such that the voltage levels at the end of the feeders have a suitable voltage level all year round. The effective turns ratio is set to 10,450/400 V to satisfy the limits defined in the Enemalta Network Code [21].

RECs at a LV network level can take various shapes and sizes. In this study, it is assumed that the energy community (EC) configurations are a direct result of how the consumers are presently connected to the LV network. Each LV feeder was considered as a separate REC. In this way, RECs can be formed without the need to reconfigure the present distribution network. Therefore, each EC consists of different combinations of single-phase and three-phase consumers/prosumers.

The distribution of the customers connected at each feeder is given in Table 1. The net-demand profiles for each of the single-phase and three-phase consumers were obtained from the local DSO in the form of a single spreadsheet for a one-year period (from 1 May 2020 to 31 April 2021) with a 15-min time resolution. Table 1 additionally shows the subset of consumers that also have a functional residential-scale PV system installed (i.e., the prosumers in the RECs). The PV generation profiles for each of the residential-scale PV systems were also obtained in the same format from the local DSO. An important limitation is that most of the single-phase main meters did not have any logs for the exported active power. Therefore, the relationship between the PV generation and exported active power had to be analyzed and approximated for each individual consumer. The single-phase PV systems installed capacities range between systems of 1.38 kWp and 4.38 kWp, while the three-phase PV systems range between 1.84 kWp and 10.8 kWp. The total installed PV capacity in each of the RECs are as follows: 52.72 kW (REC 1), 48.56 kW (REC 2), 37.87 kW (REC 3), 29.07 kW (REC 4) and 12.78 kW (REC 5).

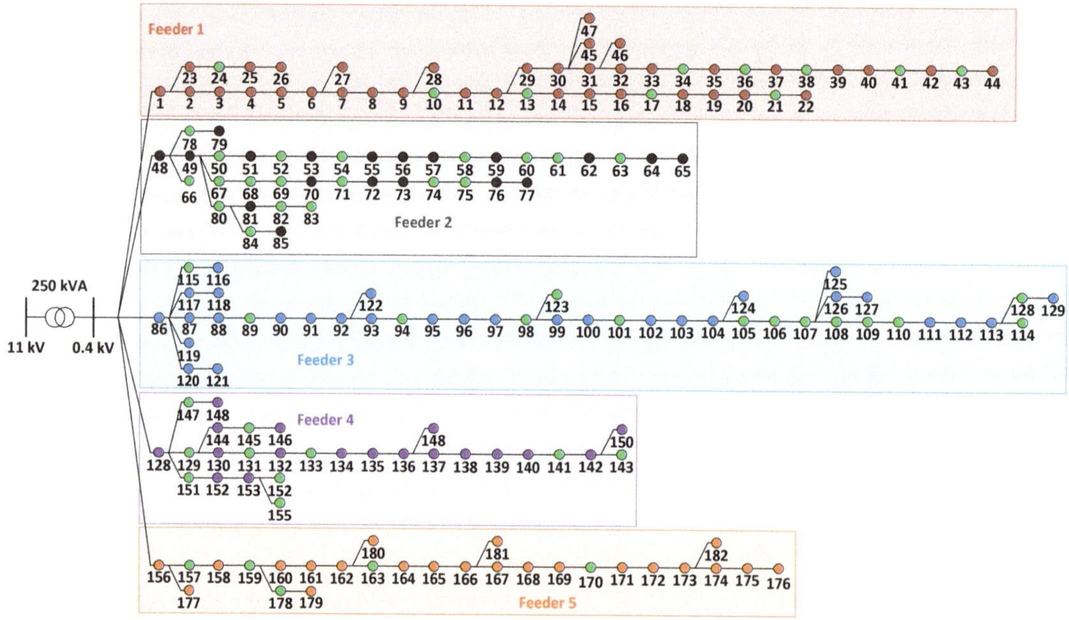

Figure 4. Simplified line diagram of the LV network at secondary substation. Feeder nodes are color coded for ease of reference (Red—Feeder 1, Black—Feeder 2, Blue—Feeder 3, Purple—Feeder 4, Orange—Feeder 5). Nodes where PV systems are connected are shown in green. Substation transformer is showing the nominal ratio without the off-load tap changer settings. The numbers in the figure identify the individual nodes of the LV network.

Analysis of the Net-Demand

Figure 5 illustrates the daily net-demand curves for each of the RECs. The estimated REC net-demand characteristics were determined from the data measured by the smart meters installed at the respective consumers/prosumers. This estimate of the total net-demand per REC gives a good approximation but it does not consider the distribution losses. Detailed analysis of the daily net-demands of all five RECs shows that oversupply occurs on more than 340 days for RECs 1 to 4, while for REC 5, oversupply occurs on 265 days. Oversupply was always observed to occur during the middle of the day as this coincides with the peak of PV generation.

One of the primary concerns resulting from the reverse power flow is the significant voltage rise that can occur along the feeder. The feeder voltage could potentially increase enough to violate the ±10% steady state tolerance defined in the Enemalta network code [21]. The voltage levels of the other feeders in the LV network can also be affected by the reverse power flowing back to the substation transformer. An additional concern is that, on a larger scale, the reverse power flow could lead to dynamic stability issues. During instances of reverse power, there is a lack of system inertia, which is an issue when coupled with the intermittency risks associated with PV generation.

Table 1. Consumers and PV systems connected to the secondary substation LV network.

REC (Feeder)	Phase	Single-Phase Consumers	Single-Phase PV Systems	Three-Phase Consumers	Three-Phase PV Systems	Maximum Feeder Lengths
1	A B C	10 12 19	4 4 2	9	3	542.466 m
2	A B C	12 15 9	7 6 5	1	-	221.6 m
3	A B C	16 25 17	4 7 2	3	1	623.16 m
4	A B C	5 12 13	2 3 4	1	1	302.074 m
5	A B C	1 13 13	- 4 1	-	-	621.852 m

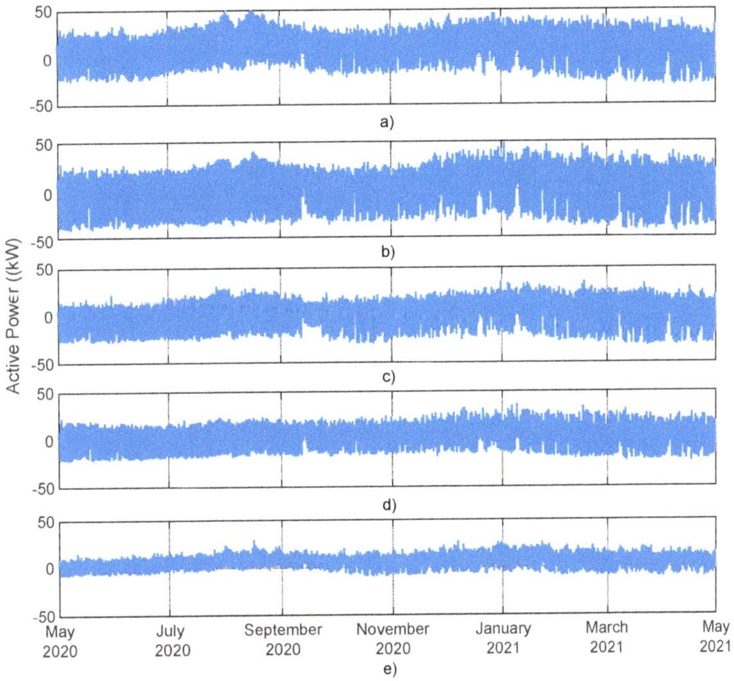

Figure 5. Daily Net-Demand Curve for each of the eCs during the considered period. (**a**) EC1 (Feeder 1). (**b**) EC2 (Feeder 2). (**c**) EC3 (Feeder 3). (**d**) EC4 (Feeder 4). (**e**) EC5 (Feeder 5). Positive active power values for the net-demand represent the power consumption by the consumer loads while negative active power values represent the reverse power flow due to the local PV generation.

One must note that this study was performed during the COVID-19 pandemic. The COVID-19 pandemic has affected every aspect of life, including the operation of the utility grid due to changes in the energy usage patterns of residential, commercial, and public entities. According to a study published by the Malta National Statistics Office [19], the total electricity supplied in 2020 amounted to 2496.4 GWh, which is a decrease of 5.4% when compared to the previous year.

4. Community BESSs for the RECs

Community battery energy storage systems (BESSs) have a typical storage capacity ranging between tens of kilowatt-hours (kWh) and a few megawatt-hours (MWh). BESSs are an effective and energy efficient solution to limit reverse power flows in the LV distribution network. The community-scale BESSs can be sized according to a wide variety of site performance requirements. In this paper, the community-scale BESSs are deployed at specific nodes in the REC and controlled to reduce the power exchanges with the medium voltage network (11 kV). Two locations for the integration of BESS into the RECs were investigated in this study: at the start of the REC and at the end of the REC. Figure 6 shows an example of the possible locations that were identified for EC5. These siting constraints were imposed by the typical densely populated Maltese towns/villages that do not enable the deployment of central community BESSs at the intermediate nodes.

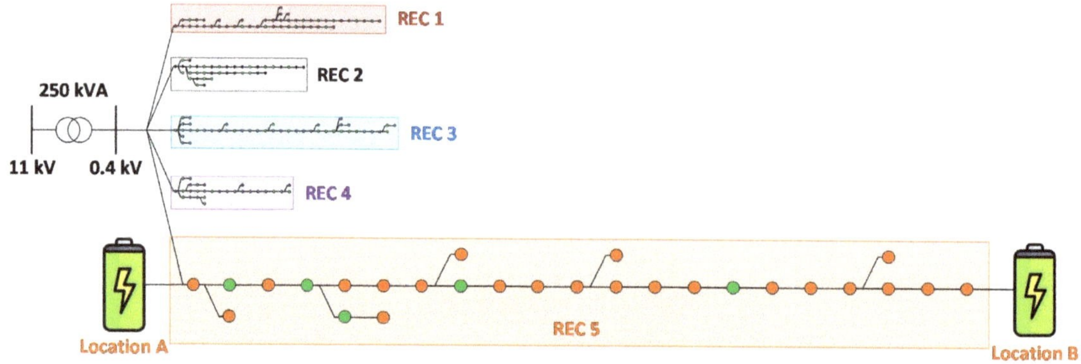

Figure 6. Possible locations for Centralized Community Storage Systems for EC5. Location A is the start of the REC. Location B is at the end of the REC.

4.1. *Community BESS Sizing Strategy*

Sizing strategies that are available in the literature typically result in a trade-off between the near-term cost and long-term technical complexity. In this study, the battery utilization factor (BUF), previously defined in [20], was used to size the BESSs at each respective REC. The BUF is determined by the daily variations in the state of charge (SoC) (i.e., cycling of the BESSs) as the BESS operates daily to minimize the power exchanged with the grid. The daily variations in the SoC of the BESSs is affected by the operating conditions, including the charge and discharge rates, depth of discharge, cycle duration, and length of time in the standby mode [20]. The BUF for the community-scale BESS at each respective REC can be defined by:

$$\mathrm{BUF} = \frac{\sum_{n=1}^{365}\left(\frac{W_1 T_{Charge_n}}{T_{ReversePower_n}}\right)\left(\frac{W_2 T_{Discharge_n}}{T_{PeakDemand_n}}\right)}{\text{Number of days}} \quad (9)$$

where n = 1, … , 365 are the days over the considered period; W_1 is the weighting during the charging periods based on the daily SoC variations during charging; W_2 is the weighting during the discharging periods based on the daily SoC variations during discharging. The daily charging duty cycle for the nth day is defined as the ratio of the actual charging time (T_{Charge} in minutes) to the reverse power flow duration ($T_{ReveresePower}$ in minutes). T_{Charge} is defined as the time duration required to charge the battery to full capacity each day while $T_{ReveresePower}$ is the daily time interval during which reverse power flows back to the substation transformer. The daily discharging duty cycle for the nth day is defined as the ratio of the actual discharging time ($T_{Discharge}$ in minutes) to the duration of the EC evening peak ($T_{PeakDemand}$ in minutes). $T_{Discharge}$ is the daily time duration required to discharge the battery to the minimum SoC while $T_{PeakDemand}$ is the time interval during which the maximum power at the EC exceeds the pre-set daily limit.

4.2. *BUF and Optimal Community BESS (Present PV Penetrations)*

The flowchart in Figure 7 describes the BESS sizing procedure. The BUF calculation, as defined in (9), was used as a discrete computational procedure to determine the optimal BESS solution for each of the RECs. The BESS optimization process for the RECs is independent of the objective function being optimized. This was conducted by initially defining for each REC: $P_{MAX}^{charge,bess}$; the lower BESS capacity bound, C_{MIN}^{bess}; an upper BESS capacity bound, C_{MAX}^{bess}; and the number of iterations by the step size, C_{step}^{bess}. The evaluated BESS capacity, C_x^{bess}, is initially set to the lower capacity bound and the BUF is determined by the methodology shown in Figure 7. The C_x^{bess} is then incremented with each iteration of the BUF algorithm. Each iteration gives a BUF point on the curve that can be plotted on a BUF vs. BESS capacity curve. An example of the BUF vs BESS capacity curve is shown in Figure 8. The maxima resulting from this curve yields the BESS capacity with the maximum utilization factor. Due to the weighted formulation of (9), there can only be one maximum for each BUF vs BESS capacity curve. The BUF was determined for BESS systems rated between 1 kWh and 200 kWh (in steps of 1 kWh) at a $P_{MAX}^{charge,bess}$ of 0.1C up to 0.5C (in steps of 0.1C). The resulting BUF vs BESS capacity curves are shown in Figure 8. The $P_{MAX}^{demand,REC}$ daily maximum power limits for each respective REC were set as a constant throughout the entire year according to the pre-determined thresholds for the daily peak demands at each EC. These were determined to be 10 kW (EC1 to EC4) and 7 kW (EC5), respectively.

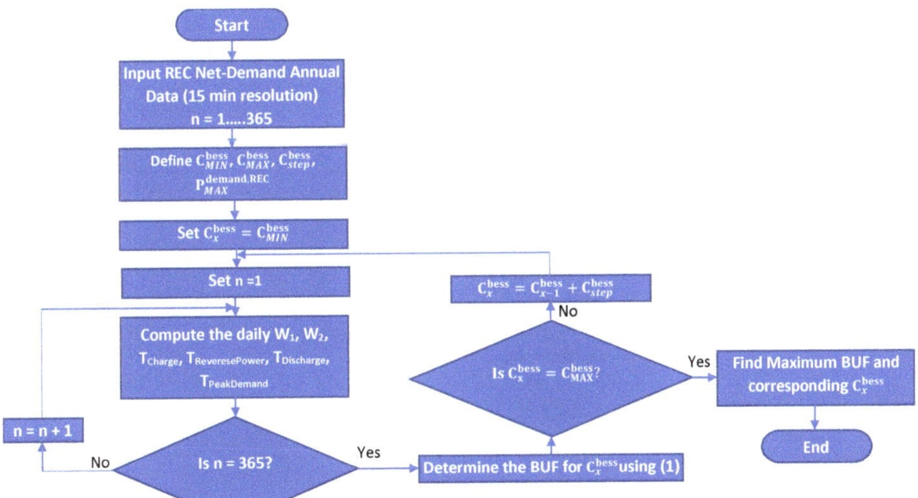

Figure 7. Flowchart of the BESS sizing procedure using the BUF Strategy (n = 1, . . . , 365 is the day number.

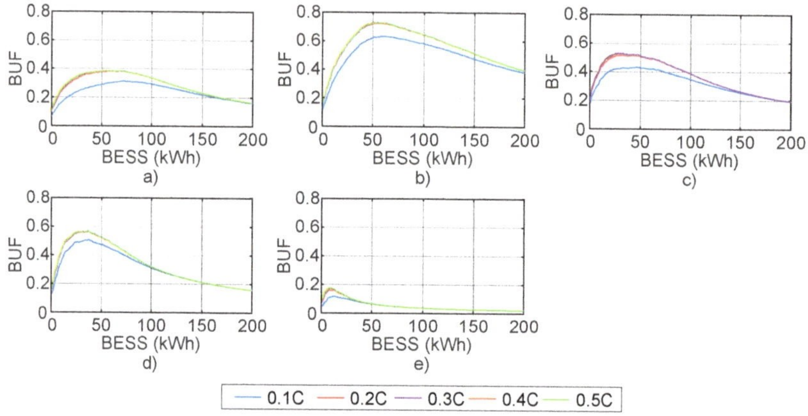

Figure 8. Battery Utilization Factor for BESS systems rated from 1 kWh to 200 kWh at charging rates of 0.1C up to 0.5C. (**a**) BUF for EC 1. (**b**) BUF for EC 2. (**c**) BUF for EC 3. (**d**) BUF for EC 4. (**e**) BUF for EC 5.

An analysis of Figure 8 reveals that the BUF characteristic are directly dependent on the daily net-demand characteristics (including the reverse power flows and peak evening load demands). In all of the RECs, the BUF was observed to be affected by low values of $P_{MAX}^{charge,bess}$, with the worst performance occurring at 0.1C. However, the improvements observed in the BUF for $P_{MAX}^{charge,bess}$ above 0.3C were negligible. The lowest BUF from all ECs can be observed at EC5 as the BESS can only charge during the reverse power periods. If the peak load demand does not exceed the threshold, the BESS does not discharge, resulting in extended periods of idle time. The highest BUF was observed for EC2 that has high reverse power flows and peak demands that are of similar magnitudes. The optimal sizes for communal BESSs (at a charging rate of 0.3–0.5C) were determined to be of 57 kWh (EC 1), 55 kWh (EC 2), 31 kWh (EC 3), 37 kWh (EC 4) and 10 kWh (EC 5), respectively.

4.3. *Energy Community Net Demand with BESSs*

The hourly net-demand curves for the each of the RECs with the addition of community storage are considered in this section. Only the results obtained with the determined $BESS_{BUF}$ are given in this section for each of the RECs. For EC1, additional results were included for a BESS twice the optimal value to compare the operational performance. The simulations were performed to evaluate the impact of the $BESS_{BUF}$ on the reverse power flow and total energy demand at each REC. In addition, the daily SoC variations for the community scale BESS are also given.

4.3.1. Energy Community 1

Figure 9a shows the hourly net-demand curves with and without the $BESS_{BUF}$ of 57 kWh (showing the average power over 15-min intervals in kW), together with the daily SoC for EC1 over the considered year. The total energy demand in the EC without storage is of 111.3 MWh, while the total energy flowing back to the substation amounts to 17.5 MWh. The total energy demand in the EC with $BESS_{BUF}$ is of 102.98 MWh, while the total energy flowing back to the substation amounts to 8.19 MWh. The BESS is not large enough to eliminate the reverse power flows on most days. One can observe that the BESS is cycled nearly daily between the predefined SoC limits. An exception occurs in the period between mid-July and mid-September due to lower magnitudes of reverse power flow resulting from higher electricity demands. In addition, one can also observe that the BESS does not always have sufficient energy to limit the evening peak demand to the pre-defined set point.

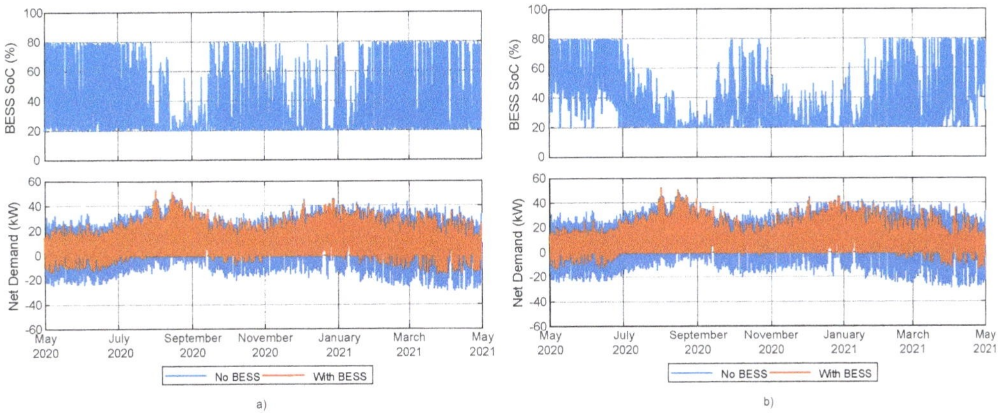

Figure 9. Net demand with and without storage and daily variations in the State of Charge for EC1. (**a**) With a 57 kWh BESS. (**b**) With a 114 kWh BESS.

Increasing the size of the BESS to 114 kWh results in further reductions of the reverse power flow. With this larger BESS, the total energy flowing back to the substation reduces to 4.66 MWh. The total energy demand reduces to 99.7 MWh as more energy was supplied by the BESS during peak shaving operation on most days where the demand exceeded the defined set-point. Figure 9b shows the hourly net-demand curves with and without storage, together with the SoC variations for EC1. The BESS works in daily microcycles on days where: (a) the absorbed reverse power flow exceeds the discharged energy during the evening peak demand periods; and (b) the absorbed reverse power flow is much less than the evening peak demand periods. While these microcycles have the advantage of increasing the lifetime of the battery, this comes at significantly higher capex and spatial footprint requirements. The latter is an extremely critical aspect due to the densely built Maltese environment, which limits the deployment of distributed large-scale BESSs.

4.3.2. Energy Community 2

Figure 10a shows the hourly net-demand curves with and without the $BESS_{BUF}$ of 55 kWh (average power over 15-min intervals in kW), together with the SoC variations for EC2, over the entire year. The total energy demand in the EC without storage is of 77.23 MWh, while the total energy flowing back to the substation amounts to 52.3 MWh. The total energy demand in the EC with $BESS_{BUF}$ is of 67.47 MWh (reduction of 12.64%), while the total energy flowing back to the substation amounts to 41.47 MWh (reduction of 20.7%). One can observe that the BESS is cycled nearly daily between the predefined SoC limits, except in the period between May and July, when the electricity demand was low and the PV generation was high. However, the BESS is not large enough to eliminate the reverse power flows on all days. In addition, the BESS does not always have sufficient energy to limit the evening peak demand to the pre-defined set point.

4.3.3. Energy Community 3

Figure 10b shows the hourly net-demand curves with and without a 31 kWh BESS (average power over 15-min intervals in kW), together with the SoC variations for EC3, over the considered period. The total energy demand in the EC without storage is of 59.65 MWh, while the total energy flowing back to the substation amounts to 32.78 MWh. The total energy demand in the EC with $BESS_{BUF}$ is of 55.2 MWh (reduction of 7.46%), while the total energy flowing back to the substation amounts to 27.8 MWh (reduction of 15.19%). One can observe that the BESS is cycled nearly daily between the predefined SoC limits, except in the period between early-May and the end of June, when the daily maximum electricity demand is very close to the defined setpoint. However, the BESS is not large enough to eliminate the reverse power flows on all days. In addition, the BESS does not always have sufficient energy to limit the evening peak demand to the pre-defined set point.

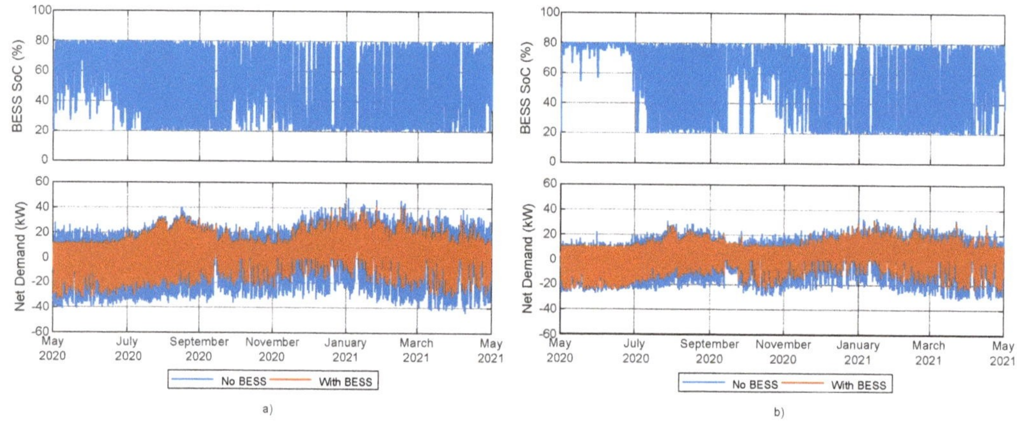

Figure 10. Net demand with and without storage and daily variations in the State of Charge. (**a**) EC2 with a 55 kWh BESS. (**b**) EC3 with a 31 kWh BESS.

4.3.4. Energy Community 4

Figure 11a shows the hourly net-demand curves with and without a 37 kWh BESS (average power over 15-min intervals in kW), together with the SoC variations for EC4, over the considered period. The total energy demand in the EC without storage is of 56.01 MWh, while the total energy flowing back to the substation amounts to 23.42 MWh. The total energy demand in the EC with $BESS_{BUF}$ is of 50.75 MWh (reduction of 9.39%), while the total energy flowing back to the substation amounts to 17.57 MWh (reduction of 25%). One can observe that the BESS is cycled nearly daily between the predefined SoC limits, except in the period between the beginning of May and mid-July. From Figure 11a, one can observe that the BESS is effective in limiting the maximum demand to the required reference of 10 kW for all days in this period. In this EC, the BESS is not large enough to eliminate the reverse power flows on all days. In addition, the BESS does not always have sufficient energy to limit the evening peak demand to the pre-defined set point, except in the period between the beginning of May and mid-July.

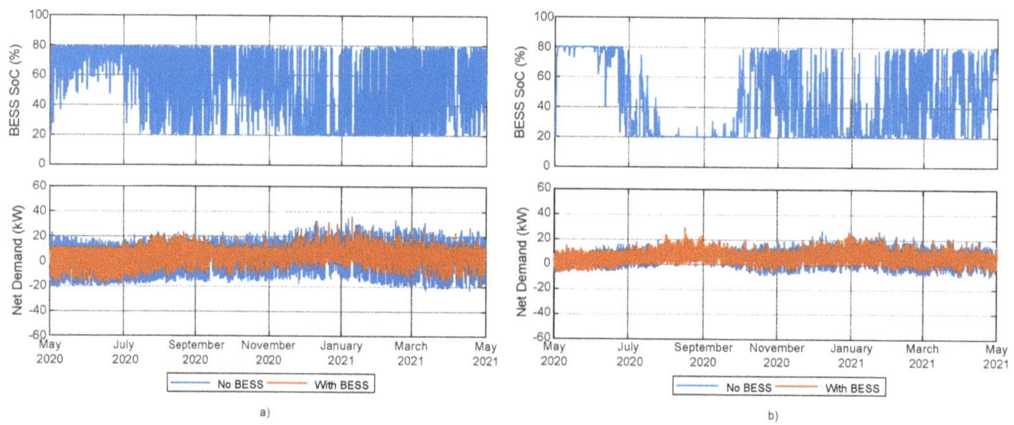

Figure 11. Net demand with and without storage and daily variations in the State of Charge. (**a**) EC4 with a 37 kWh BESS. (**b**) EC5 with a 10 kWh BESS.

4.3.5. Energy Community 5

Figure 11b shows the hourly net-demand curves with and without a 10 kWh BESS (average power over 15-min intervals in kW), together with the SoC variations for EC5, over the considered period. The total energy demand in the EC without storage is of 55.05 MWh, while the total energy flowing back to the substation is the lowest of all ECs, at 2.78 MWh. The total energy demand in the EC with $BESS_{BUF}$ is of 54.3 MWh (reduction of 1.36%), while the total energy flowing back to the substation amounts to 1.94 MWh (reduction of 30.2%). Even though the size of the BESS is small, the BESS is underused throughout the entire year. During May and June, the electricity demand is lower than the maximum threshold, resulting in the BESS operating at the maximum SoC of 80%. Hence, the BESS could not reduce the reverse power flows during this period. In the late summer months, the electricity demand increases due to the cooling requirements (air-conditioning loads). This results in no reverse power flow and, hence, the BESS operates at the minimum SoC of 20%. Therefore, one can clearly conclude that the BESS was underused between May and September 2020. Between October 2020 and April 2021, the reverse power flows and the evening peak demands were reduced as the BESS was cycled on a frequent basis. Increasing the size of the BESS to values higher than the $BESS_{BUF}$ at the present levels of PV generation would yield even further underutilization. Assuming that the electricity demand of REC5 does not increase, energy independence from the utility grid can be achieved by increasing both the PV generation and the BESS size.

5. Case Study: Typical Maltese Spring Day

The secondary substation distribution network shown in Figure 4 was modelled in MATLAB/Simulink. The cables and overhead lines were represented as distributed Pi-transmission line models. Technical data on each cable segment included the cable type per segment, segment lengths, and the electrical parameters per unit length (resistance, inductance, and line-to-ground capacitance) were provided by the local DSO to achieve a detailed schematic. These detailed cable parameters are critical to evaluating the effects of high periods of PV generation on the node voltage profiles of each energy community.

Electrical parameters of the secondary substation transformer were also included for these simulations. The primary side of the substation transformer acts as the slack bus to balance the active power and reactive power in the modelled network. This bus acts as a reference to the simulation model and is the only known voltage at the start of the simulation. Single-phase and three-phase load buses were implemented at each node to model the respective consumer loads and local PV generation.

Power flow simulations were carried out for two grid placement scenarios to evaluate the impact of communal BESS placement on the node voltages of all five energy communities. The first location is at the start of the feeder (Location 1) as the BESSs can be located within the substation or in its vicinity (i.e., the most accessible location). The second location is at the end of the feeder (Location 2), where the reduction in the energy community node rms voltages is expected to be more significant. However, in practice, there might be practical limitations that could not allow the deployment of communal BESSs at Location 2 (refer to Section 5).

5.1. Spring Net-Demand Characteristics

Spring is the best performing season for PVs in Malta due to a combination of cooler ambient temperatures, low electricity demand for heating/cooling, high irradiation levels and high levels of sun-hours [20]. On the other hand, winter has a higher electricity demand resulting from space/water heating and the lowest levels of sun hours, when compared to the other seasons.

The hourly net demand curves without community BESS for 8 May 2020 are shown in Figure 12. One can observe that the net demand curves for all ECs on this day follow the duck curve characteristic. Oversupply occurs during the middle of the day, as this coincides with the peak of PV generation. This negative net-demand occurs between, approximately,

8 am and 6 pm for all ECs. EC2 has the highest reverse power of all the ECs at −23.8 kW, with the reverse peak being twice the evening peak demand.

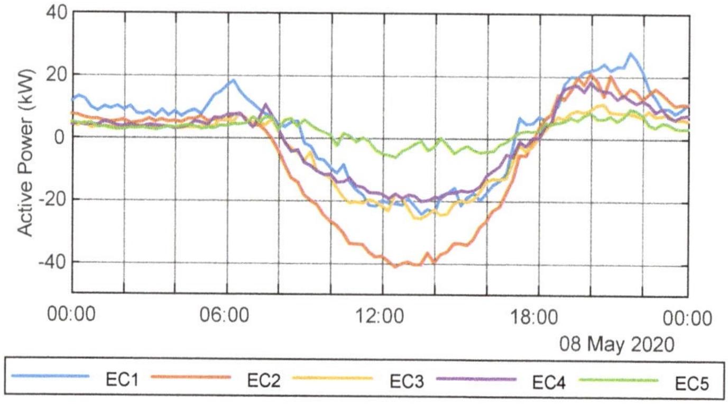

Figure 12. The hourly net demand curves for all ECs on 8 May 2020.

5.2. EC Net-Demand Characteristics with Community BESS

The hourly net demand curves with community BESS for 8 May 2020, together with the SoC variations for each EC, are shown in Figure 13. The net-demand curves are obtained from a snapshot taken of the simulations carried out in Section 4.3. Therefore, the initial SoC for each EC is determined from the operation of the community BESS on the previous days. The initial SoC values for EC1 to EC5 are 20%, 45%, 79.2%, 58.2%, and 76.6%, respectively. One can immediately deduce that, for EC3 and EC5, the difference between the maximum SoC and the initial SoC significantly limits the capacity to reduce the reverse power flows for the considered day.

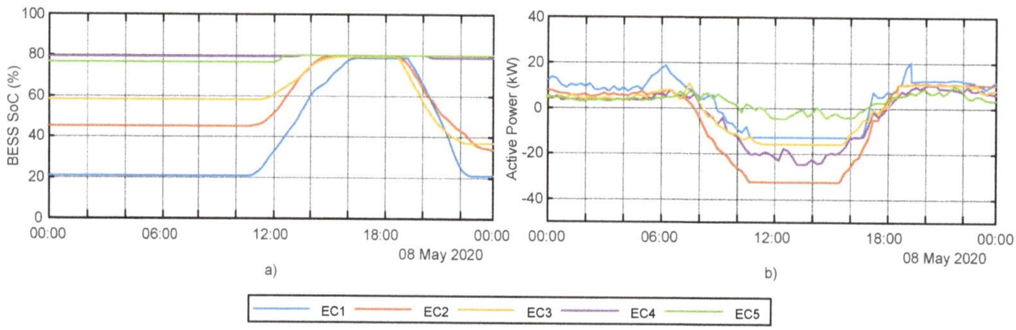

Figure 13. Net demand of the ECs with community storage for the selected case scenario. (**a**) The SoC variations of the BESS at the respective ECs. (**b**) The resulting net-demand at each EC.

Figure 13a shows the SoC variations for each BESS, whereby all of the community batteries reach the predefined limit of 80% during instances of reverse power flows. Figure 13b shows that EC2 has the highest reverse power of all the ECs, at −13.1 kW. For all of the ECs, the peak demands are limited to the predefined maximum power setpoints. An exception to this result occurs for EC1, where part of the peak demand is not compensated by the BESS as there was not enough capacity available. As expected, EC3 and EC5 showed small reductions in the reverse power flows due to the high initial SoC. These results clearly show that there is a compromise between the consumer's objective to maximizing self-consumption and the utility's objectives of maintaining a minimum base load and reducing the reverse power flows.

5.3. Impact of Communal BESS Placement on the EC Nodes

Simulations were then performed to verify the effect of the communal BESS placement on the voltage profile at each EC. Three case studies were simulated: (a) No communal BESS; (b) Communal BESS placed at the start of EC (substation secondary side); and (c) Communal BESS placed at the end of EC (furthest node from substation secondary side). Figure 14a shows a swarm plot of the phase (line-to-neutral) RMS voltages of all the nodes present in REC 1, at 15 min intervals throughout the day, without the communal BESS. From the figure, one can observe that there are overvoltage events due to the reverse power flow between 2:00 p.m. and 1:30 p.m. This corresponds to the period where solar PV generation reached a maximum. Violations of the 230 V +10% maximum limit were observed on multiple nodes in the REC.

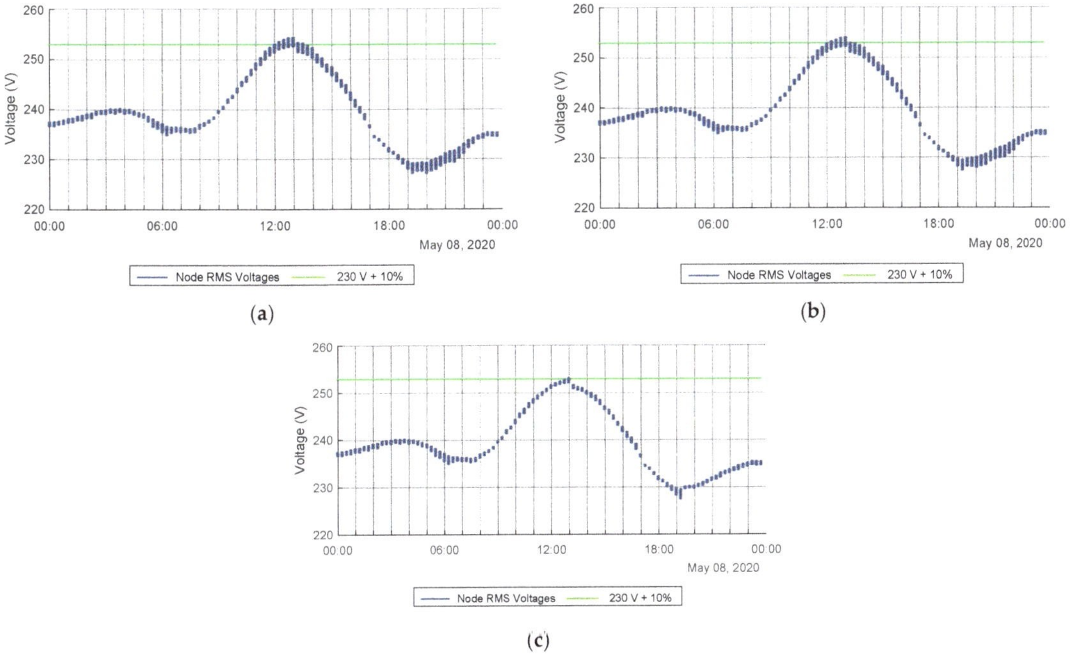

Figure 14. Swarm plot of the node phase voltages in REC 1 at 15-min intervals throughout the day. The green line represents the 230 V + 10% maximum limit (line-to-neutral). Each point is a resultant phase voltage at one of the nodes in the LV network. (**a**) No communal storage. (**b**) Communal storage placed at start of REC. (**c**) Communal Storage placed at the end of the REC.

Figure 14b shows a swarm plot of the phase (line-to-neutral) RMS voltages of all the nodes present in REC 1, at 15 min intervals throughout the day, with the communal BESS at the start of the REC. From the figure, one can observe that there are no significant changes in the voltage distributions over the whole day, as violations of the 230 V + 10% maximum limit were observed on multiple nodes in the REC. Figure 14c shows a swarm plot of the phase (line-to-neutral) RMS voltages of all the nodes present in REC 1, at 15 min intervals throughout the day, with the communal BESS at the end of the REC. From the figure, one can observe that there are reductions in the maximum voltages at the nodes, with violations of the 230 V + 10% maximum limit only occurring at 1 pm. In addition, there were also reduced voltage variations across all nodes in the network over the entire day (This is shown by smaller clusters in the swarm plot).

Similar plots were obtained for all five of the RECs and are not included here for the purpose of clarity. However, the resulting voltage variations at each node of the five RECs that can be determined from these swarm plots are summarized in the bar graph of Figure 15. The horizontal line (purple) represents the 230 V + 10% maximum voltage limit for each of the five ECs. Without the communal BESS, this maximum voltage limit is exceeded in multiple nodes of nearly all of the ECs, with the exception of EC2. The most severe voltage magnitudes occurred in EC 3, where all the nodes exceeded the maximum voltage threshold. The resulting maximum voltages for EC1 to EC5 are 254.45 V, 253.25 V, 255.17 V, 253.75 V and 253.54 V, respectively. The minimum voltages that result during the peak power demand for EC1 to EC5 are 227.19 V, 228.87 V, 228.11 V, 228.23 V and 226.45 V, respectively.

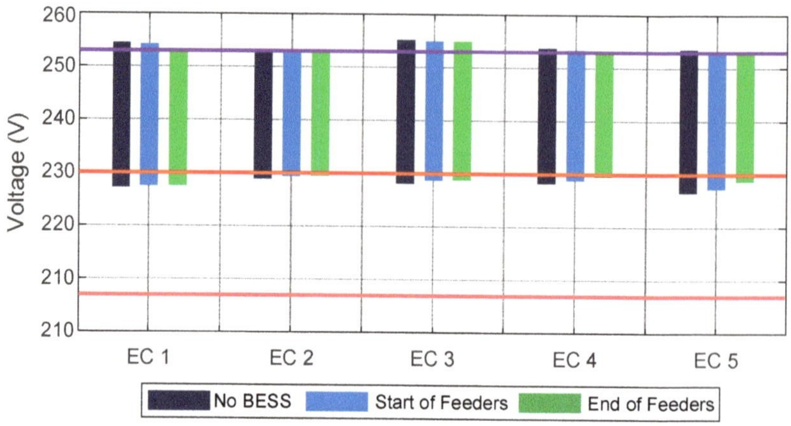

Figure 15. Bar graph of node voltage variations for each EC with and without the community BESS. The horizontal lines represent the 230 V ± 10% nominal voltage range for the Maltese LV network: Maximum limit (Purple), Nominal voltage (Red) and minimum voltage (Pink).

With the communal BESSs installed at the start of the EC, a reduction in the magnitude and duration of the reverse power flow at the substation was observed. EC 3 and EC 5 show the least reductions in reverse power flow as the initial SoC was close to the defined maximum threshold. in ECs where the initial SoC was low (e.g., EC 1 and EC 2), the operation of the communal BESS reduced the voltage magnitudes, however the maximum voltage limit was still exceeded in nearly all ECs, with the exception of EC 2. The resulting maximum voltages for EC1 to EC5 are 254.16 V (−0.14%), 253 V (−0.1%), 254.9 V (−0.11%), 253.37 V (−0.15%) and 253.05 V (−0.19%), respectively. Due to the reduction in power consumption during the peak demand, the minimum voltages for EC1 to EC5 increased to 227.51 V (+0.14%), 229.4 V (+0.23%), 228.71 V (+0.26%), 228.74 V (+0.22%) and 227.22 V (+0.34%), respectively.

With the communal BESSs installed at the end of the EC, the reductions in the magnitude and duration of the reverse power flow resulted in the lowest maximum voltages in the ECs. Similarly to the previous case scenario, EC 3 and EC 5 show the least reductions in reverse power flow as the initial SoC was close to the defined maximum threshold. in ECs where the initial SoC was low (e.g., EC 1 and EC 2), the operation of the communal BESS reduced the voltage magnitudes. In this scenario, the maximum voltage limit was only exceeded in EC 1 and EC 3. The resulting maximum voltages for EC1 to EC5 are 253.24 V (−0.48%), 252.76 V (−0.19%), 254.9 V (−0.11%), 252.9 V (−0.33%) and 252.67 V (−0.34%), respectively. Due to the reduction in power consumption during the peak demand, the minimum voltages for EC1 to EC5 increased to 227.53 V (+0.15%), 229.43 V (+0.25%), 228.77 V (+0.29%), 229.47 V (+0.54%) and 228.62 V (+0.96%), respectively.

6. Discussion

As mentioned in the previous sections, the challenges of achieving high penetrations of renewables in the Maltese distribution network, and in island power systems in general, should not be taken lightly. The formation of RECs provides an attractive solution to efficiently use any available assets by locally consuming the energy generated from renewables. The analysis for the five hypothetical RECs carried out in this study shows that, with the present penetration of renewables, the substation transformer sees an annual net-demand of 359.24 MWh, with the total reverse power injected in the grid amounting to 128.78 MWh. The maximum evening peak electricity demand at the transformer reaches 153 kW, while the reverse peak reaches a maximum of −125 kW. Deploying the energy storage units sized according to the BUF strategy ($BESS_{BUF}$) within the RECs resulted in a 7.95% reduction in the total energy demand from the grid during peak demand hours, and 16.9% reduction in the total energy supplied back to the grid during midday. The maximum evening peak electricity demand at the transformer was reduced to 78.2 kW, while the reverse peak reaches a maximum of −53.5 kW. Therefore, the installed PV capacity within the five RECs can be increased by up to 7.4% from the present installed capacity, while reaching the present reverse energy injection levels into the grid. This assumes a typical specific yield of 1626 kWh/kWp for PV systems in the Maltese islands [22].

By reducing their daily peak electricity demand and the reverse power flow into the utility grid, RECs can minimize their impact on the utility grid through increased self-sufficiency. The case study for a typical Maltese spring day showed that the communal BESS can also provide advantages within the same RECs. When installed at the end of the RECs, the communal BESS reduced the node rms voltages magnitudes by up to 0.48% by absorbing the reverse power flow. The degree of reduction in reverse power flow depends on the initial SoC of the batteries. Larger community storage systems could be deployed if additional ancillary services are provided to the grid in order to minimize the payback period and maximize profits. The ancillary services could include power quality improvement, regulation, and flexibility reserve.

Limitations

One should note that modelling assumptions were performed that might affect the results. As already mentioned in Section 3, most single-phase main meters in households where PV systems were installed did not have any logs for the exported active power. Therefore, the relationship between the PV generation and the exported active power for this study had to be analyzed and approximated for each individual consumer. This might affect the potential energy reductions for the modelled RECs.

Another assumption of this study concerns the controllability of the storage units within the REC. BESSs control the real power injection/absorption symmetrically on all three phases. However, in practice, all the three phases might not be available at the last node of the REC. An individual phase can directly feed several consumers as it is significantly cheaper to deploy single phase conductors than three phases (+neutral). In this context, only the partial load on that phase of the REC can be compensated by the BESS. Normally, there are no connections among phases (through power electronic converters) to allow for intra-phase power transfer. In addition, the charging/discharging rates must be limited to avoid overloading the conductors.

7. Conclusions

Renewable energy communities in the Maltese context can provide significant advantages to the citizens and the DSO. This study investigated the coordinated operation of the energy storage assets within the community, aiming at reducing the peak power exchanged between the REC and the main grid. Each LV feeder in the considered secondary substation was considered as an REC, such that RECs can be formed without the need to reconfigure the present distribution network. Each REC consists of different combinations of single-phase and three-phase consumers/prosumers. An analysis of the daily net-demand curves for each REC revealed that oversupply occurs on more than 340 days for RECs 1 to 4, while for REC 5 oversupply occurs on 265 days. Community storage was proposed in this study as a means to reduce the peak power exchanged by each REC with the grid. The proposed sizing strategy for the community BESS is based on the battery utilization factor, which uses historical data for both demand and PV generation and ensures maximum utilization of the storage assets. The optimal sizes for the communal BESSs were determined to be of 57 kWh (EC 1), 55 kWh (EC 2), 31 kWh (EC 3), 37 kWh (EC 4) and 10 kWh (EC 5), respectively. Detailed analysis of the daily net-demand curves with and without storage for the each of the RECs showed reductions in the energy demand and reverse power flow during peak PV hours. The communal BESSs in the RECs are cycled nearly daily between the predefined SoC limits, except in certain periods of the year where the electricity demand is higher than the norm or when it is very close to the defined reference value.

Finally, power flow simulations were carried out on a typical spring day for two grid placement scenarios (start and end of the feeder, respectively). The impact of the communal BESS placement on the node voltages of all five energy communities was determined through a detailed analysis of the node phase rms voltages of each REC. The battery placement was seen to play a part in the benefits to the energy community itself. When the BESS was placed at the end of the EC, the voltage violations of the maximum limit were observed only for two communities out of the five modelled (EC 1 and EC 5), with resulting maximum voltages for EC1 to EC5 of 253.24 V (−0.48%), 252.76 V (−0.19%), 254.9 V (−0.11%), 252.9 V (−0.33%) and 252.67 V (−0.34%), respectively.

Author Contributions: Conceptualization, A.M.; methodology, A.M., C.S.S. and J.L.; software, A.M.; validation, C.S.S. and J.L.; formal analysis, A.M.; investigation, A.M., C.S.S. and J.L.; resources, A.M.; data curation, A.M.; writing—original draft preparation, A.M.; writing—review and editing, C.S.S. and J.L.; visualization, A.M.; project administration, A.M. and C.S.S. All authors have read and agreed to the published version of the manuscript.

Funding: This research was financed by the Energy and Water Agency under the National Strategy for Research and Innovation in Energy and Water (2021–2030), grant agreement reference number ESTELLE EWA 110/20/2/001-C.

Data Availability Statement: The original data used for the simulations are not publicly available.

Acknowledgments: The authors would like to acknowledge Enemalta PLC for providing the data required to conduct this study.

Conflicts of Interest: The authors declare no conflict of interest. The funders had no role in the design of the study; in the collection, analyses, or interpretation of data; in the writing of the manuscript, or in the decision to publish the results.

Nomenclature

BESS	Battery Energy Storage System
$BESS_{BUF}$	Battery Energy Storage System sized according to BUF
BUF	Battery Utilization Factor
CAPEX	Capital Expenditure
DSOs	Distribution System Operators
EC	Energy Community
EU	European Union
ICT	Information and Communications Technology
KPI	Key Performance Indicators
kWh	kilowatt-hours
LV	Low Voltage
MWh	megawatt-hours
P2P	Peer-to-Peer
PV	Photovoltaic

RECs	Renewable Energy Communities
SoC	State of Charge
TOU	Time-of-Use

References

1. EC Directive (EU) 2018/2001 of the European Parliament and of the Council of 11 December 2018 on the promotion of the use of energy from renewable sources. *Off. J. Eur. Union* **2018**, *L 328*, 82–209.
2. Menniti, D.; Pinnarelli, A.; Sorrentino, N.; Vizza, P.; Barone, G.; Brusco, G.; Mendicino, S.; Mendicino, L.; Polizzi, G. Enabling Technologies for Energy Communities: Some Experimental Use Cases. *Energies* **2022**, *15*, 6374. [CrossRef]
3. Di Silvestre, M.L.; Ippolito, M.G.; Sanseverino, E.R.; Sciumè, G.; Vasile, A. Energy self-consumers and renewable energy communities in Italy: New actors of the electric power systems. *Renew. Sustain. Energy Rev.* **2021**, *151*, 111565. [CrossRef]
4. Ceglia, F.; Esposito, P.; Faraudello, A.; Marrasso, E.; Rossi, P.; Sasso, M. An energy, environmental, management and economic analysis of energy efficient system towards renewable energy community: The case study of multi-purpose energy community. *J. Clean. Prod.* **2022**, *369*, 133269. [CrossRef]
5. Katsaprakakis, D.A.; Proka, A.; Zafirakis, D.; Damasiotis, M.; Kotsampopoulos, P.; Hatziargyriou, N.; Dakanali, E.; Arnaoutakis, G.; Xevgenos, D. Greek Islands' Energy Transition: From Lighthouse Projects to the Emergence of Energy Communities. *Energies* **2022**, *15*, 5996. [CrossRef]
6. O'Neill-Carrillo, E.; Mercado, E.; Luhring, O.; Jordan, I.; Irizarry-Rivera, A. Community Energy Projects in the Caribbean: Advancing Socio-Economic Development and Energy Transitions. *IEEE Technol. Soc. Mag.* **2019**, *38*, 44–55. [CrossRef]
7. Chamorro, H.R.; Yanine, F.F.; Peric, V.; Diaz-Casas, M.; Bressan, M.; Guerrero, J.M.; Sood, V.K.; Gonzalez-Longatt, F. Smart Renewable Energy Communities–Existing and Future Prospects. In Proceedings of the IEEE 22nd Workshop on Control and Modelling of Power Electronics (COMPEL), Cartagena, Colombia, 2–5 November 2021; pp. 1–6.
8. Felice, A.; Rakocevic, L.; Peeters, L.; Messagie, M.; Coosemans, T.; Camargo, L.R. Renewable energy communities: Do they have a business case in Flanders? *Appl. Energy* **2022**, *322*, 119419. [CrossRef]
9. Barone, G.; Buonomano, A.; Forzano, C.; Giuzio, G.F.; Palombo, A. Increasing renewable energy penetration and energy independence of island communities: A novel dynamic simulation approach for energy, economic, and environmental analysis, and optimization. *J. Clean. Prod.* **2021**, *311*, 127558. [CrossRef]
10. Doroudchi, E.; Khajeh, H.; Laaksonen, H. Increasing Self-Sufficiency of Energy Community by Common Thermal Energy Storage. *IEEE Access* **2022**, *10*, 85106–85113. [CrossRef]
11. Cielo, A.; Margiaria, P.; Lazzeroni, P.; Mariuzzo, I.; Repetto, M. Renewable Energy Communities business models under the 2020 Italian regulation. *J. Clean. Prod.* **2021**, *316*, 128217. [CrossRef]
12. Herenčić, L.; Kirac, M.; Keko, H.; Kuzle, I.; Rajšl, I. Automated energy sharing in MV and LV distribution grids within an energy community: A case for Croatian city of Križevci with a hybrid renewable system. *Renew. Energy* **2022**, *191*, 176–194. [CrossRef]
13. Dolatabadi, M.; Siano, P.; Soroudi, A. Assessing the Scalability and Privacy of Energy Communities by Using a Large-Scale Distributed and Parallel Real-Time Optimization. *IEEE Access* **2022**, *10*, 69771–69787. [CrossRef]
14. Zhu, H.; Ouahada, K.; Abu-Mahfouz, A.M. Peer-to-Peer Energy Trading in Smart Energy Communities: A Lyapunov-Based Energy Control and Trading System. *IEEE Access* **2022**, *10*, 42916–42932. [CrossRef]
15. Frisch, C.; Donohoo-Vallett, P.; Murphy, C.; Hodson, E.; Horner, N. An Electrified Nation: A Review of Study Scenarios and Future Analysis Needs for the United States. *IEEE Power Energy Mag.* **2018**, *16*, 90–98. [CrossRef]
16. Zhong, W.; Xie, K.; Liu, Y.; Yang, C.; Xie, S. Multi-resource allocation of shared energy storage: A distributed combinatorial auction approach. *IEEE Trans. Smart Grid* **2020**, *11*, 4105–4115. [CrossRef]
17. Aziz, M.; Dagdougui, H.; Elhallaoui, I. A Decentralized Game Theoretic Approach for Virtual Storage System Aggregation in a Residential Community. *IEEE Access* **2022**, *10*, 34846–34857. [CrossRef]
18. Sudhoff, R.; Schreck, S.; Thiem, S.; Niessen, S. Operating Renewable Energy Communities to Reduce Power Peaks in the Distribution Grid: An Analysis on Grid-Friendliness, Different Shares of Participants, and Economic Benefits. *Energies* **2022**, *15*, 5468. [CrossRef]
19. Regional, Geospatial, Energy and Transport Statistics Unit, Renewable Energy from Photovoltaic Panels: 2021. *Natl. Stat. Off. Malta Tech. Rep.* **2022**. Available online: https://nso.gov.mt/en/NewsReleases/Documents/2022/06/News2022106.pdf (accessed on 1 October 2022).
20. Micallef, A.; Spiteri Staines, C.; Cassar, A. Utility-Scale Storage Integration in the Maltese Medium-Voltage Distribution Network. *Energies* **2022**, *15*, 2724. [CrossRef]
21. Enemalta Corporation. The Network Code. Malta. 2013. Available online: https://www.enemalta.com.mt/wp-content/uploads/2018/05/Network-Code-EMC-Approved-October-2013-1.pdf (accessed on 1 October 2022).
22. Micallef, A.; Spiteri Staines, C. Generation Performance Analysis for Installed Photovoltaic Systems on the Maltese Islands. In Proceedings of the 1st Workshop on Blockchain for Renewables Integration (BLORIN 2022), Palermo, Italy, 2–3 September 2022.

energies

MDPI

Article

Technical and Economic Assessment of Battery Storage and Vehicle-to-Grid Systems in Building Microgrids

Alexandre F. M. Correia, Pedro Moura * and Aníbal T. de Almeida

Institute of System and Robotics, Department of Electrical and Computer Engineering, University of Coimbra, 3030-290 Coimbra, Portugal
* Correspondence: pmoura@isr.uc.pt

Abstract: In an electrical microgrid, distributed renewable generation is one of the main tools used to achieve energy sustainability, cost efficiency and autonomy from the grid. However, reliance on intermittent power sources will lead to a mismatch between generation and demand, causing problems for microgrid management. Flexibility is key to reducing the mismatch and providing a stable operation. In such a context, demand response and energy storage systems are the main factors that contribute to flexibility in a microgrid. This paper provides an assessment of the technical and economic impacts of a microgrid at the building level, considering photovoltaic generation, battery energy storage and the use of electric vehicles in a vehicle-to-building system. The main novel contributions of this work are the quantification of system efficiencies and the provision of insights into the design and implementation of microgrids using real on-site data. Several tests were conducted using real on-site data to calculate the overall efficiencies of the different assets during their operation. An economic assessment was carried out to evaluate the potential benefits of coordinating battery storage with a vehicle-to-building system regarding the flexibility and cost-efficient operation of the microgrid. The results show that these two systems effectively increase the levels of self-consumption and available flexibility, but the usefulness of private electric vehicles in public buildings is constrained by the schedules and parking times of the users. Furthermore, economic benefits are highly dependent on the variability of tariffs and the costs of energy storage systems and their degradation, as well as the efficiency of the equipment used in the conversion chain.

Keywords: microgrids; electric vehicles; vehicle-to-building; battery energy storage; distributed generation

Citation: Correia, A.F.M.; Moura, P.; de Almeida, A.T. Technical and Economic Assessment of Battery Storage and Vehicle-to-Grid Systems in Building Microgrids. *Energies* **2022**, *15*, 8905. https://doi.org/10.3390/en15238905

Academic Editors: Alexander Micallef and Zhaoxia Xiao

Received: 31 October 2022
Accepted: 22 November 2022
Published: 25 November 2022

1. Introduction

1.1. Motivation

Following the recent challenges related to climate change mitigation and reducing dependency on fossil fuels, new technologies have been developed to increase end-use energy efficiency and improve the harvest of renewable energies cost-effectively. The increased penetration of distributed energy resources, such as wind, solar photovoltaic (PV) and biomass sources, and energy storage systems has rekindled interest in the development of electric microgrids. According to the Microgrid Exchange Group, "A microgrid is a group of interconnected loads and distributed energy resources within clearly defined electrical boundaries that acts as a single controllable entity with respect to the grid" [1]. Electric microgrids have the ability to work independently of the main utility grid and can be used to promote the installation of distributed energy generation, reduce the costs of energy transportation, improve local power quality, increase renewable source self-consumption levels and provide auxiliary services to the main grid. Microgrids are also highly desirable in situations in which the high reliability of an energy supply service is required for critical facilities, such as hospitals, airports and military centers [2]. Microgrids can also be used to maintain energy supply in remote locations and to supply areas isolated during catastrophic events.

Microgrids can be classified into different types, each with its own purpose and application [3]. Isolated microgrids are mostly used on islands and in off-grid locations, being frequently reliant on diesel generators for supply. Commercial or industrial microgrids have large generation capacities and are focused on economic benefits via reduced costs or by providing the utility grid with ancillary services. Military microgrids are used for military and naval operations and have high redundancy levels. Community microgrids are centered around small villages or neighborhoods and have high levels of participants focused on reducing electricity costs and increasing autonomy and self-consumption. Lastly, building/campus microgrids are focused on aggregating loads and generation in order to increase building/campus efficiency. They differ from community microgrids in that, in most cases, generation and demand assets have the same owner.

The microgrid tested in this work is a campus microgrid located in the Department of Electrotechnical Engineering at the University of Coimbra. The characteristics of the microgrid are explained in Section 3. Microgrids rely on high levels of distributed generation, mostly renewable sources, which are inherently intermittent and non-dispatchable. Such dependency commonly leads to a mismatch between generation and demand, which is mitigated by either curtailing generation or injecting surplus energy into the grid, thus reducing cost efficiency and being particularly harmful in utility grids not prepared for bidirectional power flows [4]. As the share of intermittent power sources in a grid increases, the more difficult it will be to strike a balance between renewable generation and demand. The most visible symptom of this problem is the famous "duck curve", which has become prevalent in countries with high solar generation penetration [5]. The "duck curve" is the result of the solar output rising during the day, when demand is lower, and a steep ramp-up occurring during the late afternoon when the sun goes down. The consequences are the curtailment of generation, potentially negative energy prices and grid stress caused by turning the generation assets on and off following the quick ramp-up and -down.

For a microgrid to work efficiently, it is desirable to reduce as much as possible the mismatch between supply and demand. Therefore, flexibility is paramount in maximizing the use of existing resources in order to increase cost efficiency and maintain the stability of the grid or microgrid. The level of flexibility present in a microgrid is highly correlated with its demand-response capability and energy storage systems, such as batteries and vehicle-to-building (V2B) systems [6,7].

Electric mobility has also seen significant technological breakthroughs, not only due to the decreasing costs of batteries but also due to the higher availability of charging infrastructures and the incentives provided by most countries for the acquisition of electric vehicles (EVs). During the first half of 2022, 4.3 million battery and plug-in hybrid EVs were delivered worldwide—an increase of 62% compared to the same period in 2021. By the end of 2022, nearly 27 million EVs are expected to be in operation [8]. The number of sales is predicted to reach 26 million EVs per year by 2030, corresponding to a share of 28% of total motor sales [9]. EVs can have a stronger impact when considering smaller sub-grids, such as community or campus microgrids. For the standard EV client, the concept of vehicle-to-building/home is easier to understand and hence more attractive [10]. Alongside government policies and incentives promoting the transition to EVs, it is important to direct efforts towards developing the infrastructure needed for standardizing vehicle-to-grid interaction and simplifying the technologies required for it to become conventional for social and market acceptability [11].

Regarding energy storage, almost all of the existing options for microgrid-level energy storage are focused on stationary battery systems, due to the recent developments in terms of both price and performance which have made the technology viable. However, it is also important to take into account the increasing penetration of EVs, which will play a huge role in microgrid management, since EVs can be used both for demand response and as energy storage resources using vehicle-to-building systems. The massification of battery energy storage systems (BESSs) will also open up new possibilities for grid management, including microgrids, building energy management and vehicle-to-grid interaction.

This work aims to contribute a technical and economic assessment of the usage of battery energy storage systems and vehicle-to-building systems integrated into a campus microgrid to quantify system efficiencies and provide insights into the design and implementation of microgrids. All the assets were already installed inside the building, and data were collected during real operations, providing more realistic data than simulated results. The operational efficiencies were assessed and used to evaluate round-trip efficiency and cycle economies, which information is to be used as a reference by the energy management system. Furthermore, an economic assessment with different scenarios was made to highlight the impacts that flexibility can have in terms of increasing a building's self-consumption and reducing overall electricity needs. In summary, the novel contributions to the field of study are:

a. An analysis of round-trip efficiencies of the most commonly used distributed energy systems in a microgrid using real on-site data;
b. A technical comparison of conventional bidirectional EV chargers and novel silicon carbide (SiC) bidirectional EV chargers;
c. Insights regarding the economic aspects of using BESSs and vehicle-to-building for load shifting and peak shaving in buildings.

1.2. Related Works

In Europe, there is a current policy trend to work towards near-zero energy buildings (nZEBs). One of the main research subjects is the implementation of better energy storage facilities in buildings. Battery systems and, more recently, vehicle-to-building systems have been subjects of increasing research in the past decade due to more accessible costs, more variety in the technology available on the market, the potential for flexibility and policies that encourage nZEB renovations.

The way in which EVs are used is also evolving. Increasing numbers of EV models are being built with an increased focus on the bidirectional transfer of power. This feature presents a viable solution for increasing flexibility by adding a new storage asset to microgrids. The functionality of EVs will not be limited to mobility only; they will provide various services to the user and the grid through the vehicles' unused battery capacities being taken advantage of. The work presented in [12] offers a comparison between vehicle-to-grid, vehicle-to-building and vehicle-to-home (V2H). It is stated that V2G is the most complex and ambitious concept to apply, being heavily reliant on the penetration of EV technology to be successful. V2B and V2H are easier to implement, neither requiring large infrastructures or large numbers of EVs to be available. Another advantage is that these two methods prove to be more beneficial for the individual user. In [13], an economic evaluation of EV charging integrated into microgrids is presented. A campus microgrid is used as a test bench and several business models based on the self-consumption of electricity and smart charging are simulated. It concludes that using PV systems coupled with smart charging presents the highest profit for the operator. The installation of a battery storage system is less profitable, but increases flexibility and reduces dependency on the grid, particularly when PV is not being generated. In [14], the authors simulate a V2B system with six electric vehicles and evaluate the impact of battery degradation. The results prove that V2B has relatively significant profit potential, even considering the costs associated with battery degradation.

The work in [15] addresses energy sharing through a transactive energy market in community microgrids, using stationary vehicles and EVs as flexibility resources. The proposed method indicates that, with the management of energy storage resources, it is possible to reduce the mismatch between demand and local generation as well as operating costs. The method also highlights the benefits of aggregating stationary or mobile assets to increase impacts on the grid. Similarly, Ref. [16] simulates a method for energy communities in which EVs are used as flexibility resources between several buildings. The results show that the flexibility provided by EVs is more impactful in terms of providing reduced costs and increased self-consumption for buildings integrated into energy communities as compared with the individual management of buildings. In [17], a concept of collaborative charging for vehicle-to-building is proposed. The study demonstrates that it is economically viable for a building manager to provide free EV charging to users in exchange for control of the storage capacity during allotted time periods. The user benefits from the free energy and the building benefits from the increased storage reserve that can be used to reduce power peaks or increase self-consumption of intermittent renewable sources. In [18], a coordination scheme for EV charging in office buildings is proposed. The method is focused on taking advantage of integrated distributed resources, such as PV generation and battery storage systems, to achieve an energy cost reduction by coordinating with the known schedules of the building employees and loads. This method illustrates an energy cost reduction of 14% when compared with a first-come-first-served approach. In [19], the authors study the impacts that plug-in EVs can have regarding contributions to a nearly zero-energy building. The results show that, by sharing even a fraction of their battery capacities, plug-in EVs can reduce the amount of energy supplied by the grid by up to 40% in the considered scenarios. The authors also conclude that stationary storage systems have increased benefits when used to compensate for the intermittent presence of charging EVs. Similarly, the works [20,21] analyze vehicle-to-building-to-vehicle in different management schemes to reduce a building's energy signature and its electric consumption. On the topic of integration into microgrids, Ref. [22] focuses on the system and protocols needed for microgrids to integrate EVs. The authors suggest utilizing a blockchain-based method to provide adequate security and protection for users' data.

Most of the aforementioned works conclude upon the effectiveness, value or profitability of V2G as a resource to be integrated into buildings or cities. However, there are few works exploring the efficiencies and losses of energy transactions between assets, such as V2G or BESS, integrated into the same building/microgrid. The work in [23] is claimed to be the first to experiment on the round-trip efficiency of V2G systems. The results achieved for V2G efficiency are very close to those achieved for the experiments in this paper, but the scope of the research is narrow and does not include analysis of other common operations inside buildings, such as charging from PV or from/to BESS. In [24], the relationship between energy cost and efficiency is studied. The authors present data regarding V2G efficiency using EVs and plug-in hybrid EVs and make remarks about the economic impacts and benefits of V2G for system operators. In [25], a thorough study is presented regarding several aspects of V2G in order to propose regulations to maximize profit and determine optimal operating points. The work mentions the efficiency of V2G but does not present any experimental results that demonstrate or can be used by system operators to account for losses during energy transactions in microgrids or buildings. All these works approach V2G efficiencies in different ways but fail to provide a panoramic view of the expected efficiencies in a microgrid.

On the subject of operational stability, in a microgrid environment, the enclosed and limited resources mean that power fluctuations pose higher risks to operational stability [26]. The high reliance on intermittent resources heightens the probability of unwanted power fluctuations. Due to their ability to store energy and schedule charges, electric vehicles, as well as energy storage systems, present new opportunities for applying demand-side measures, such as demand response [27]. Demand response has been proven to dramatically increase microgrid reliability by minimizing peak demand, increasing the match between generation and demand, and improving cost efficiency [28–32]. Grid operators also rely on other demand-side measures, such as tariffs and monetary incentives, to smooth the demand curve and modify consumption habits to better match the existing energy availability [33]. All these works conclude that EVs will have a major role in achieving nZEB building status, whether from the perspective of supplying extra storage capacity to buildings, improving charging services for users or providing ancillary services, such as demand response. As such, this work focuses on analyzing the efficiency of several methods for integrating EVs in microgrids, the main one being vehicle-to-building.

Battery energy storage systems have been subject to considerable advances in the past decade, and the battery pack price has fallen from over 1000 €/kWh in 2010 to 132 €/kWh in 2021 [34], and the forecasts point to a 100 €/kWh mark by the middle of this decade. Such a reduction in costs will lead to higher incentives for energy storage systems in stationary applications, as well as the electrification of the transportation sector. The main limitation of such battery systems has been their limited numbers of cycles, which cause the capacity to degrade both with time and usage. Daily charge–discharge cycles will gradually degrade a battery's health and reduce its efficiency and useful capacity. Therefore, it is important to address the issue of battery degradation and its impacts on the economic and technical operations of the microgrid.

The work in [35] presents two experiments designed to assess the battery degradation caused by V2G and calendar aging due to temperature and state of charge (SoC). For these experiments, batteries were discharged twice a day at the maximum power. With this usage profile, it was shown that the lifetime of the battery pack decreased to nearly half its predicted lifetime. These experiments also led to a new method to track the battery's state of health. In [36], a methodology is proposed to quantify EV battery degradation from driving only versus driving and vehicle-to-grid services. It was concluded that interactions, such as peak load shaving and frequency regulation, at a typical power rate, do not significantly accelerate battery degradation when compared with degradation due to driving or calendar aging. When correctly used, vehicle-to-grid/building impacts on battery degradation are insignificant. In [37], a thorough battery degradation model is suggested that considers calendar aging, capacity throughput, temperature, state of charge, depth of discharge and current rate. The results of such work indicate that this model can extend the life of the EV battery beyond the situation in which there is no V2G by optimizing state of charge and power transfer. In [38], measurements made to study the power losses during the charge and discharge of an EV are presented. One-way losses varied from 12% to 36%, with most losses occurring inside power electronics. Based on these results, the authors underline the importance of choosing adequate charging stations for increasing efficiency.

According to the European Environment Agency [39], reducing system losses and therefore being able to achieve more efficient distribution systems, whether they are centralized systems or microgrids, is crucial to achieving the targets for decarbonization and decentralization of the electric grid. The aforementioned works are all in agreement that vehicle-to-building is crucial to achieving reduced emissions in buildings and improving energy costs and efficiency, in which battery storage systems and electric vehicles have a natural synergy. Therefore, this work aims to complement such conclusions by contributing real-use data on vehicle to building system efficiency and strategies for the implementation of such systems in microgrid environments.

1.3. Paper Organization

The remainder of the paper is structured as follows: Section 2 presents the characteristics of the microgrid, as well as the specifications of the equipment used in the experimental work. Section 3 presents the results of the experimental tests, as well as an economic assessment of several different scenarios in which various levels of storage capacity are available to the building, which results are discussed in Section 4. Lastly, Section 5 summarizes the paper, highlighting its main conclusions.

2. Materials and Methods

2.1. Microgrid Characteristics

The experiments were conducted at the Department of Electrical Engineering of the University of Coimbra. The building has an area of about 10,000 m^2 and electricity consumption of 500 MWh/year. The building comprises several classrooms, laboratories, administrative services and three research institutes. The highest energy consumption of the building is in the periods from 10 am to noon and from 5:30 pm to 7 pm. This partially coincides with peak tariff hours: 9:30 am to noon and 6:30 pm to 9 pm. The distribution transformer that services the building has a maximum load of 630 kVA and operates with a load factor of around 30–40%. The building is equipped with 292 PV panels corresponding to about 70 kW with AC injection. A diagram of the microgrid installed in the building is presented in Figure 1.

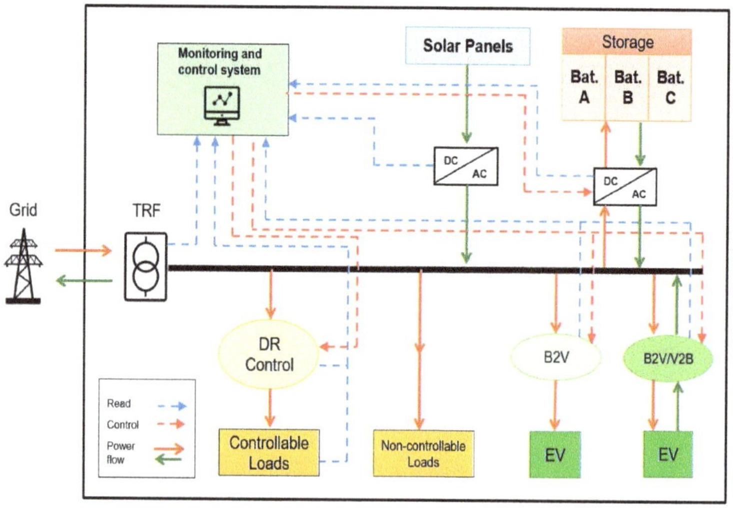

Figure 1. DEEC microgrid schematic.

An energy storage system consisting of three LG Chem RESU10H 400V Li-ion [40] batteries is operational, with a total capacity of about 29.4 kWh (27.9 kWh of usable energy). The batteries are connected to three Sunny Boy Storage 5.0 [41] single-phase inverters, each with a rated power of 5 kW and a rated max efficiency of 97.5%.

The building is also equipped with the ability to ensure V2G/V2B electric vehicle charging with several conventional DC chargers and one prototype of a silicon carbide charger, all using the CHAdeMO protocol. The selected DC charger is a Magnum CAP three-phase bidirectional DC charger [42] with a maximum charging output of 10 kW and an efficiency of 93% at rated voltage and power. The power circuit of the chargers is composed of an AC/DC and a DC/DC converter and transformers to deliver the correct power either to the vehicle or to the residence outlet. Due to the high switching frequency of the semiconductors for power conversions and the conduction losses, large amounts of heat are produced during the operation of the charger, which requires the usage of a forced ventilation system. Along with the onboard CPU, all these components and power conversions accrue losses during the charging operation, further decreasing the efficiency. The segregation of these losses is beyond the scope of this paper, with the charger considered as a single entity. A prototype of a fully functional three-phase SiC technology DC charger is also available and was used for further testing in parallel with the conventional DC charger. Due to its novel technology, the SiC charger is smaller and produces less heat and noise, having higher efficiency when compared to the conventional DC charger [43,44].

The building is also equipped with several heating, ventilation and air conditioning (HVAC) systems for temperature regulation in classrooms, offices and laboratories. During the summer and winter months, nearly 40% of the demand is due to the HVAC systems. These conditions are ideal to test and apply demand-control measures. For this, smart thermostats were developed and used to regulate the part of the HVAC load inside classrooms. The results of these tests are not within the scope of this paper, since they were previously studied in [45]. Lastly, monitoring software has been developed to display the data for and the status of all the microgrid's assets in real time. The monitoring system is used to make decisions based on the current and predicted parameters of the microgrid, considering electricity prices, predicted PV production, building occupancy, weekdays and seasonal changes. A screenshot of the display is presented in Figure 2.

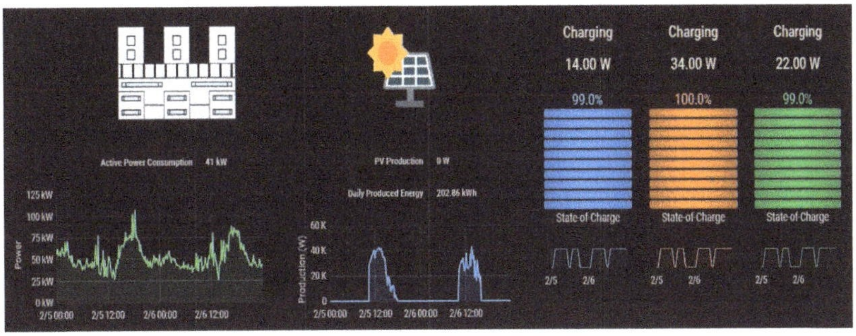

Figure 2. Microgrid monitoring system.

Figure 3 presents a daily net load diagram and PV production for two days in the Department of Electrical and Computer Engineering during the late summer (September 2020). The difference between the net load from the weekend (Sunday) to the weekday (Monday) is clearly visible, as well as the grid injection due to excess solar generation. As can be seen, there is clearly a mismatch between PV generation and demand. During weekends and summertime, the demand level is lower than the generation, leading to a generation surplus (negative net demand), and some energy needs to be injected into the grid, which is paid with a very low tariff. Similarly, during weekdays, the peak in PV generation coincides with the lunch break.

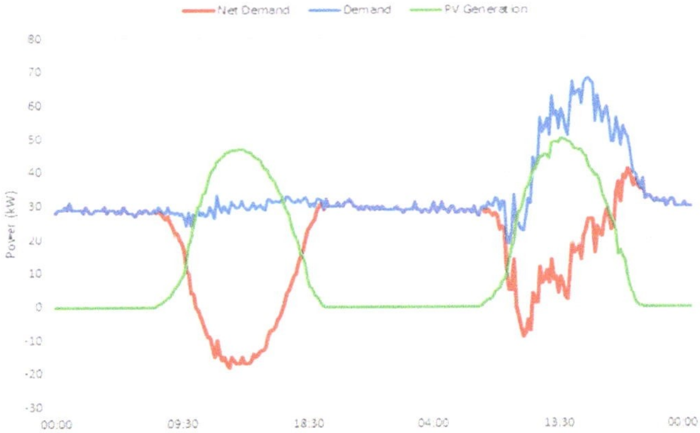

Figure 3. Net Demand, Demand and PV Generation in the DEEC Building for 2 days (From Sunday to Monday, 6–7 September).

It can be observed that there is a need for energy flexibility to improve the usage of the energy resources. By coordinating the PV production with the available storage capacity, it is possible to reduce the energy injected into the grid and use it later in the day when electricity tariffs are higher. This situation will ensure a double benefit to the microgrid by increasing self-consumption and reducing peak demand. EVs can be used to mitigate this issue. By leaving a vehicle connected to the building during lunch breaks and work hours, the building management system may opt to use the EV's available storage to supplement the building's BESS and artificially increase demand when there is a renewable generation surplus. This energy can either be paid for by the EV owner at a lower price or returned to the building later in the day during peak hours. Additionally, by increasing the levels of flexibility in the building, it will be possible to increase the autonomy of the microgrid from the utility grid and further increase the existing photovoltaic generation, thereby approaching the goal of a near-zero emissions building.

2.2. Microgrid Efficiency

The interconnection between different sources of energy generation, energy storage systems and a variety of loads implies many stages of power conversion for voltage level compatibility in the power exchange. PV systems generate DC power, but most building loads are in AC. Even when storing solar energy in a battery, a DC-to-DC converter has to be used to make the voltages and currents compatible with a BESS. The need for all these energy conversions adds up to infrastructural costs and reduces the overall efficiency of the microgrid. Hence, it is desirable to look for opportunities to reduce the number of power conversions needed and to increase the efficiency of the power conversion equipment. An example of a solution is the direct charging of EVs or BESSs using PV energy [46]. Currently, the solar energy produced is converted to AC to enter the grid and then it is converted back to DC to charge the electric vehicle. By having a direct DC link from the PV generation to the EV charger, it is possible to avoid one stage of power conversion and reduce the cycle losses by about 10% while still maintaining the AC connection and the possibility of V2B, as was assessed in the experimental trials.

EV chargers are another source of cycle inefficiency. Since most of the charging is carried out directly from the grid, conversion from AC to DC is needed. This power conversion is performed using high-speed semiconductors (MOSFETs) with high switching frequencies which generate switching and conduction losses, leading to large amounts of heat that must be managed. A new semiconductor technology, silicon carbide (SiC), is being introduced in electronics, with considerable benefits [47]. SiC semiconductors have much lower drain–source resistance and are much faster, which translates into lower conduction and lower switching losses. These characteristics allow the construction of lighter and smaller charging stations with higher efficiency, fewer ventilation needs and even potentially lower costs with mass production [48].

Regarding the efficiency of vehicle-to-building systems, there is a trade-off, since, compared with traditional unidirectional charging, it leads to an increased number of charging and discharging cycles and hence to increased battery degradation and operational losses. There are several factors and variables that affect battery degradation, ranging from chemistry to driving habits to charging conditions, among others. As such, studies regarding the impact of vehicle-to-grid/building on battery degradation have reached different conclusions, from neglectable to relevant impacts [36,49–52]. Battery degradation reduces the usability of the vehicle, and operational losses diminish economical returns and overall system efficiency. These two constraints limit the profitability of V2B and are a barrier to convincing the user to commit their vehicle to the building. In order to be attractive to the user, V2B must have relevant economic benefits and a proper assessment of the costs and benefits must be provided.

Regarding BESSs, they are composed of a power conversion system connected to the grid and a battery management system (BMS) to ensure even charge distribution in the battery cells and reliable battery operation. Energy losses present in a BESS are due to conversion losses, ohmic resistance losses and battery losses, with the battery accounting for the majority of these losses. Battery losses occur due to internal resistance, heat produced during the electrochemical reactions, cell voltage imbalance, charging profile, BMS self-consumption, aging and temperature. For example, frequently using fast and ultra-fast charging (>1 C) will rapidly decrease battery capacity, whereas freezing temperatures slow the electrochemical reactions, reducing performance, and high temperatures accelerate aging due to stress. Each one of these components will contribute a small amount to the overall losses of the system, and careless operation will accelerate the natural capacity degradation of the battery, reducing the number of working cycles. Since BESSs are still relatively expensive, it is desirable to take maximum advantage of available capacities to ensure cost-effectiveness.

3. Results

In this section, the experimental trials for the BESS and V2G systems are presented. The results are for several tests with different working conditions. For the BESS, full charge–discharge cycles were made, and, for V2G, different power levels were used to charge the battery for a given time period and SoC. The conducted experiments also intended to ensure an overview of the overall power losses and efficiencies present in the BESS during a full charge–discharge cycle. The round-trip efficiency (RTE) could then be calculated and used to evaluate the operating efficiencies of the overall system. The round-trip efficiency of an energy storage system is a term used to describe how much useful energy the system can provide versus the amount of energy inputted into the system.

3.1. Solar PV Monitoring

The PV system has been monitored daily since the start of its operation. It was identified that there was a surplus of generation during weekends from 10 am to 4 pm and on clear sunny days on workdays during lunch hours. About 70 kWh need to be injected into the grid each day of the weekend and about 40 kWh during weekdays in the summer. During winter, there is very little injection of generation into the grid. Over the last 4 years, the system has ensured an average generation of about 75 MWh/year, of which around 3 to 4 MWh/year are estimated to be generation surplus to be injected into the grid. The inverters have a calculated average efficiency of around 95%, which means that only 71.25 MWh/year are used in the building.

3.2. Li-on Batteries

The objective of the experiments with the Li-on batteries was to collect data from different locations of the power circuit to assess the efficiencies of the charging and discharging processes. Figure 4 presents a diagram of the circuit along with the different powers measured, P1 to P4. The points that were chosen to take measurements from were: Point A—between the main grid and the converter; Point B—between the converter and the terminals of the batteries before the BMS. Each battery has a total energy capacity of 9.8 kWh and 9.3 kWh of usable energy. Therefore, in the charging–discharging tests, only 9.3 kWh of energy was cycled. The values for the SoC use as a reference the usable energy capacity, e.g., 0% SoC in the results corresponds to 5% real SoC (500 Wh) of the battery. The tests consisted of ensuring a full charge–discharge cycle, from 0% to 100% to 0% of the usable SoC, and measuring the currents and voltages at Points A and B. The energy was then calculated in each stage, as well as the respective efficiencies. The efficiencies were calculated using Equations (1) to (4). Equation (1) presents the round-trip efficiency (η_{RTE}), Equation (2) the converter charging efficiency (η_{char}), Equation (3) the converter discharging efficiency (η_{dischar}) and Equation (4) the battery efficiency (η_{bat}).

$$\eta_{\text{RTE}} = \frac{P4}{P1} \tag{1}$$

$$\eta_{\text{char}} = \frac{P2}{P1} \tag{2}$$

$$\eta_{\text{dischar}} = \frac{P4}{P3} \tag{3}$$

$$\eta_{\text{bat}} = \frac{P3}{P2} \tag{4}$$

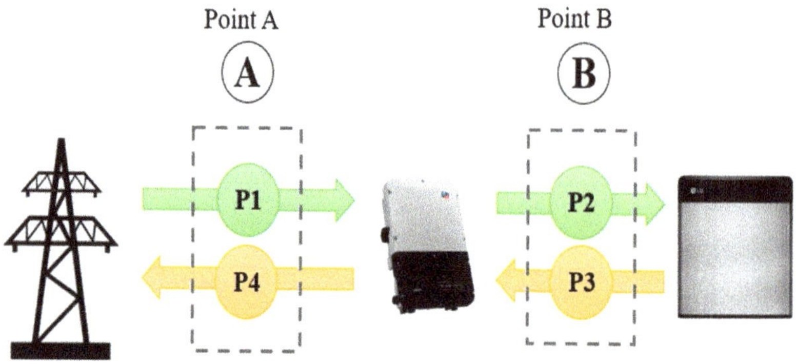

Figure 4. Electric circuit schematic and measuring points in the bidirectional grid–inverter–battery power flow.

A full charge and discharge cycle was tested for each of the three batteries with the maximum charging power allowed by the inverter, which is 4.8 kW, corresponding to around 0.5 C. The second round of tests was then performed with one of the batteries, where the same measurements were repeated for 0.1 C and 0.3 C. At the time of these tests, the BESS had around one year of lifetime and had been working at around a charge–discharge cycle every two days. The state of health of all the batteries was around 95%. The results of the first round of tests are presented in Tables 1 and 2.

Table 1. Energy measurements for Points A and B, in Wh.

Battery	Operation	Point A	Point B	Battery Losses	Converter Losses
A	Charge Discharge	9158 7809	8885 8164	720.4	273.6 355.5
B	Charge Discharge	9138 7799	8855 8079	676.1	282.4 280.3
C	Charge Discharge	9149 7852	8832 8146	686.1	317.3 294.2

Table 2. Calculated efficiencies.

Battery	η_{char}	$\eta_{dischar}$	η_{Bat}	η_{RTE}
A	97.01%	95.64%	91.89%	85.26%
B	96.90%	96.52%	91.23%	85.34%
C	96.53%	96.38%	92.23%	85.81%

From these results, it can be seen that the efficiency of the converters is around the rated value of 97%. Most of the system losses occur inside the battery due to its electrochemical reactions, internal resistance and temperature. Part of the energy lost inside the battery is used for the operation of the battery management system, which is integrated directly inside the battery. In sum, the round-trip efficiency for this BESS is around 85%, with 5% of the energy lost in conversion and nearly 10% of it lost while being stored inside the battery. Ohmic losses due to cabling are also present, but since they are highly dependent on the system's physical architecture and the value is too small to have a real economic impact, they were not considered. Nevertheless, in this system, ohmic losses account for 0.5% of the total losses, this percentage being diluted inside the losses throughout the various cycle stages.

The third round of tests was made to ascertain the differences in efficiency caused by a different charging power level. One battery was chosen, and the full charge–discharge tests were repeated. Table 3 presents the achieved results. As expected, the charging power affected the efficiencies.

Table 3. Calculated efficiencies for different C values.

C	η_{char}	$\eta_{dischar}$	η_{Bat}	η_{RTE}
0.1 C	94.1%	94.43%	95.03%	84.45%
0.3 C	94.8%	95.01%	95.04%	85.60%
0.5 C	96.6%	96.52%	91.23%	85.34%

With lower currents, the inverters will work farther from the nominal operating point, thus reducing efficiency. Oppositely, less current will benefit the battery by producing less heat and reducing electrochemical stress. Even though at lower Cs the battery efficiency increases, the overall round-trip efficiency will be higher when working with lower values of power and at the inverter's rated power. Higher C values were not possible to test due to the inverter power ratings.

A brief analysis was also made regarding battery degradation by measuring the state of health (SoH), which is the metric used to measure the condition of a battery, this being the ratio between the battery's current capacity and the specified rated capacity. Li-ion batteries will lose capacity and gain internal resistance over time due to the wear and tear of the cathodes and anodes. Common factors that affect battery aging are calendar time, cell chemistry, temperature, average SoC, C rate, cycle number and depth of discharge (DoD). In this case, the batteries are installed in an underground garage, where the temperature is mostly constant all year round. At the time of these measurements, they had an estimated age of 3 years and had been operating with 5 full charge–discharge cycles per week (on weekdays), counting around 625 cycles since the start of operation. In normal conditions, the C rate was kept at 0.5 C, and the DoD was from 100% to 5%, which is larger than what is normally advised. The SoH was then measured with external equipment. The measured values and also the SoHs calculated by each of the integrated battery management systems are shown in Table 4.

Table 4. Measured states of health of the batteries.

Battery	BMS SoH	Measured SoH
A	96%	94.96%
B	96%	95.21%
C	97%	97.18%

Two of the batteries indicated a decrease in storage capacity of around 4–5% and the other a decrease of 3%. The lifetime expectancy stated by the manufacturer is 10 years or 6000 cycles for the maintenance of at least 60% of their initial rated energy storage capacities. This means that, in all cases, capacity degradation is within or below the expected levels of −4%, with around 300 cycles per annum. Further analysis of battery degradation and its costs was beyond the scope of this paper but is planned for future research.

3.3. V2G Chargers

For the V2G assessment, two three-phase bidirectional DC chargers were tested. One charger used conventional silicon MOSFETs, and another charger was built using high-efficiency silicon carbide technology. The purpose of these experiments was to evaluate the efficiencies and to identify the different types of losses occurring in the electric circuit comprising the grid, the charger and the EV. A Nissan Leaf with a 40 kWh battery and V2G capacity was used for the tests.

An approach similar to the one applied in the battery tests was taken. Strategic points in the power circuit were identified, and at each of these points the voltage and current were measured to calculate the power and energy. The measuring points were as follows: Point A—AC measurement between the grid and the charger; Point B—DC measurement at the output of the charger; Point C—DC measurement at the terminal ends of the vehicle's battery. The considered circuit and the points are illustrated in Figure 5.

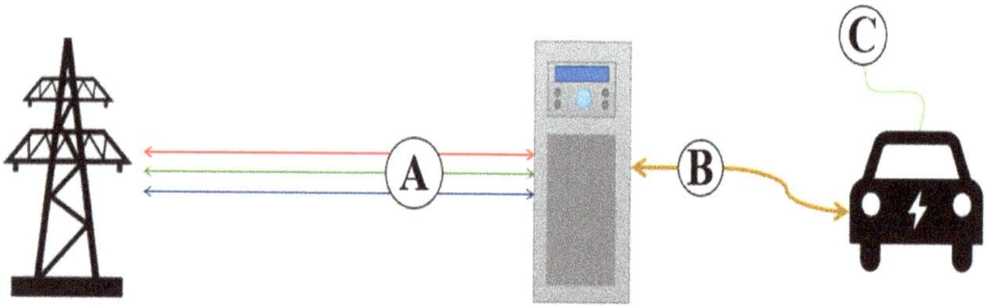

Figure 5. Measuring points of the charging–discharging process.

Several tests were performed for each charger with power levels of 2.5 kW, 5 kW, 7.5 kW and 10 kW, during both charging and discharging. All tests were conducted with a battery SoC of about 50%, as at this level the losses due to battery energy storage are reduced. For each power level, the vehicle was put on charge for the time necessary time for a change of 2% in the SoC, following a discharge, at the same power level, for a similar amount of time. The SoC of the battery was monitored using the onboard software. A test was also performed in which an order to charge the vehicle at 0 kW of power was placed on the charger to study its standalone consumption (only Point A was monitored), as well as to evaluate the standby consumption. The efficiencies were calculated using Equations (5) to (6). Equation (5) presents the charger charging efficiency (η_{char}), Equation (6) the discharging efficiency ($\eta_{dischar}$) and Equation (7) the round-trip efficiency of the charger (η_{cyc}).

$$\eta_{char} = \frac{Eb}{Ea} \tag{5}$$

$$\eta_{dischar} = \frac{Ea}{Eb} \tag{6}$$

$$\eta_{cyc} = \eta_{char} \times \eta_{dischar} \tag{7}$$

The values of the calculated efficiencies for the charger are presented in Table 5, with negative values of power for discharge. One of the main conclusions that can be derived from these tests is that for higher power values there will be an increase in charger efficiency (leading to higher cycle efficiency). This was expected, mainly due to the operation being near the rated conditions of the chargers. Another reason is that by increasing power, the constant losses present in the chargers become less impactful than the delivered power.

Besides the losses caused by the Joule effect in the copper wirings and charger cable and the semiconductors losses (switching and conduction), there are other electronic components, such as transformers, relays and other passive elements, that contribute to the overall losses during operation (leakage currents, hysteresis losses, etc.). These losses remain mostly constant during the operation and are independent of the charging power. Thus, when working with higher power levels, their impact on the overall efficiency is less noticeable.

The results also indicate a relevant disparity between charging and discharging efficiencies. This can be explained by the asymmetrical path that is needed for power conversion. Whereas in AC to DC a rectifier is used, the reverse operation is performed by an inverter, which will have different levels of loss, and the current flows through a different circuit. Figure 6 presents a comparison of the charger efficiencies.

Table 5. Chargers and Cycle Efficiencies.

Charger	Power	Charger Efficiency	Cycle Efficiency
Conventional DC Charger	2.5 kW −2.5 kW	84.80% 76.65%	64.60%
	5 kW −5 kW	89.56% 85.49%	76.56%
	7.5 kW −7.5 kW	90.80% 87.62%	79.55%
	10 kW −10 kW	91.10% 88.25%	80.39%
Sillicon Carbide Charger	2.5 kW −2.5 kW	94.01% 96.73%	90.93%
	5 kW −5 kW	94.66% 96.37%	91.22%
	7.5 kW −7.5 kW	95.30% 96.12%	91.60%
	10 kW −10 kW	95.46% 96.23%	91.86%

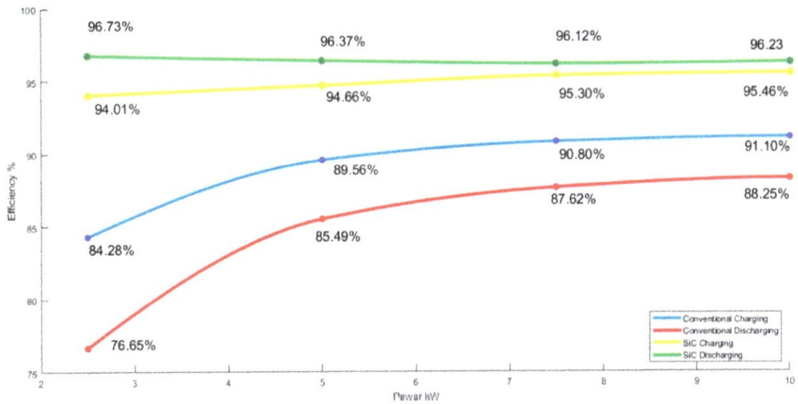

Figure 6. Charger Efficiency Comparison: SiC vs. conventional chargers.

Temperature is also an impactful factor regarding overall efficiency, and, during the developed tests, both chargers were situated inside a cool and dry place, where the air temperature was mostly constant. The SiC charger was noticeably cooler during operation than the conventional IGBT charger. The first charger was also noticeably quieter due to its requiring less ventilation, and therefore less power was needed for temperature control. This is one of the main reasons for the increased efficiency levels of the SiC charger. By using SiC semiconductors, the overall losses in power conversion are reduced. There will be smaller switching and conduction losses, less heat produced and therefore less power used for ventilation. The increased bandgap in the SiC charger also allows the semiconductors to work at higher voltages and with higher frequencies, with higher efficiency.

Regarding charging, it can be concluded that higher levels of power will lead, in most cases, to higher efficiencies. Nevertheless, the impact of charging with high levels of current on the efficiency and health of the batteries must also be taken into account. For the conducted experiments, charging and discharging powers were kept at about 0.5 C, which was considered safe and not high enough to hinder the batteries' health. These results also show the advantages of SiC technology in terms of achieving a boost of more than

10% of the cycle efficiency. The SiC charger was also much less affected by the different charging powers.

The charging of EVs not only results in changes to the electricity demands of buildings but can also have a significant impact on power quality. An EV charger is a nonlinear load that can produce large current harmonics that will flow through an electric grid, distorting voltage. Residential EV owners could face problems with power quality and, in larger buildings, when considering a high number of chargers and several EVs charging simultaneously, the harmonics could have a relevant impact on the voltage levels. Previous works [53,54] regarding the harmonic impact of multiple electric vehicle charging have concluded that harmonic limitations may be a greater barrier than power limitations regarding EV charging. The IEEE 519–2022 [55] is the current active standard that dictates the maximum distortion allowed for electric power systems in the USA.

During the charging and discharging tests, the harmonic current components were monitored. The main purpose was to assess the level of harmonics produced by the used chargers. As can be observed in Figure 7, the total current harmonic distortion is situated at around 3.6%, and the following harmonics are all under the requirements of the IEEE 519–2022 standard. Specifically, in the tested case, the building has three EV chargers installed, but since the transformer is underutilized (with a load factor of around 30–40%) the current harmonics created by the EV charging were not expected to have a significant impact on the voltage levels of the building.

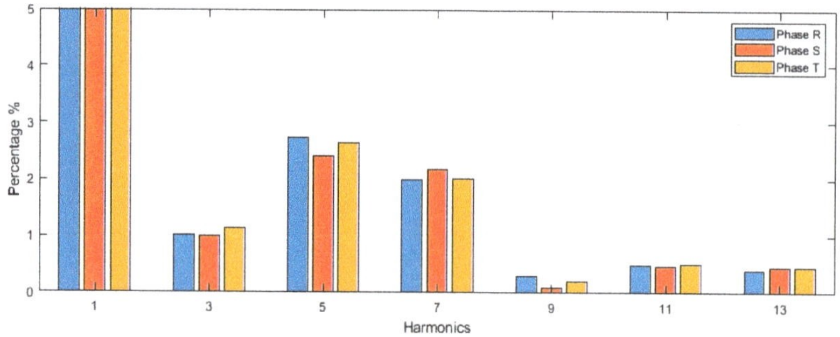

Figure 7. Current harmonics with a conventional charger.

3.4. Economic Assessment

Using the aforementioned results, an economic assessment was carried out, considering several different scenarios for the DEEC building. Net load diagrams for a weekday and a weekend day are presented in Figure 8a,b. In these load diagrams, the impact of PV generation is visible. The steep peaks are explained by the sample rate, of 10 min, which means that the load diagram shows drastic changes in consumption, particularly during the late afternoon (19:30 h). The building is also prone to heavy loads due to the use of electric motors in laboratories, which originate the various peaks during the workday. Since the peak load of the building is low when compared to the installed power (292 kW), there is little advantage in aiming for a peak shaving strategy, although the use of such a method is feasible in other buildings. Instead, the main purpose of the BESS and V2B systems with respect to cost efficiency is to absorb the surplus generation and take advantage of the tariff variation.

For the economic analysis of the scenarios, it was first considered to group the available storage capacities for both the BESS and V2B and take into account the sum of the storage capacities as a whole. However, this was not possible due to: (1) the difference in the efficiencies of the charging cycles when using the BESS or V2B and (2) the fact that the periods of availability for the BESS and V2B are different, since, whereas BESS storage capacity is available 24/7, the availability of V2B is constrained by user habits. If the EV is privately owned, then the only hours when it is available for the building microgrid are during work hours. Inversely, if the EV is owned by the company using the building, the period in which it is guaranteed to be available for V2B is during off-work hours. Since only private EV users currently utilize the building chargers, the first case for V2B availability was considered, that is, storage capacity from V2B being only available during work hours (9 am to 6 pm). For these reasons, the storage capacities for the BESS and V2B were treated as separate, and several scenarios were studied with different amounts of storage.

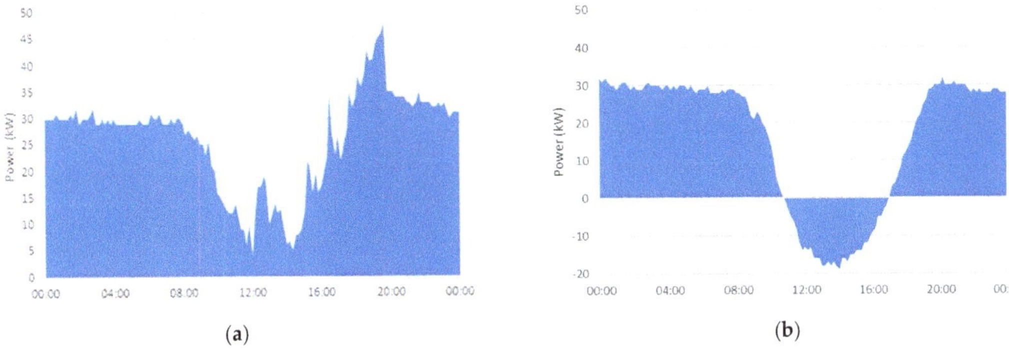

Figure 8. Twenty-Four-Hour Typical net Load Diagrams: (**a**) Weekday; (**b**) Weekend.

The base scenario is one in which where there is no energy storage capacity whatsoever to be used by the building. This scenario will be used for comparison and for calculating economic and technical benefits. In Scenario A, the existing and operational BESS, with a total of 27.9 kWh of battery storage, was considered. For the analysis of the scenarios with V2B, some considerations have to be made: for Scenario B, 30 kWh of V2B storage capacity was added. This value was chosen based on the existing two V2G chargers, and the assumption was made that each user allotted 15 kWh of the EV battery for V2B operations and that the capacity injected into the grid had to be recovered by the end of the day. This is a limitation and, in some cases, may lead to an increase in demand during peak hours and a reduction in the overall usable storage capacity. The considered available time, for a standard Portuguese working schedule, is from 09:00 to 18:00; given this, V2B capacities are only available during that time. Due to these limitations, the economic benefits from the building's point of view are limited. There is almost no generation surplus to be considered during weekdays, and during weekends the users are not typically present in the building. Instead, the plausible benefit for the building is related to using smart charging to avoid charging EVs during peak hours.

By calculating the energy consumed during peak hours on weekdays and the generation surplus during the peak periods, it was concluded that there is a margin to upgrade the building storage capacity by 70 kWh, which is under consideration. Since there is a need for more storage capacity and the DEEC building has plans for increasing the storage capacity of the DEEC building, scenarios C and D were considered, in which the storage capacities for the BESS and V2B were increased to double their existing values to assess the economic impacts. The different scenarios are summarized in Table 6.

Table 6. Tested Scenarios.

Scenario	BESS Capacity (kWh)	V2B Capacity (kWh)
Base	0	0
A	27.9	0
B	27.9	30
C	70	0
D	70	60

When assessing the scenarios, it is important to take into account the weekly tariff cycle currently in use by the DEEC building: during weekdays, the peak periods are from 09:30 h to 12 h and from 18:30 h to 21 h, and there is no peak tariff during weekends. The rest of the weekly tariffs and their prices are presented in Appendix A. During injection, the value paid for by the grid was calculated to be EUR 0.18 per injected kWh. The cost for the base scenario was calculated to be 76.6 €/day during weekdays and 27.64 €/day for the weekends. These were the reference costs to be used to calculate savings. Other assumptions were made. For the BESS, it was considered that the SoC at the start of the day was always zero, and for V2B it was assumed that the vehicles all had at least 15 kWh available for the grid at the start of the day and that each vehicle was charged with equal amounts of power and energy. The results for the economic scenarios are presented in Figures 9–12.

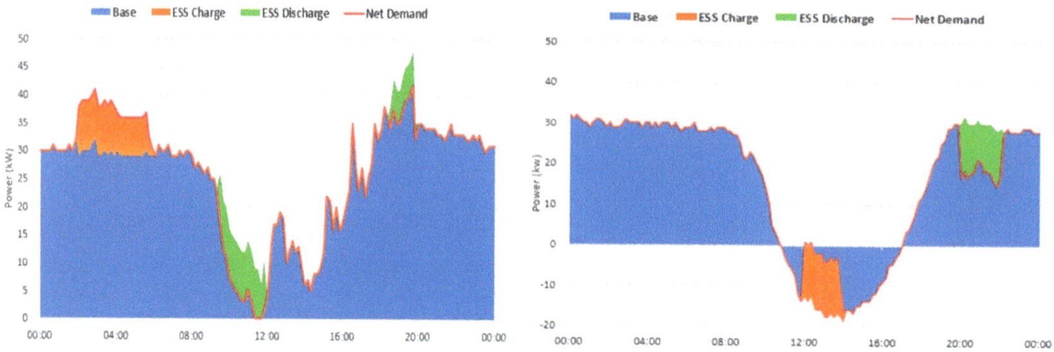

(a) (b)

Figure 9. Scenario A: weekday (**a**); weekend (**b**).

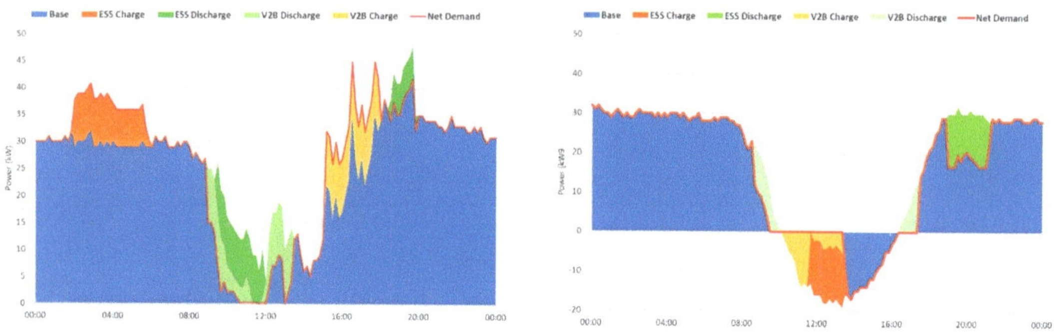

(a) (b)

Figure 10. Scenario B: weekday (**a**); weekend (**b**).

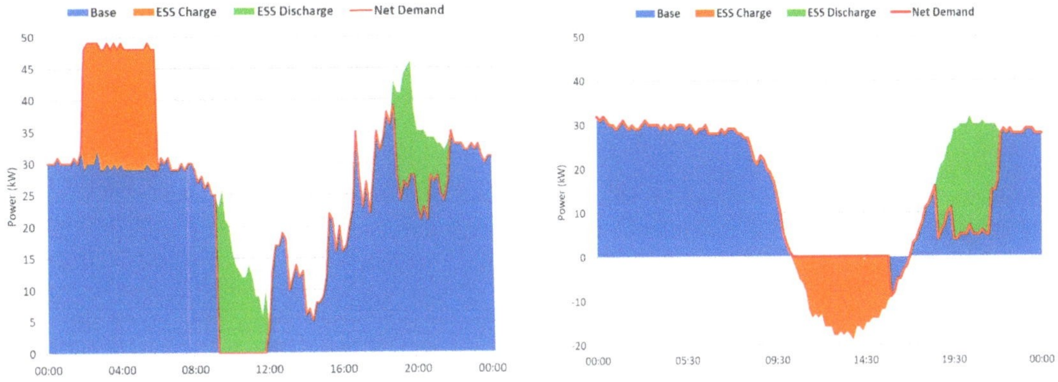

(**a**) (**b**)

Figure 11. Scenario C: weekday (**a**); weekend (**b**).

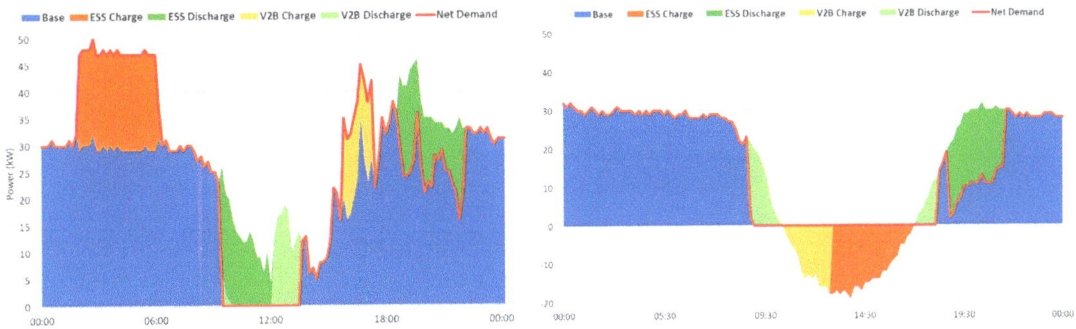

(**a**) (**b**)

Figure 12. Scenario D: weekday (**a**); weekend (**b**).

4. Discussion

4.1. Microgrid Efficiency Analysis

Following the experimental results, the obtained data were gathered and Table 7 was created to summarize the efficiencies of all the tested assets as well as to evaluate the final operational efficiency. All the values for efficiencies presented in the table are the values achieved through the developed experiments, except those for the grid transformer and the efficiency of the EV battery (marked with an asterisk (*) in Table 7), for which the datasheet value for the asset was used.

In the 'Operation' column, several possible energy-transfer actions are presented. For example, the 'PV-Grid' line is the efficiency related to transferring photovoltaic energy directly to the grid, whereas 'G2V' refers to a grid-to-vehicle operation. When SiC is mentioned, this means that the electric vehicle charging was made using the SiC charger. For the efficiencies of the EV chargers, the values at maximum power were used. The efficiency of the operation was then calculated by multiplying the different efficiencies at each stage for each operation. For instance, for the ESS to EV operation, power must be converted from DC to AC, then AC to DC in the charger; lastly, both the storage efficiencies of the ESS battery and the EV battery have to be considered as well for calculating the round-trip efficiency. The real values of round-trip efficiencies have small variations of +−1% due to differences in the battery storage systems, the EV batteries and the operating conditions. Nevertheless, the stated values provide a distinct overview of the efficiencies present for each operation.

Table 7. Microgrid stage efficiencies.

Operation	Conversion		Grid	BESS	EV	EV	EV	Total
	DC-AC	AC-DC	TRF *		Char.	Disc.	Total	
PV-Grid	95%	-	99%	-	-	-	94%	94%
Grid-BESS	-	96.9%	99%	91.23%	-	-	87.51%	87.51%
BESS-Grid	96.52%	-	99%	91.23%	-	-	87.17%	87.17%
BESS-EV (B2V)	96.52%	-	-	91.23%	91.10%	-	76.20%	76.20%
BESS-EV (SiC)	96.52%	-	-	91.23%	95.46%	-	79.85%	79.85%
G2V	-	-	99%	-	91.10%	-	85.68%	85.68%
G2V (SiC)	-	-	99%	-	95.46%	-	89.78%	89.78%
V2G	-	-	99%	-	-	88.25%	83.00%	83.00%
V2G (SiC)	-	-	99%	-	-	96.23%	90.50%	90.50%
V2B (EV-BESS)	-	96.9%	-	91.23%	-	88.25%	74.11%	74.11%
V2B SiC(EV-BESS)	-	96.9%	-	91.23%	-	96.23%	80.82%	80.82%

The purpose of the table is to provide an overview of the expected efficiencies in the assessed microgrid. The table can then be integrated into a decision-making algorithm that takes into account the efficiencies of each microgrid asset, allowing the prediction of energy generation, available capacity and user limitations relevant to choosing when and how to store energy or maximize economic benefit. By analyzing the table, it can be concluded that opting for SiC technology chargers will significantly improve the overall efficiency of operation. Grid operators or building management systems should give priority to charging–discharging using this type of equipment, when available, over conventional chargers. With SiC chargers, storing energy in EVs is more efficient than storing it in BESSs. This decision is also heavily dependent on the number of available SiC charges, user limitations and EV available capacities. The least efficient operations for the microgrid are charging or discharging an electric vehicle directly to or from another BESS. In these types of operations, a significant amount of energy is lost during power conversion. Since there is no direct DC-to-DC charging in this system, power has to be converted twice: once when going into the grid and again when injected either into a BESS or EV. Therefore, charging an EV directly from a BESS should be avoided when taking into account efficiency factors. These results indicate that the installation of a DC bus connecting solar, battery and DC charger systems could be an interesting solution to increase the efficiency of the system even more. This would allow the charging of EVs directly from PV generation (a direct solar charging solution), avoiding the conversion of this energy to AC and back to DC. In the following section, these efficiencies will be used for an assessment of the economic impacts.

4.2. Economic Analysis

Following the economic analysis, the main results are summarized in Table 8.

In scenario A, only the operating BESS is used (27.9 kWh). During the weekday, the batteries store energy during super off-peak hours and discharge during peak hours at lunch and late afternoon. Since there is no PV generation surplus and the peak power is not troublesome to the building, the benefit of operating the BESS during weekdays is restricted to the profit achieved due to the reduction in consumption during peak hours. The calculated daily cost was EUR 72.33 for weekdays, which represents a saving of EUR 426 in electricity consumption per day or 1114 €/annum (for the calculations of yearly savings, 261 weekdays and 104 weekend days were considered). At the weekends, due to the PV generation surplus, the batteries charge during lunch hours. In this case, the battery capacity is not sufficient to absorb all the generation surplus, and about 40 kWh of PV generation is still injected into the grid. The stored energy is then used by the building during the normal tariff period at night (since there is no peak period during weekend days). Without the BESS, the weekend daily electricity cost is EUR 38.06, whereas with the BESS the daily cost decreases to EUR 35.9.

Table 8. Results for the economic scenarios.

Scenario	Grid Injection (kWh)		Energy Consumed in Peak (kWh)		Cost (€/Annum)	
	Weekday	Weekend	Weekday	Weekend	Weekday	Weekend
Base	0	74.83	133.33	N/A	19,993	3958
A	0	48.05	107.29	N/A	18,879 (−5.56%)	3733 (−5.66%)
B	0	28.98	93.37	N/A	18,648 (−6.72%)	3606 (−8.86%)
C	0	5.33	61.83	N/A	15.865 (−20.64%)	3179 (−19.67%)
D	0	0	61.16	N/A	15,980 (−20.07%)	3231 (−18.36%)

Scenario B takes advantage of both the BESS and V2B storage capacity. During the weekday, the BESS charges at night and discharges during daily peak hours. At the weekend, the BESS waits until noon to start charging and until the late afternoon for the discharge. The cost on weekdays is reduced to EUR 71.45 per day and on weekends to EUR 34.68 per day. For V2B, during weekend days, the same schedule of availability was used as that on weekdays (9 am to 6 pm). Therefore, the V2B system starts discharging the battery of an EV as soon as the client enters the workplace. (Note: V2B power is capped at 10 kW.) Since it was considered a restriction that the energy used by the building be returned to the EV by the end of the day to the V2B client, it was only possible to utilize 18.8 kWh of the storage capacity during weekends. This value is the sum of the energy discharged from the vehicle in the morning period (5.3 kWh) and the predicted late-afternoon discharged energy (11.5 kWh) before 6 pm. Even though there is still leftover capacity in the EVs, if the building overcharges the vehicles during the period in which there is an excess of solar generation it will not be able to recover that charge due to the time limit imposed by the EV clients, for whom the day ends at 6 pm. If a client leaves earlier, before 5 pm, then their vehicle would have charged more energy than the amount supplied to the building. This is not necessarily a problem, and benefits/consequences will vary depending on the economic model adopted by the building operator for V2B, but for this work the aim was to ensure flexibility and the maintenance of an energy equilibrium; if the building operator wishes to prevent this issue, V2B charging would have to be halted in the afternoon, as soon as the charged value was equal to the discharged value in the morning. In both cases, the usable storage capacity for V2B will be lower than the initially allotted capacity of 30 kWh. If there were no schedule limitations, then it would be possible to recoup all the charged energy in the EVs later in the afternoon. This is a barrier to the efficacy of V2B and highlights the advantage of 24/7 BESS availability. Due to the constraint of energy equilibrium, there will be an increase in demand late in the afternoon on workdays. The system operator must therefore plan accordingly, so that neither the client is left lacking energy nor is the electricity bill increased due to charging during peak hours.

Due to the generation surplus during weekends, a scenario in which the BESS is upgraded to a larger capacity was considered. In scenario C, the scheduling for charging/discharging is the same as in scenario A: during weekdays, the BESS charges during super off-peak hours (2 am to 6 am) and discharges during peak hours. At weekends, it waits for the PV generation surplus to charge and discharges during the evening. Using this method, the energy injected into the grid decreases from 74.83 kWh to 5.33 kWh—a reduction of more than 90%. With a BESS with 70 kWh capacity, the daily costs are reduced to EUR 60.78 per weekday and the energy injected into the grid decreases from around 7 MWh/year to only 0.5 MWh/year. The cost for weekends is reduced to EUR 30.56—a reduction of 19.6% from the base cost.

Finally, in the last scenario, 60 kWh of V2B storage capacity was added to the BESS. The schedules are similar to those in scenario C, and during weekdays the batteries charge at night, during the super off-peak period, and discharge at peak hours. As soon as the EV clients enter the workplace, the system starts discharging the batteries using the V2B system. According to the defined constraints, the system must try to maintain a zero sum between the charged and discharged energy in the EVs. Therefore, during the afternoon, the amount of energy charged to the parked vehicles is limited by the amount of energy discharged into the grid during the morning period and the predicted late-afternoon energy discharge. When the maximum usable storage from V2B is reached, the system initiates the BESS charging cycle. As soon as the energy surplus from PV generation is over, the V2B system starts discharging until 6 pm or when the maximum energy limit is reached. The gap between V2B and battery discharge is due to clients leaving at 6 pm, but batteries only start discharging at 6:30 pm, when the peak tariff period starts. Once again, due to the constraints of the V2B schedule and energy equilibrium, the usable storage capacity is reduced. From the available 70 kWh allocated to the EV users, only a maximum of 25.83 kWh was used at once. In comparison, for the BESS, 49 kWh of storage capacity was used. In this scenario, the morning peak period consumption is eliminated and all the excess surplus energy is used by the system. This scenario leads to savings of EUR 61.23 per weekday and EUR 31.06 at weekends.

As can be seen from the comparisons presented in Table 8, it was possible to significantly reduce the energy injected into the grid and the energy consumed during peak hours. Expanding the existing BESS to 70 kWh (Scenario C) is more cost-effective than trying to complement the existing battery storage capacity with V2B (Scenario B). However, if the building has available EV users, it may be faster and easier to create additional storage using V2B than to invest in upgrading the existing system.

The usability of V2B is heavily limited by its availability. The small difference in the reduction in energy consumed in peak periods between scenarios C and D is caused by the schedule limitation for V2B, since it is not possible to use that storage capacity to cover the late-afternoon peak. During peak periods, priority is given to the BESS, due to the system being more efficient (Table 7). As such, if the BESS is capable of soaking up all the energy consumed during peak, it is not economically viable to rely on V2B, because it will not be possible to take advantage of a difference in price from the tariff, leading to reduced savings. This was demonstrated in scenario C during weekdays, in the time periods where V2B charges and discharges had the same cost, having no direct economic benefit. Building operators may still opt to use V2B in similar scenarios to reap other technical benefits.

5. Conclusions

This work has presented a technical and economic analysis of a battery energy storage system and a vehicle-to-building system integrated into a campus microgrid. The results were gathered using the distribution systems already operating in a campus microgrid. For the technical analysis, full charge–discharge cycles were considered to calculate the round-trip efficiencies of several common interactions in the microgrid. A prototype SiC bidirectional charger was analyzed and compared with a conventional V2G charger, and a power quality analysis of vehicle chargers was also conducted. The results showed that the SiC technology is superior to conventional IGBT bidirectional chargers, with operating efficiency increases ranging from 10 to 26% depending on the power used. The results also showed that, in most cases, using energy stored in batteries is more efficient than using the same amount of energy stored in vehicles. The more stages and transitions of power there are in a microgrid, the less efficient the overall performance will be. The installation of a DC bus connecting PV generation, batteries and DC charging systems could significantly increase the round-trip efficiency of all energy transfers.

For the economic analysis, the impacts of the BESS and V2B regarding the reduction of excess grid injection and peak-hour energy consumption were explored. Four different scenarios with different levels of storage capacity were simulated. It can be concluded that, by using both a BESS and V2B, it is possible to significantly reduce the peak energy consumption and the grid energy injection. However, a BESS is more effective in reducing peak energy consumption and more cost-effective than V2B. The usability of V2B is hindered by the availability of EV clients and tariff periods. If it is not possible to take advantage of different tariff prices when V2B is available, it will not be possible to obtain a cost advantage. The benefits of V2B are also limited by the conditions agreed upon between the microgrid operator and the EV users. In this analysis, it was decided that the quantity of energy charged into the vehicle by the microgrid must be recovered by the end of the same day. This led to situations in which consumption shifted but without originating any economic benefit. In comparison, the BESS was more efficient, and the 24/7 availability provided the most flexibility for the microgrid. As for V2B, due to the volatility and relative unpredictability of the users' schedules, it should be seen as a way to complement existing storage when available and not as the main means of achieving flexibility.

Electric vehicles and energy storage systems have been shown to be primary agents with respect to achieving the objective of nZEB status. The results of this work provide insights and a technical table to be used by building and microgrid designers or operators for the development of better system architectures and better planning for existing resources.

Author Contributions: A.F.M.C.: conceptualization, methodology, data collection and assessment, writing—original draft, writing—review and editing. P.M.: conceptualization, writing—review and editing, supervision. A.T.d.A.: project funding, writing—review and editing, supervision. All authors have read and agreed to the published version of the manuscript.

Funding: This research was supported by the Fundação para a Ciência e a Tecnologia (FCT), through the projects UIDB/00048/2020, ResiMicrogrid (EXPL/EEI-EEE/1611/2021) and ML@GridEdge (UTAP-EXPL/CA/0065/2021), and by the European Regional Development Fund and national funds, through the project EVAICharge (CENTRO-01-0247-FEDER-047196).

Data Availability Statement: Not applicable.

Conflicts of Interest: The authors declare no conflict of interest. The funders had no role in the design of the study; in the collection, analyses, or interpretation of data; in the writing of the manuscript; or in the decision to publish the results.

Appendix A

The electricity tariffs and prices per kWh paid by the DEEC building are presented in Table A1. A decision was made to use the prices for 2021 so as to avoid the high volatility of prices due to the socio-economic situation in Europe in 2022. The price paid per kWh injected into the grid is the median value of the daily OMIE—the Iberian Electricity Market Operator—for March 2021.

Table A1. Weekly tariff cycle for the DEEC building.

Period	Weekdays	Weekends	Price
Peak	09:30–12:00 18:30–21:00	N/A	0.205932 €/kWh
Normal	07:00–09:30 12:00–18:30 21:00–24:00	09:30–13:00 18:30–22:00	0.115202 €/kWh
Off-Peak	00:00–02:00 06:00–07:00	00:00–02:00 06:00–09:30 13:00–18:30 22:00–24:00	0.090208 €/kWh
Super Off-Peak	02:00–06:00	02:00–06:00	0.062263 €/kWh
Grid Injection	N/A	N/A	0.040842 €/kWh

References

1. U.S. Department of Energy. *DOE Microgrid Workshop Report*; U.S. Department of Energy: San Diego, CA, USA, 2011; p. 32.
2. Pullins, S. Why Microgrids Are Becoming an Important Part of the Energy Infrastructure. *Electr. J.* **2019**, *32*, 17–21. [CrossRef]
3. Hayden, E. *Introduction to Microgrids*; Securicon: Caracas, Venezuela, 2013.
4. Gao, D.W. (Ed.) *Energy Storage for Sustainable Microgrid*; Academic Press: Oxford, UK, 2015; p. IV.
5. California Independent System Operator, Flexible Resources Fast Facts 2016. Available online: https://www.caiso.com/documents/flexibleresourceshelprenewables_fastfacts.pdf (accessed on 16 October 2022).
6. Söder, L.; Lund, P.D.; Koduvere, H.; Bolkesjø, T.F.; Rossebø, G.H.; Rosenlund-Soysal, E.; Skytte, K.; Katz, J.; Blumberga, D. A Review of Demand Side Flexibility Potential in Northern Europe. *Renew. Sustain. Energy Rev.* **2018**, *91*, 654–664. [CrossRef]
7. Good, N.; Ellis, K.A.; Mancarella, P. Review and Classification of Barriers and Enablers of Demand Response in the Smart Grid. *Renew. Sustain. Energy Rev.* **2017**, *72*, 57–72. [CrossRef]
8. EV-Volumes—The Electric Vehicle World Sales Database. Available online: https://www.ev-volumes.com/ (accessed on 21 October 2022).
9. EVO Report 2022 | BloombergNEF | Bloomberg Finance LP. BloombergNEF. Available online: https://about.bnef.com/electric-vehicle-outlook/_fastfacts.pdf (accessed on 16 October 2022).
10. Kern, T.; Dossow, P.; Morlock, E. Revenue Opportunities by Integrating Combined Vehicle-to-Home and Vehicle-to-Grid Applications in Smart Homes. *Appl. Energy* **2022**, *307*, 118187. [CrossRef]
11. Ravi, S.S.; Aziz, M. Utilization of Electric Vehicles for Vehicle-to-Grid Services: Progress and Perspectives. *Energies* **2022**, *15*, 589. [CrossRef]
12. Garcia-Villalobos, J.; Zamora, I.; Martín, J.; Junquera Martínez, I.; Eguia, P. Delivering Energy from PEV Batteries: V2G, V2B and V2H Approaches. In Proceedings of the International Conference on Renewable Energies and Power Quality (ICREPQ'15), La Coruna, Spain, 25–17 March 2015.
13. Popova, R.; Nguyen, H.Q.; Sprengeler, M.; Feizi, T.; la Rotta, M.R.; Stryi-Hipp, G. Economic Evaluation of Electric Vehicle Charging Infrastructure Integration in Microgrids. In Proceedings of the 2020 Forum on Integrated and Sustainable Transportation Systems (FISTS), Online, 3–5 November 2020; pp. 45–50.
14. Ghaderi, A.; Nassiraei, A.A.F. The Economics of Using Electric Vehicles for Vehicle to Building Applications Considering the Effect of Battery Degradation. In Proceedings of the IECON 2015—41st Annual Conference of the IEEE Industrial Electronics Society, Yokohama, Japan, 9–15 November 2015; pp. 003567–003572.
15. Moura, P.; Sriram, U.; Mohammadi, J. Sharing Mobile and Stationary Energy Storage Resources in Transactive Energy Communities. In Proceedings of the 2021 IEEE Madrid PowerTech, Madrid, Spain, 28 June–2 July 2021; p. 6.
16. Moura, P.; Yu, G.K.W.; Mohammadi, J. Management of Electric Vehicles as Flexibility Resource for Optimized Integration of Renewable Energy with Large Buildings. In Proceedings of the 2020 IEEE PES Innovative Smart Grid Technologies Europe (ISGT-Europe), Delft, The Netherlands, 26–28 October 2020; pp. 474–478.
17. Tanguy, K.; Dubois, M.R.; Lopez, K.L.; Gagné, C. Optimization Model and Economic Assessment of Collaborative Charging Using Vehicle-to-Building. *Sustain. Cities Soc.* **2016**, *26*, 496–506. [CrossRef]
18. Yoon, S.; Park, K.; Hwang, E. Connected Electric Vehicles for Flexible Vehicle-to-Grid (V2G) Services. In Proceedings of the 2017 International Conference on Information Networking (ICOIN), Da Nang, Vietnam, 11–13 January 2017; pp. 411–413.
19. Ancillotti, E.; Bruno, R.; Crisostomi, E.; Tucci, M. Using Electric Vehicles to Improve Building Energy Sustainability. In Proceedings of the 2014 IEEE International Electric Vehicle Conference (IEVC), Graz, Austria, 17–19 December 2014; pp. 1–7.
20. Barone, G.; Buonomano, A.; Calise, F.; Forzano, C.; Palombo, A. Building to Vehicle to Building Concept toward a Novel Zero Energy Paradigm: Modelling and Case Studies. *Renew. Sustain. Energy Rev.* **2019**, *101*, 625–648. [CrossRef]
21. Barone, G.; Buonomano, A.; Forzano, C.; Giuzio, G.F.; Palombo, A. Increasing Self-Consumption of Renewable Energy through the Building to Vehicle to Building Approach Applied to Multiple Users Connected in a Virtual Micro-Grid. *Renew. Energy* **2020**, *159*, 1165–1176. [CrossRef]
22. Umoren, I.A.; Shakir, M.Z.; Tabassum, H. Resource Efficient Vehicle-to-Grid (V2G) Communication Systems for Electric Vehicle Enabled Microgrids. *IEEE Trans. Intell. Transp. Syst.* **2021**, *22*, 4171–4180. [CrossRef]
23. Schram, W.; Brinkel, N.; Smink, G.; van Wijk, T.; van Sark, W. Empirical Evaluation of V2G Round-Trip Efficiency. In Proceedings of the 2020 International Conference on Smart Energy Systems and Technologies (SEST), Istanbul, Turkey, 7–9 September 2020; pp. 1–6.
24. Kriukov, A.; Gavrilas, M. Energy/Cost Efficiency Study on V2G Operating Mode for EVs and PREVs. In Proceedings of the 2019 8th International Conference on Modern Power Systems (MPS), Shanghai, China, 21–23 May 2019; pp. 1–6.
25. Coban, H.H.; Lewicki, W.; Sendek-Matysiak, E.; Łosiewicz, Z.; Drożdż, W.; Miśkiewicz, R. Electric Vehicles and Vehicle–Grid Interaction in the Turkish Electricity System. *Energies* **2022**, *15*, 8218. [CrossRef]
26. Koul, B.; Singh, K.; Brar, Y.S. Chapter 4—An Introduction to Smart Grid and Demand-Side Management with Its Integration with Renewable Energy. In *Advances in Smart Grid Power System*; Tomar, A., Kandari, R., Eds.; Academic Press: Oxford, UK, 2021; pp. 73–101.
27. Zheng, D.; Zhang, W.; Netsanet Alemu, S.; Wang, P.; Bitew, G.T.; Wei, D.; Yue, J. Chapter 9—Tertiary Control of Microgrid. In *Microgrid Protection and Control*; Zheng, D., Zhang, W., Netsanet Alemu, S., Wang, P., Bitew, G.T., Wei, D., Yue, J., Eds.; Academic Press: Oxford, UK, 2021; pp. 255–295.

28. Imani, M.H.; Ghadi, M.J.; Ghavidel, S.; Li, L. Demand Response Modeling in Microgrid Operation: A Review and Application for Incentive-Based and Time-Based Programs. *Renew. Sustain. Energy Rev.* **2018**, *94*, 486–499. [CrossRef]
29. Mazidi, M.; Zakariazadeh, A.; Jadid, S.; Siano, P. Integrated Scheduling of Renewable Generation and Demand Response Programs in a Microgrid. *Energy Convers. Manag.* **2014**, *86*, 1118–1127. [CrossRef]
30. Gomes, I.; Melicio, R.; Mendes, V.M.F. Assessing the Value of Demand Response in Microgrids. *Sustainability* **2021**, *13*, 5848. [CrossRef]
31. Beyazıt, M.A.; Taşcıkaraoğlu, A.; Catalão, J.P.S. Cost Optimization of a Microgrid Considering Vehicle-to-Grid Technology and Demand Response. *Sustain. Energy Grids Netw.* **2022**, *32*, 100924. [CrossRef]
32. Harsh, P.; Das, D. Energy Management in Microgrid Using Incentive-Based Demand Response and Reconfigured Network Considering Uncertainties in Renewable Energy Sources. *Sustain. Energy Technol. Assess.* **2021**, *46*, 101225. [CrossRef]
33. Understanding Demand-Side Management in Microgrid Systems. Available online: https://www.hivepower.tech//blog/understanding-demand-side-management-in-microgrid-systems (accessed on 16 October 2022).
34. Battery Pack Prices Fall to an Average of $132/KWh, but Rising Commodity Prices Start to Bite. *BloombergNEF*, 30 November 2021. Available online: https://about.bnef.com/blog/battery-pack-prices-fall-to-an-average-of-132-kwh-but-rising-commodity-prices-start-to-bite/ (accessed on 16 October 2022).
35. Dubarry, M. *Electric Vehicle Battery Durability and Reliability Under Electric Utility Grid Operations*; Hawaii Natural Energy Institute: Honolulu, HI, USA, 2017.
36. Wang, D.; Coignard, J.; Zeng, T.; Zhang, C.; Saxena, S. Quantifying Electric Vehicle Battery Degradation from Driving vs. Vehicle-to-Grid Services. *J. Power Sources* **2016**, *332*, 193–203. [CrossRef]
37. Shinzaki, S.; Sadano, H.; Maruyama, Y.; Kempton, W. *Deployment of Vehicle-to-Grid Technology and Related Issues*; SAE International: Warrendale, PA, USA, 2015.
38. Apostolaki-Iosifidou, E.; Codani, P.; Kempton, W. Measurement of Power Loss during Electric Vehicle Charging and Discharging. *Energy* **2017**, *127*, 730–742. [CrossRef]
39. A Policy Framework for Climate and Energy in the Period from 2020 to 2030—European Environment Agency. Available online: https://www.eea.europa.eu/policy-documents/com-2014-15-final (accessed on 16 October 2022).
40. LG RESU10H Installation Manual. Available online: https://www.manualslib.com/manual/1313452/Lg-Resu10h.html (accessed on 16 October 2022).
41. SMA Solar Technology. *User Manual—Sunny Boy Storage 3.8-US/5.0-US/6.0-US. 88*; SMA Solar Technology: Niestetal, Germany, 2019.
42. V2G—Magnum Cap. Available online: https://magnumcap.com/v2g/ (accessed on 16 October 2022).
43. Loncarski, J.; Ricco, M.; Monteiro, V.; Monopoli, V.G. Efficiency Comparison of a Dc-Dc Interleaved Converter Based on SiC-MOSFET and Si-IGBT Devices for EV Chargers. In Proceedings of the 2020 IEEE 14th International Conference on Compatibility, Power Electronics and Power Engineering (CPE-POWERENG), Setúbal, Portugal, 8–10 July 2020; Volume 1, pp. 517–522.
44. Ucer, E.; Buckreus, R.; Haque, M.E.; Kisacikoglu, M.; Sozer, Y.; Harasis, S.; Guven, M.; Giubbolini, L. Analysis, Design, and Comparison of V2V Chargers for Flexible Grid Integration. *IEEE Trans. Ind. Appl.* **2021**, *57*, 4143–4154. [CrossRef]
45. Correia, A.; Ferreira, L.M.; Coimbra, P.; Moura, P.; de Almeida, A.T. Smart Thermostats for a Campus Microgrid: Demand Control and Improving Air Quality. *Energies* **2022**, *15*, 1359. [CrossRef]
46. Bauer, P.; Ram, G. *Direct Solar EV Charging with V2G and Intelligent EMS. 21*; TU Delft: Delft, The Netherlands, 2020.
47. Benefits of Using SiC Power in EV Fast Chargers | Wolfspeed. Available online: https://www.wolfspeed.com/knowledge-center/article/sic-drives-cost-performance-benefits-in-ev-fast-chargers/ (accessed on 16 October 2022).
48. Ditze, S.; Ehrlich, S.; Weitz, N.; Sauer, M.; Aßmus, F.; Sacher, A.; Joffe, C.; Seßler, C.; Meißner, P. A High-Efficiency High-Power-Density SiC-Based Portable Charger for Electric Vehicles. *Electronics* **2022**, *11*, 1818. [CrossRef]
49. Peterson, S.B.; Apt, J.; Whitacre, J.F. Lithium-Ion Battery Cell Degradation Resulting from Realistic Vehicle and Vehicle-to-Grid Utilization. *J. Power Sources* **2010**, *195*, 2385–2392. [CrossRef]
50. Bishop, J.D.K.; Axon, C.J.; Bonilla, D.; Tran, M.; Banister, D.; McCulloch, M.D. Evaluating the Impact of V2G Services on the Degradation of Batteries in PHEV and EV. *Appl. Energy* **2013**, *111*, 206–218. [CrossRef]
51. Petit, M.; Prada, E.; Sauvant-Moynot, V. Development of an Empirical Aging Model for Li-Ion Batteries and Application to Assess the Impact of Vehicle-to-Grid Strategies on Battery Lifetime. *Appl. Energy* **2016**, *172*, 398–407. [CrossRef]
52. Guo, J.; Yang, J.; Lin, Z.; Serrano, C.; Cortes, A.M. Impact Analysis of V2G Services on EV Battery Degradation—A Review. In Proceedings of the 2019 IEEE Milan PowerTech, Milano, Italy, 23–27 June 2019; pp. 1–6.
53. Lucas, A.; Bonavitacola, F.; Kotsakis, E.; Fulli, G. Grid Harmonic Impact of Multiple Electric Vehicle Fast Charging. *Electr. Power Syst. Res.* **2015**, *127*, 13–21. [CrossRef]
54. Megha, A.; Mahendran, N.; Elizabeth, R. Analysis of Harmonic Contamination in Electrical Grid Due to Electric Vehicle Charging. In Proceedings of the 2020 Third International Conference on Smart Systems and Inventive Technology (ICSSIT), Tirunelveli, India, 20–22 August 2020; pp. 608–614.
55. IEEE SA—IEEE 519-2022. Available online: https://standards.ieee.org/ieee/519/10677/ (accessed on 29 October 2022).

energies

MDPI

Article

Optimizing Fuel Efficiency on an Islanded Microgrid under Varying Loads

Joo Won Lee [1], Emily Craparo [1], Giovanna Oriti [2,*] and Arthur Krener [3]

1 Department of Operations Research, Naval Postgraduate School, Monterey, CA 93943, USA
2 Department of Electrical and Computer Engineering, Naval Postgraduate School, Monterey, CA 93943, USA
3 Department of Applied Mathematics, Naval Postgraduate School, Monterey, CA 93943, USA
* Correspondence: goriti@nps.edu

Abstract: Past studies of microgrids have been based on measurements of fuel consumption by generators under static loads. There is little information on the fuel efficiency of generators under time-varying loads. To help analyze the impact of time-varying loads on optimal generator operation and fuel consumption, we formulate a mixed-integer linear optimization model to plan generator and energy storage system (ESS) operation to satisfy known demands. Our model includes fuel consumption penalty terms on time-varying loads. We exercise the model on various scenarios and compare the resulting optimal fuel consumption and generator operation profiles. Our results show that the change in fuel efficiency between scenarios with the integration of ESS is minimal regardless of the imposed penalty placed on the generator. However, without the assistance of the ESS, the fuel consumption increases dramatically with the penalty imposed on the generator. The integration of an ESS improves fuel consumption because the ESS allows the generator to minimize power output fluctuation. While the presence of a penalty term has a clear impact on generator operation and fuel consumption, the exact type and weight of the penalty appears insignificant; this may provide useful insight for future studies in developing a real-time controller.

Keywords: microgrid optimization; energy storage system; time-varying loads; fuel efficiency; energy management; generator

Citation: Lee, J.W.; Craparo, E.; Oriti, G.; Krener, A. Optimizing Fuel Efficiency on an Islanded Microgrid under Varying Loads. *Energies* **2022**, *15*, 7943. https://doi.org/10.3390/en15217943

Academic Editor: Alexander Micallef

Received: 15 September 2022
Accepted: 21 October 2022
Published: 26 October 2022

1. Introduction

Demand for energy is increasing significantly worldwide. This is particularly true in the United States Department of Defense (DoD), which is the largest consumer of energy in the federal government [1]. As warfighting transitions into the cyber and space domains, energy "has been and will remain a fundamental enabler of military capability" according to the United States (US) Office of Assistant Secretary of Defense for Energy, Installations, and Environment [2]. Military organizations need reliable and sustainable microgrid power systems to effectively generate energy required for deployed military operating bases.

Microgrid Operations

When temporary military bases are built in forward operating areas such as the Middle East, self-sustainability is one of the most important features to take into account, as one cannot rely on local power grids. An islanded microgrid is formed when an electrical grid is capable of operating in isolation from the main grid [3]. Islanded microgrid operations are very common in military settings.

An islanded microgrid typically consists of generators, an energy storage system (ESS), and loads that need to be met. A centralized controller calculates the optimal power flow balance between the generators and the ESSs [4], as shown in Figure 1. This ensures that all critical loads are met during all microgrid operating modes [5,6]. As a result, the controller improves reliability, reduces cost, and diversifies power generation. As microgrids are intended to be self-sufficient, the ESS is one of the primary ways to ensure that the system is stabilized against fluctuating loads and unmanaged energy sources [5].

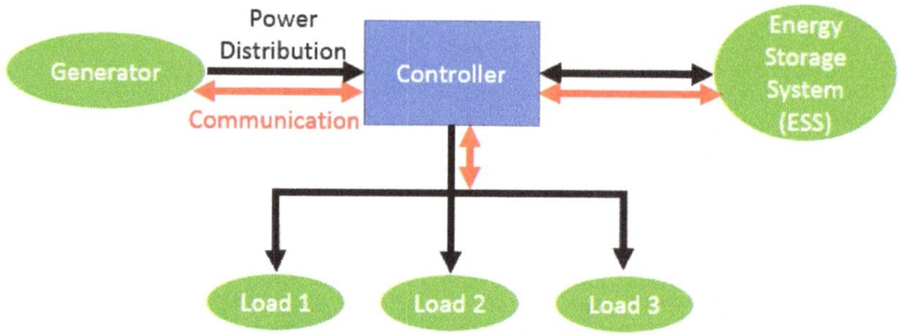

Figure 1. Notional microgrid controller flowchart with power distribution and communication flow. Adapted from Craparo and Sprague [7].

A key function of the controller is to determine the optimal power flow among the generators, ESSs, and loads in such a way as to satisfy demand while minimizing fuel consumption. To accomplish this, the controller must model fuel consumption as a function of generator output. Existing studies primarily utilize fuel efficiency (or consumption) profiles based on static generator output (e.g., [8–11]). To build these profiles, generator manufacturers run generators at various constant loads for long periods of time and measure the resulting fuel consumption [12]. In practice, loads encountered on islanded microgrids are generally not constant but instead vary over time, often quite rapidly. To our knowledge, no data exist quantifying generator efficiency under time-varying loads. However, data from other settings (e.g., automotive fuel efficiency) suggest that fluctuating loads may significantly degrade generator efficiency [13].

In this study, we develop an optimization-based approach to model loss of efficiency with fluctuating generator output and to optimally plan generator and ESS operations in such a way as to minimize fuel consumption while satisfying demand. To our knowledge, this is the first work to consider efficiency losses due to fluctuating generator output. In addition to potential fuel savings, reducing generator output fluctuations can improve generator stability as well as operational lifetime [14,15]. Additionally, any potential fuel savings at an islanded microgrid are generally magnified when one considers the fully burdened cost of fuel, accounting for the monetary and energy costs associated with transporting fuel to the location at which it will be consumed. Ref. [8] provides an overview of this cost in military settings, including both the energy cost and cost in human life to transport fuel to a forward operating theater.

2. Background and Literature Review

Previous studies explore different approaches to vary the architecture of microgrids, find unique ways to optimally schedule energy distribution, and formulate optimization models. It is imperative to optimize microgrids to be cost effective and reliable to deploy based on measure of fuel consumption.

2.1. Microgrid Architecture Variation

The two main energy resources that make up an islanded microgrid architecture are the generator(s) and the ESS. It is important to determine the correct size of these resources in a microgrid to meet the power demand at any given time [16,17]. One way to properly size microgrid components is in accordance to the peak-load demand criteria [18]. In addition to using peak-load demand data, research has been conducted on integrating an optimal hybrid photovoltaic (PV)/wind/diesel generating system combination identified by a sizing algorithm based on power supply availability and power system design specifications [19].

Building off of a rudimentary sizing algorithm using past daily and seasonal load fluctuation, a more sophisticated dynamic energy flow model is used to determine the appropriate size of the microgrid components [20]. Then, past weather information such as solar and wind data are utilized to maximize the microgrid's "islanding time", which is defined as the time the microgrid can operate in a self-sufficient manner without external fuel supplies or connection to a main power grid [21].

An ESS sizing methodology uses a mixed-integer linear program (MILP) to choose an optimal storage size by taking initial investment costs and microgrid operating costs into account [22]. Similarly, some researchers have taken a more economic approach by introducing a cost–benefit analysis using forecast data to minimize total microgrid costs while maximizing the benefits from the respective storage size through an MILP technique [23].

2.2. Optimal Energy Scheduling Techniques

Optimal energy scheduling improves energy utilization while minimizing generator fuel consumption costs in a microgrid. An initial optimal energy scheduling technique was used to show that microgrid with ESS saves a significant amount of fuel compared to a microgrid without an energy storage unit [24]. Indeed, multiple studies have identified the ESS as a key component to ensure cost reduction [24,25]. Energy scheduling is especially important when ESSs are introduced into the microgrid architecture, as the system has the ability to store additional power for future use [22,23]. To reduce fuel cost and increase fuel efficiency, load scheduling becomes a crucial addition to the controller algorithm.

Another way to minimize fuel cost is to implement a day-ahead scheduling optimization technique for various microgrid operation modes such as utility grid-connected mode and off-grid operation mode [25]. Some work in day-ahead scheduling incorporates weather forecast data and models its impact on renewable production [26,27].

Instead of modeling a microgrid as a single entity, the particle swarm optimization (PSO) technique models the energy resources as individual particles that are parallel with each other, which creates a large network made up of many entities [28,29]. The parallel modeling technique enables shorter computation time when compared to a general MILP approach. The PSO optimization approach reduces computation time while minimizing energy production expense [29].

Ref. [30] considers the optimal scheduling of a diesel generator and an ESS, where the ESS is modeled using a detailed nonlinear model. They address the resulting planning problem with a combination of MILP and PSO. Ref. [31] models an islanded microgrid using a Markov decision process framework and then uses approximate dynamic programming to solve the resulting schedule optimization problem, while [32] considers the problem of placing distributed generators in such a way as to maximize efficiency and reliability.

2.3. Fuel Consumption Minimization

The main goal in most islanded microgrid research is to minimize fuel consumption, which often corresponds to maximizing generator efficiency. A common approach to quantify fuel consumption is using a cost optimization scheme of various power-sharing techniques to share the load between energy resources in a microgrid [33]. On highly nonlinear power-sharing schemes, linear, nonlinear, and dynamic strategies have sought to maximize microgrid efficiency [20,22,24,25,33].

Power-sharing techniques have been used to determine if additional electrical loads with expensive operating costs would effect the generator fuel consumption [10]. Other studies have minimized fuel consumption by integrating additional and various combinations of energy resources such as PV cells, batteries, and generator cycling in a microgrid architecture [34]. As fuel consumption is an important measure of performance, the goal of this paper is to minimize fuel consumption while ensuring that the power demand is efficiently met.

Some researchers have considered microgrids in settings other than strictly land-based. For instance, Ref. [35] considers shipboard microgrids and propose a hierarchical coordinated control approach to effectively manage the distinct operating modes inherent in these systems; Ref. [36] studies the hybridization of railway vehicles as a potential means to improve efficiency while satisfying various operational constraints. They demonstrate the potential for significant emmissions reduction by incorporating a properly sized ESS.

2.4. Novel Contribution and Organization

To our knowledge, all previous studies, including those in our literature review, have modeled generator fuel efficiency based on fuel curves derived by measuring fuel efficiency under static loading conditions (e.g., [12]). Experience from other domains (e.g., automotive) suggests that generator efficiency may be significantly degraded under dynamic load conditions such as those observed in microgrid operations [13]. To study the potential impact of this degradation, we introduce a generator penalty term designed to model a loss of efficiency as the generator output fluctuates. We include an ESS in our notional microgrid to allow the generator to produce a more constant output, despite fluctuating demand.

We first consider a linear approach where the fractional change in generator power is multiplied by a predefined scalar penalty coefficient and added to the cumulative fuel consumption. Second, we use a piecewise linear approach to vary the penalty coefficient based on the fractional change in generator power.

This paper is organized as follows. Section 3 describes our microgrid architecture and MILP optimization model. Section 4 exercises the model on a case study and highlights important insights. Section 5 presents the conclusion and final thoughts from this study as well as various recommendations for future research areas.

3. Methodology

This section describes our MILP optimization model of a forward operating base (FOB) microgrid, where the primary objective is to minimize the overall generator fuel consumption while satisfying a required power demand. Our microgrid equipment consists of a fuel-based generator and an ESS, and we exercise our model using power demand data collected from a US FOB located in the Middle East [37].

The MILP acts as a rudimentary power system controller which controls the power flow between the generator and ESS to satisfy the demand. Figure 1 depicts our controller architecture, including the power flow and communication flow. The controller maintains communication between the three separate components; however, power is only produced by the generator. This power may satisfy the demand directly or charge the ESS. The ESS may discharge power to meet demand.

3.1. Microgrid Architecture

This section describes the components of our notional FOB microgrid, as shown schematically in Figure 1.

3.1.1. Fuel-Based Generator

We model a 60 kilowatt (kW) Advanced Medium Mobile Power Sources (AMMPS) generator. This is a US Army-authorized power-generating unit that replaces the second generation Tactical Quiet Generator (TQG). We obtained steady-state generator fuel consumption data for four power settings, shown in Figure 2, from the US Army Base Camp Integration Lab [38].

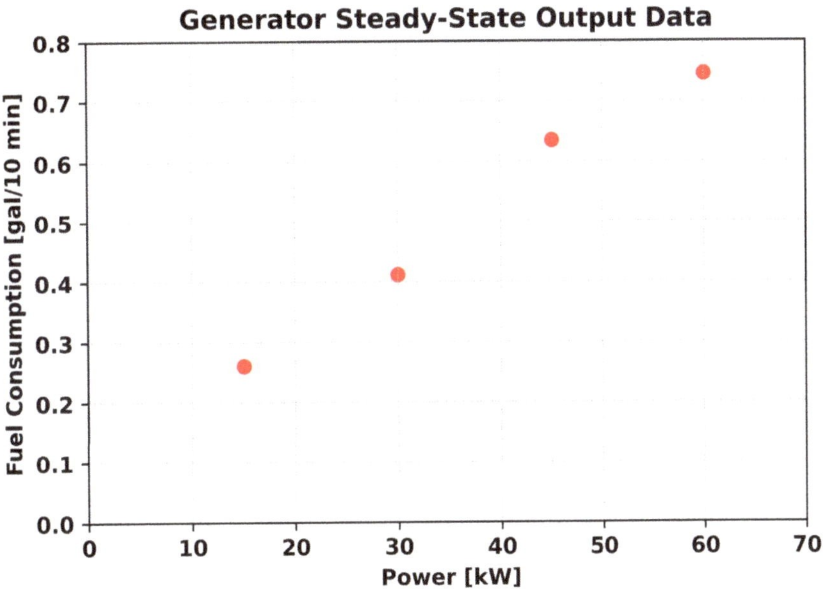

Figure 2. Steady-state fuel consumption data from a 60 kW AMMPS generator at four distinct operating regions obtained from previous studies [38].

3.1.2. Energy Storage System

We model a 25 kilowatt-hours (kWh) ESS with a maximum charge rate and discharge rate of 20 kW with 90% round-trip efficiency (RTE); these values approximate the characteristics of a lithium-ion battery [39]. We constrain the ESS to maintain a state of charge (SOC) between 20% and 80% in order to prolong the lifespan of the ESS [40,41]. We initialize the ESS to an SOC of 50% in the first time period to simulate a continuously operating power system.

3.2. Optimization Model

The overall objective of the MILP optimization model is to minimize generator fuel consumption while satisfying demand $demand_t$ in each time period t in T. Demand may be satisfied by the generator, the ESS, or some combination of the two. Additionally, the generator can produce power exceeding the demand, and the additional power charges the ESS. We denote the power produced by the generator in time step t as gen_t, the power used to charge the ESS (battery) as $cbatt_t$, and the power discharged from the battery as $dbatt_t$. Then, the power balance equation is

$$demand_t = gen_t + effd \cdot dbatt_t - cbatt_t \quad \forall\, t \in T \tag{1}$$

where $effd$ is the discharge efficiency of the battery. We express the battery SOC SOC_t as a percentage of the maximum charge $battcap$ and calculate it as

$$SOC_t = SOC_{t-1} - dbatt_t \cdot \frac{dt}{battcap} + effc \cdot cbatt_t \cdot \frac{dt}{battcap} \quad \forall\, t \in T \tag{2}$$

where $effc$ represents the charging efficiency of the battery and dt represents the length of our time step in hours.

We restrict our generator to operate between a minimum power output of $minGen = 15$ kW and a maximum power output of $maxGen = 60$ kW:

$$minGen \leq gen_t \leq maxGen \quad \forall\, t \in T. \tag{3}$$

Similarly, we define $maxCharge$ and $maxDischarge$ as the maximum charge and discharge rates of the battery, and we constrain the battery to maintain an SOC between $minSOC$ and $maxSOC$:

$$0 \leq dbatt_t \leq maxDischarge \quad \forall\, t \in T \tag{4}$$

$$0 \leq cbatt_t \leq maxCharge \quad \forall\, t \in T \tag{5}$$

$$minSOC \leq SOC_t \leq maxSOC \quad \forall\, t \in T. \tag{6}$$

To simulate ongoing operations and avoid end-of-horizon effects, we require that the final SOC is equal to the initial SOC:

$$SOC_1 = SOC_{|T|}. \tag{7}$$

The overall objective of the MILP optimization model is to minimize the fuel consumed by the generator. As shown in Figure 3, we use a least-squares linear fit to represent steady-state fuel consumption based on our data. The linear fit is defined by its slope and intercept, where sl^b is the fuel consumption slope of 0.0113 and in^b is the fuel consumption intercept of 0.0933. This steady-state fuel consumption is typically the only fuel consumption term accounted for in previous studies, as shown in Equation (8). Figure 4 shows the resulting steady-state generator efficiency curve.

$$\min \sum_{t \in T} \left[sl^b gen_t + in^b \right] \tag{8}$$

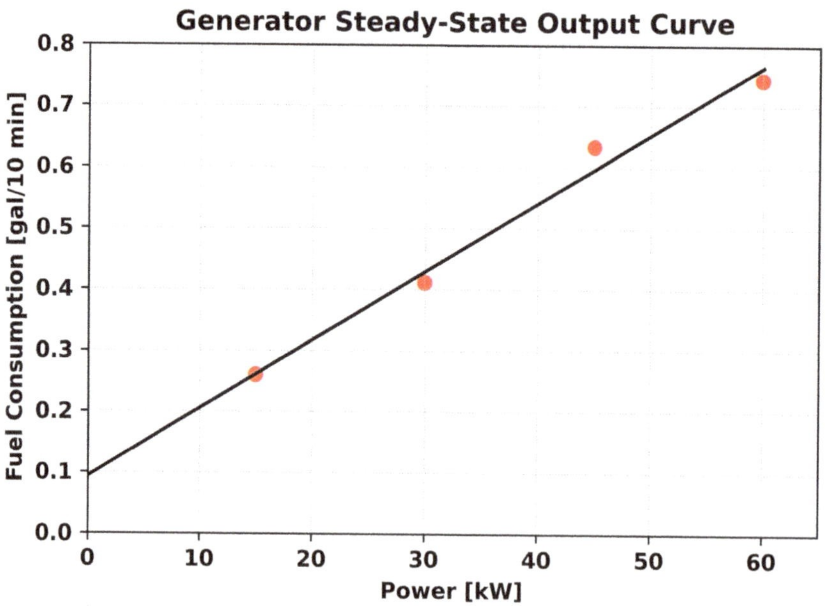

Figure 3. Fuel consumption of a 60 kW AMMPS generator with a linear fit based on generator steady-state power output data.

To address our primary study objective, we include an additional "penalty term" $penalty_t$ to capture the efficiency loss incurred when the generator output power fluctuates rapidly from time period to time period:

$$\min \sum_{t \in T} \left[sl^b gen_t + in^b + penalty_t \right]. \tag{9}$$

We now describe our methodology for calculating $penalty_t$.

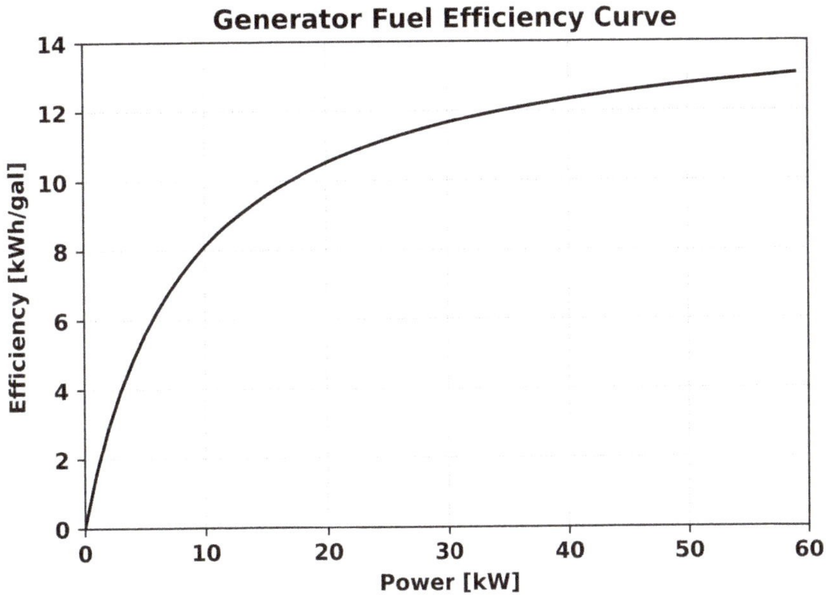

Figure 4. Fuel efficiency curve of a 60 kW AMMPS generator based on generator steady-state power output data.

3.3. Generator Penalty Concept

We consider two different approaches to calculate the "penalty term" $penalty_t$ in Equation (9). First, we calculate $penalty_t$ as a linear function of the (approximate) fractional change in generator output from time period to time period. Second, we calculate $penalty_t$ as a piecewise linear function of the fractional change in generator output.

3.3.1. Linear Penalty Approach

First, we express $penalty_t$ as a linear function of the fractional change in generator output from one time period to the next. Let $frac\_chg_t$ denote the fractional change in generator output from time period $t-1$ to time period t. Then, $frac\_chg_t$ is exactly calculated as

$$frac\_chg_t = \frac{|gen_t - gen_{t-1}|}{gen_{t-1}} \quad \forall\, t \in T. \tag{10}$$

In order to formulate a linear optimization model, we instead calculate an approximate value for $frac\_chg_t$. First, we calculate the absolute change in generator output $abs\_chg_t = |gen_t - gen_{t-1}|$ for all t in T using the linear constraints

$$abs\_chg_1 = 0 \tag{11}$$

$$gen_t - gen_{t-1} \leq abs\_chg_t \quad \forall\, t \in [2, ..., T] \tag{12}$$

$$gen_{t-1} - gen_t \leq abs\_chg_t \quad \forall\, t \in [2, ..., T] \tag{13}$$

where we rely on the fact that fluctuations are penalized in our objective function, and thus, the solver chooses the smallest feasible value for abs_chg_t, given the values of gen_{t-1} and gen_t.

After calculating the value of abs_chg_t using Equations (12) and (13), we must address the non-linearity caused by the gen_{t-1} term in the denominator of Equation (10). We do this by partitioning the generator's operating range $[minGen, maxGen]$ into a set of discrete operating regions $i \in I$. Each operating region i is defined by its lower bound l_i and upper bound u_i, where $l_1 = minGen$, $u_{|I|} = maxGen$, and $l_i = u_{i-1}$ for $i = 2, ..., |I|$. We utilize binary variable $Y_{i,t}$ to indicate that the generator is operating in region i at time t and enforce this using the following constraints:

$$\sum_{i \in I} Y_{i,t} l_i \leq gen_t \leq \sum_{i \in I} Y_{i,t} u_i \quad \forall t \in T \tag{14}$$

$$\sum_{i \in I} Y_{i,t} = 1 \quad \forall t \in T. \tag{15}$$

To obtain our linear approximation to Equation (10), we replace the gen_{t-1} term in the denominator by the midpoint of the generator's operating region at time $t-1$, i.e., $\sum_{i \in I} Y_{i,t-1} \frac{l_i + u_i}{2}$. We include the binary variable $Y_{i,t-1}$ in the numerator of our calculation and obtain the following expression for $frac\_chg_t$:

$$frac\_chg_t = \sum_{i \in I} \frac{Y_{i,t-1} abs\_chg_t}{(l_i + u_i)/2} \quad \forall t \in T. \tag{16}$$

This expression is still nonlinear due to the product of a binary variable and a continuous variable in the numerator. However, we linearize this expression by defining the continuous variable $P_{i,t}$ and using the following system of constraints to ensure that $P_{i,t} = Y_{i,t-1} abs\_chg_t$:

$$0 \leq P_{i,t} \leq Y_{i,t-1}(maxGen - minGen) \quad \forall i \in I,\ t \in T \tag{17}$$

$$abs\_chg_t - (maxGen - minGen)(1 - Y_{i,t-1}) \leq P_{i,t} \leq abs\_chg_t \quad \forall i \in I,\ t \in T. \tag{18}$$

Our expression for $frac\_chg_t$ is then

$$frac\_chg_t = \sum_{i \in I} \frac{P_{i,t}}{(l_i + u_i)/2} \quad \forall t \in T. \tag{19}$$

Lastly, we define scalar penalty coefficient $pcoe$ and express our objective function with linear fluctuation penalty as

$$\min \sum_{t \in T} \left[sl^b gen_t + in^b + pcoe \sum_{i \in I} \frac{2P_{i,t}}{l_i + u_i} \right]. \tag{20}$$

3.3.2. Piecewise Linear Penalty Approach

Next, we expand upon our linear penalty approach by constructing a piecewise linear penalty term. This allows us to model more complex penalty functions, such as a marginal penalty that increases with the fractional change in generator output $frac\_chg_t$. To construct our piecewise linear function, we first define a discrete set of regions for the value of $frac\_chg_t$ for all $t \in T$, which is defined similarly to the generator operating regions in Section 3.3.1. Denote the lower and upper bounds for fractional change region $h \in H$ as lo_h and up_h, where $lo_1 = 0$, $up_{|H|} = \frac{maxGen - minGen}{(l_1 + u_1)/2}$, and $lo_h = up_{h-1}$ for $h = 2, ..., |H|$. Then, let binary variable $W_{h,t}$ indicate that $frac\_chg_t$ lies within region h, and enforce this using the following constraints:

$$\sum_{h \in H} W_{h,t} lo_h \leq frac\_chg_t \leq \sum_{h \in H} W_{h,t} up_h \quad \forall t \in T \tag{21}$$

$$\sum_{h \in H} W_{h,t} = 1 \quad \forall t \in T. \tag{22}$$

Let slo_h and int_h denote the slope and intercept, respectively, of the piecewise linear penalty function in region h. Then, we wish to express $penalty_t$ as

$$penalty_t = \sum_{h \in H} W_{h,t}(slo_h frac\_chg_t + int_h) \quad \forall\, h \in H,\, t \in T \tag{23}$$

where, again, we have a product of a binary variable and a continuous variable $W_{h,t} frac\_chg_t$. To linearize this term, we introduce the continuous decision variable $Q_{h,t}$ and use the following system of constraints to ensure that $Q_{h,t} = W_{h,t} frac\_chg_t$:

$$0 \leq Q_{h,t} \leq up_{|H|} W_{h,t} \quad \forall\, h \in H,\, t \in T \tag{24}$$

$$frac\_chg_t - up_{|H|}(1 - W_{h,t}) \leq Q_{h,t} \leq frac\_chg_t \quad \forall\, h \in H,\, t \in T. \tag{25}$$

Thus, our objective function is

$$\min \sum_{t \in T} \left[sl^b gen_t + in^b + \sum_{h \in H} (slo_h Q_{h,t} + int_h W_{h,t}) \right]. \tag{26}$$

We experiment with four different piecewise linear penalty functions shown in Figure 5. We calculate $slo_h = h \cdot slo_1$ for each linearization region $h = 1, ..., |H|$. The intercept int_h of each segment in the piecewise linear curve is calculated so as to define a continuous piecewise linear function. The four piecewise linear functions are defined by their initial slopes (i.e., slo_1), which vary from 0.1 to 0.4 as shown in the figure legend, with $int_1 = 0$ for each function.

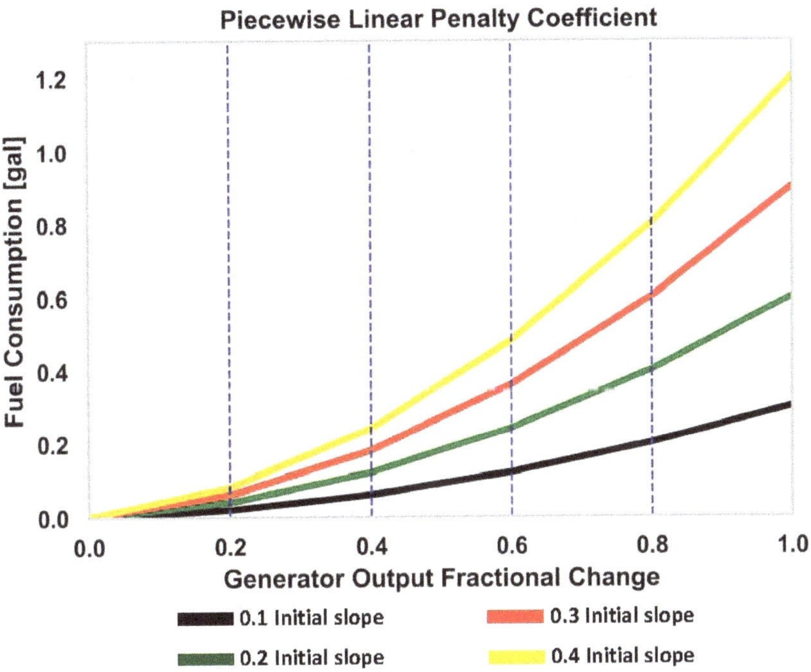

Figure 5. Fuel consumption plots for piecewise linear penalty profiles with varying initial slopes.

Our model appears in its entirety in Appendix A, which is followed by a table containing our parameter values.

4. Results and Analysis

We now exercise our optimization model on various scenarios. We first study the impact of including an ESS in our microgrid architecture by solving the model with and without an ESS; then, we quantify the impact that various penalty terms have on the microgrid's fuel consumption and the optimal generator and ESS usage.

We implement our model using Python's Pyomo package and solve it using the IBM ILOG CPLEX Interactive Optimizer 12.10.0.0 on a computer with 16 GB RAM and a 2.60 Ghertz (Hz) CPU [42]. The instances described in this section contain approximately 414–9501 constraints and 221–5761 decision variables, of which 60–2016 are binary. These instances solve to a 0–1% optimality gap in approximately 1–20 s.

We consider demand data obtained during the summer months from a US FOB located in Afghanistan and collected by the Army Logistics Innovation Agency (LIA) during the Contingency Base–Demand Data Collection (CB-DDC) project. Figure 6 depicts the power demand profile in a 48 h period, which we implement in the optimization model.

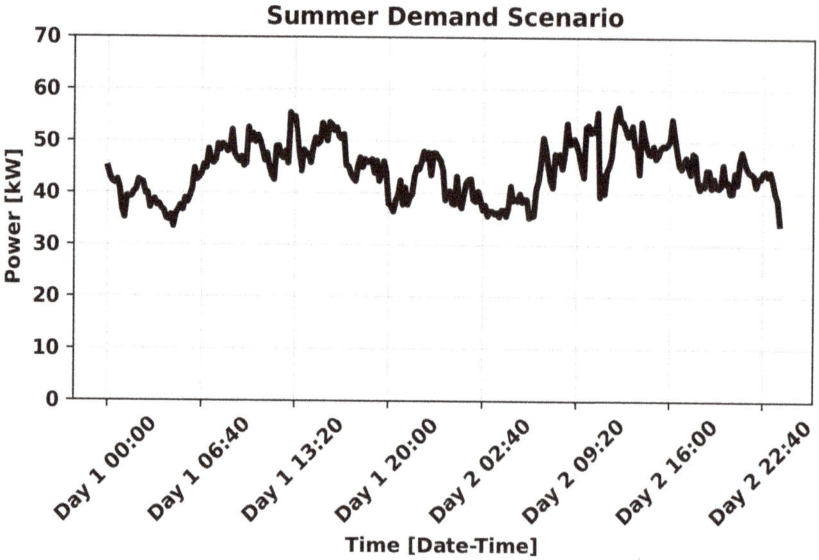

Figure 6. US FOB power demand scenario over a 48 h time frame during the summer season.

4.1. Linear Penalty

We first consider a linear penalty term with coefficients of 0.2 gallons (gal)/Δ, 0.4 gal/Δ, 0.6 gal/Δ, and 0.8 gal/Δ. Figure 7 shows the optimal power production for each of these coefficients, which is solved to a 1% optimality gap. As the figure indicates, most of the demand is satisfied by the generator. Rapid generator fluctuations decrease substantially when any of the penalties is imposed, and the decrease is larger for a higher penalty coefficient. Figure 8 shows the battery state of charge for these instances.

Mirroring Figure 7, Table 1 summarizes the results for each of the four penalty slope coefficients. We observe that the presence of a penalty significantly impacts the overall breakdown of power production, while the exact value of the penalty does not significantly affect this breakdown for the penalty values we consider. However, as the rightmost column indicates, an increasing penalty coefficient does increase fuel consumption when no ESS is present.

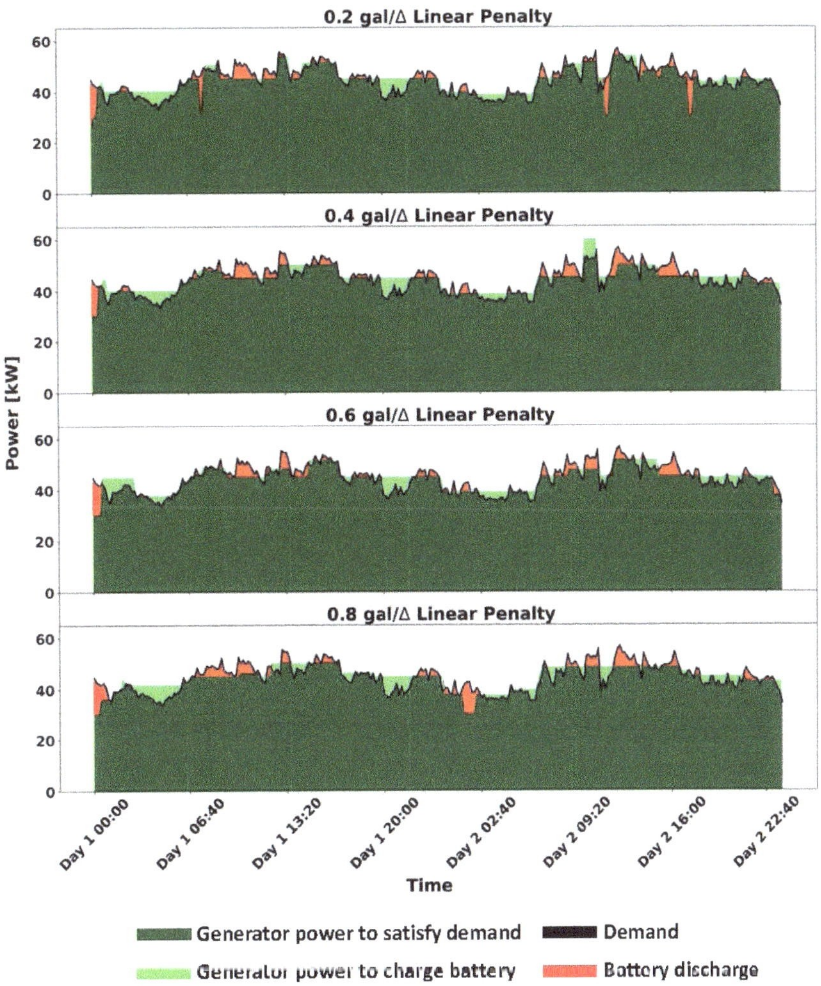

Figure 7. Power output plot based on linear penalty coefficients imposed on the generator for the summer demand scenario.

Table 1. Optimal fuel consumption and power output with various linear penalties for the summer demand scenario.

Penalty	Cumulative Fuel Consumption (gal)	Demand Met by Generator (%)	Demand Met by Energy Storage System (%)	Generator Output Used to Charge Energy Storage System (%)	Cumulative Fuel Consumption w/o Energy Storage System (gal)
No Penalty	169.97	99.85	0.15	0	170.18
Linear 0.2	171.47	97.23	2.77	2.92	171.44
Linear 0.4	171.30	97.08	2.92	3.11	175.23
Linear 0.6	171.71	97.14	2.81	3.05	177.76
Linear 0.8	171.86	96.85	3.15	3.37	180.28

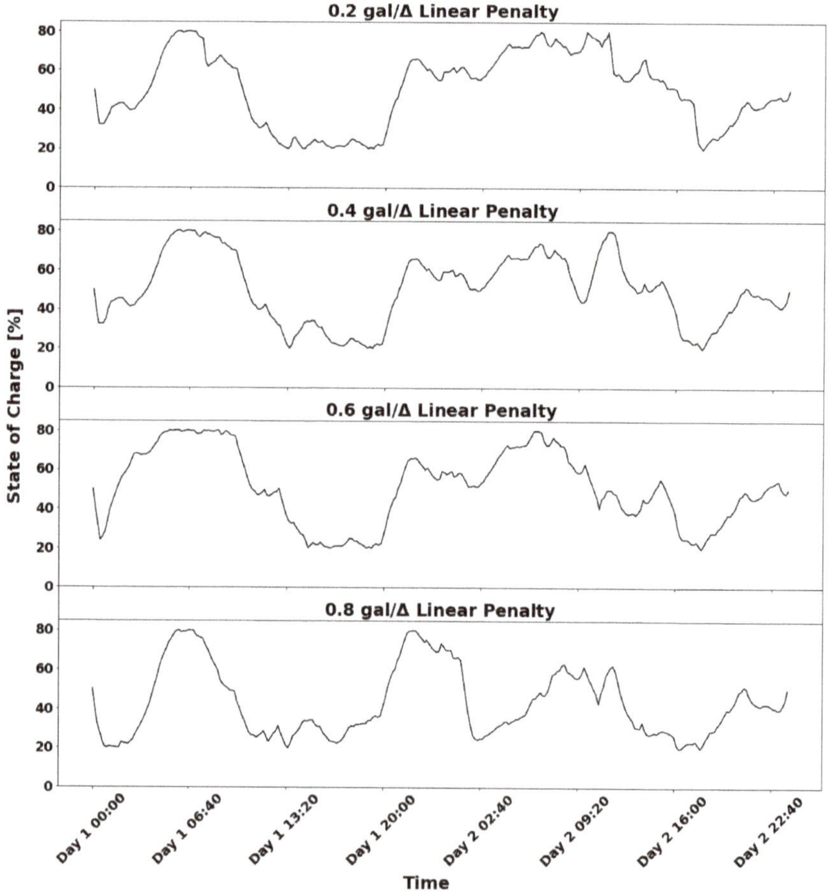

Figure 8. Battery state of charge for the instances shown in Figure 7.

4.2. Piecewise Linear Penalty

Finally, we consider the four piecewise linear penalty functions shown in Figure 5 and display the optimal solution for each of these four cases in Figure 9. The changes in generator and ESS optimal power production are minimal across the four piecewise linear penalty variations. Most of the demand is satisfied by the generator, as the dark green dominates the surface area of each plot. Additionally, the fluctuation in generator power output and ESS discharge rate decreases when the piecewise linear initial slope increases across the four penalties. When the generator exceeds the demand, the additional power charges the ESS for future use. Figure 10 shows the corresponding battery SOC.

Mirroring the graphical results, Table 2 depicts the corresponding numerical results for each of the different piecewise linear penalties. The results across the four piecewise linear penalties indicate that about 3% of the demand is satisfied by the ESS. The results display a consistent pattern of the generator supplying most of the demand. Again, the numerical results from the table reveal that the weight of the piecewise linear penalty coefficient is insignificant, as the generator and ESS scheduling power patterns are consistent across the four penalty variations. With no ESS present, we observe increasing fuel consumption with increased penalty coefficients, as expected.

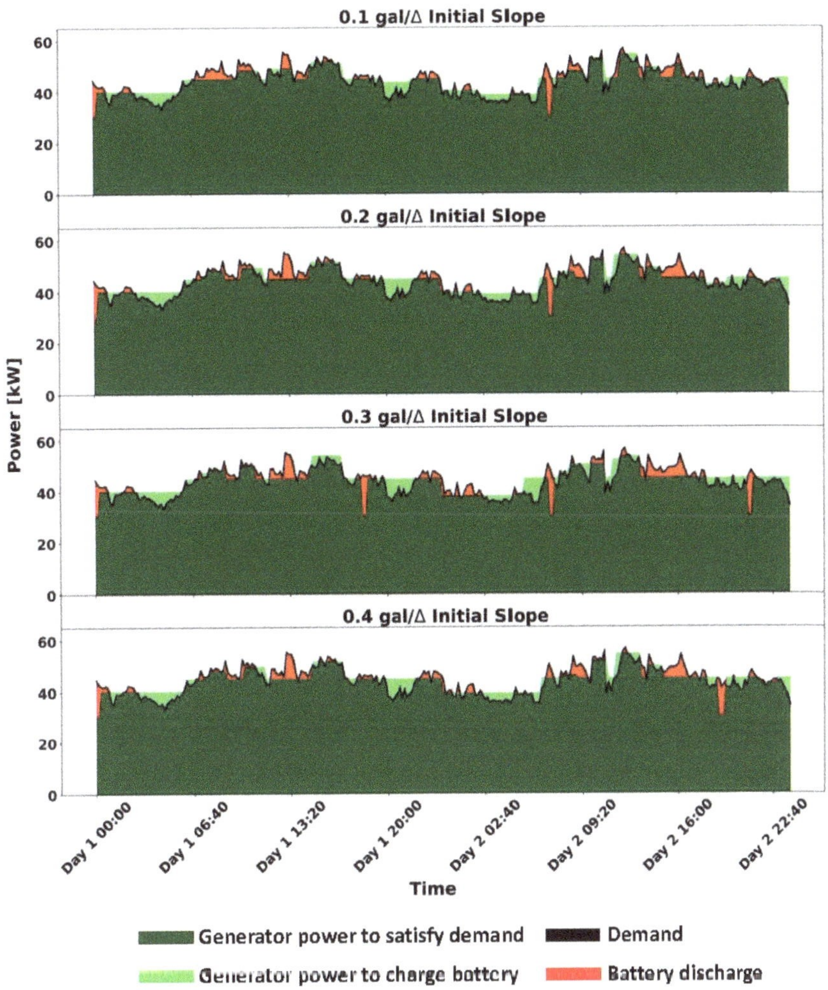

Figure 9. Power output plot based on piecewise linear penalty coefficients imposed on the generator for the summer demand scenario.

Figure 11 reiterates the importance of imposing a penalty on the generator. When an ESS is present but no penalty is imposed on generator fluctuations, the ESS is not utilized. This is not unexpected due to the fact that the ESS has an RTE less than 100%. While the ESS could, in principle, enable the generator to operate in a more efficient region of its production curve for a greater proportion of time, this apparently does not occur for this instance. The ESS is utilized more when a penalty is imposed on the generator, as optimization smooths out the generator power output to minimize cumulative fuel consumption.

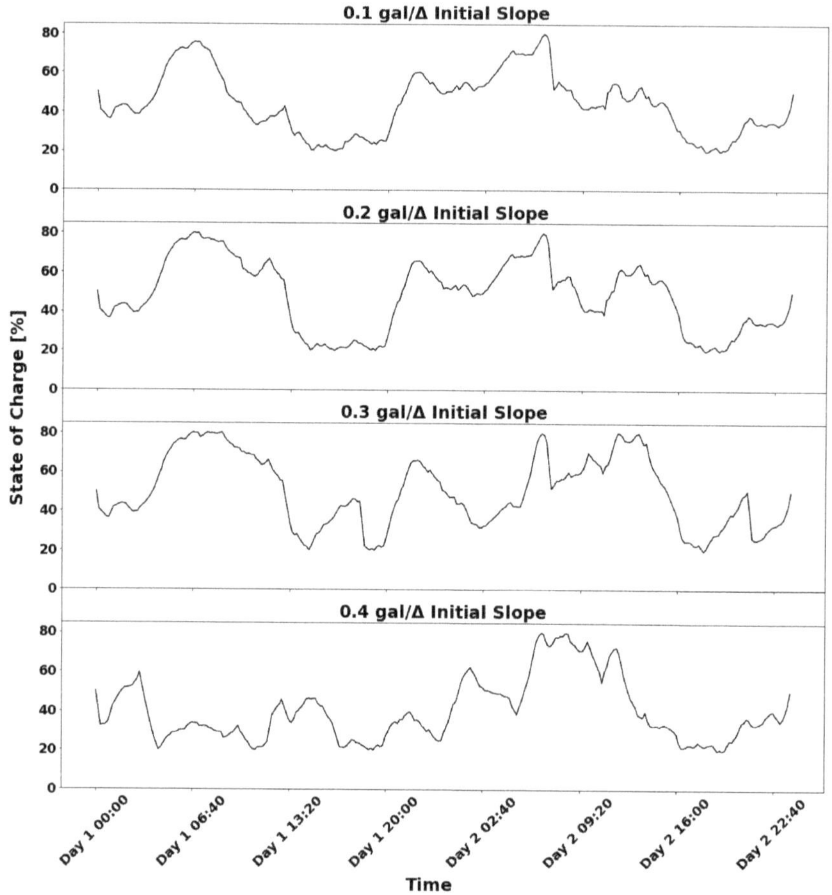

Figure 10. Battery SOC for the instances shown in Figure 9.

Table 2. Optimal fuel consumption and power output with various piecewise linear penalties for the summer demand scenario.

Penalty	Cumulative Fuel Consumption (gal)	Demand Met by Generator (%)	Demand Met by ESS (%)	Generator Output Used to Charge ESS (%)	Cumulative Fuel Consumption w/o ESS (gal)
No Penalty	169.97	99.85	0.15	0	170.18
Pi. Lin. 0.1	170.76	97.47	2.53	2.69	171.44
Pi. Lin. 0.2	171.06	97.21	2.79	2.93	172.71
Pi. Lin. 0.3	171.78	96.73	3.27	3.50	173.97
Pi. Lin. 0.4	171.69	97.12	2.88	3.07	175.24

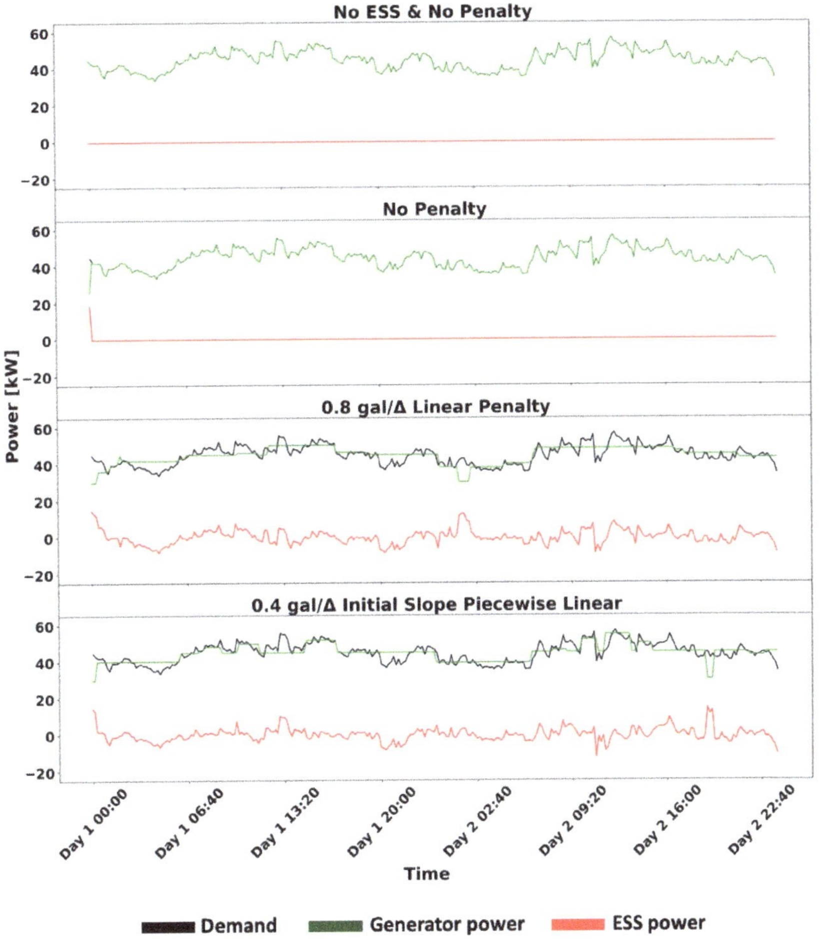

Figure 11. Optimal power distribution over various architectures for the summer demand scenario.

4.3. Impact of ESS Round-Trip Efficiency

The generator and ESS optimal power production vary drastically between the scenario with a penalty imposed on the generator and a scenario without a penalty imposed on the generator. In general, the activity of the ESS increases significantly when a penalty is imposed on the generator. In all previous scenarios, the ESS is set to a 90% RTE, which is consistent with the research outlined in Section 2. While technological advancement may increase the ESS RTE above 90%, the price of the ESS increases rapidly as the RTE increases. This is a very important factor in microgrid design.

We now study the impact of the ESS RTE on overall fuel consumption. We consider six different ESS RTEs reflective of various ESS technologies, including future technologies achieving high efficiencies: 70%, 75%, 80%, 85%, 90%, and 95%. For each ESS RTE, we run the model using four different piecewise linear penalty functions for each of ten different 24 h demand scenarios. The cumulative fuel consumption for each of these 240 configurations appears as a circular marker in Figure 12, where ESS RTE appears on the horizontal axis and the generator penalty function is denoted by color. (For readability, we introduce a small horizontal jitter at each ESS RTE.) Solid lines indicate the average fuel consumption for each ESS RTE, broken down by penalty function. As the figure indicates, there is considerable variability among scenarios for a given ESS RTE, and the average fuel consumption decreases only modestly with increasing ESS RTE. Since a lower RTE is often cheaper to obtain, this study shows possible effects and tradeoffs that designers would have to make when selecting an ESS for their respective microgrid architecture. Our results indicate that the presence of an ESS in a microgrid is of much more importance than its exact specifications for the configuration we consider.

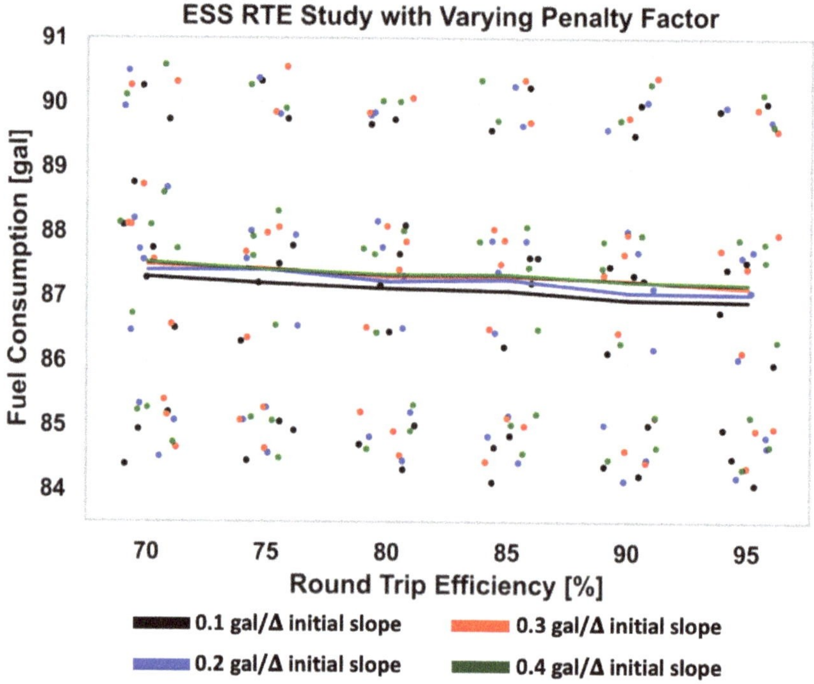

Figure 12. Total fuel consumption for varying ESS RTE and penalty functions for ten different summer demand scenarios. Circular markers indicate individual model runs, while lines indicate average fuel consumption for each ESS RTE and penalty funtion.

5. Conclusions and Future Work

This paper formulates an MILP optimization model that prescribes optimal generator and ESS usage to minimize fuel consumption while satisfying demand. A novel feature of this model is that it includes a penalty on generator output fluctuations. This penalty represents additional fuel that is consumed when the generator's output varies over time, and it has not been modeled in prior research. The inputs to the optimization model include a power demand scenario as well as the relevant characteristics of the generator, ESS, and penalty function.

We also exercise the MILP on a case study derived from actual FOB demand data. Our results indicate that an ESS is critical to achieving a smooth generator operating profile. When a penalty term is included in the objective function, we observe much smoother generator output profiles, with peak loads satisfied by the ESS and excess generator power used to charge the ESS during periods of low demand. This results in modest immediate fuel savings and can be expected to lengthen the operational lifetime of the generator, which is an important consideration in practice. We observe only minimal changes in our optimal solutions as we vary the magnitude of the penalty term, indicating that the presence of a penalty term is more important than its exact magnitude for the values we consider.

Future Work

While our MILP provides important insights on the impact of the penalty term on the optimal generator output profile and the resulting fuel consumption, it is not appropriate for implementation in a microgrid. The primary reason for this is that the MILP requires a complete demand profile as an input; it is thus "omniscient" and not suitable for real-time operation, where future demands are unknown. Thus, a natural next step for future research is to develop a real-time controller that attempts to replicate the smooth operating profiles observed in our optimal solutions. Our results indicate that the optimal load profiles are not sensitive to the exact form or magnitude of the penalty function. The implication of this is that if researchers can develop a real-time controller that produces load profiles similar to those obtained in our solutions, that controller can be expected to perform well in a variety of settings.

Our model excursions consider a simple microgrid architecture and demand data based on a US FOB in the Middle East. It is critical to explore multiple microgrid architectures when varying locations have different resources and technologies available. An appropriate next step is to alter the microgrid architectures by implementing different generators of varying sizes or even modifying the discharging and charging rates of the ESS as well as exploring additional demand scenarios.

With the initial exploration of generator penalty terms using a simple microgrid optimization model, this paper shows that a simple penalty term imposed on the generator has a significant influence on the optimal solution and overall fuel consumption. Due to the lack of empirical real-world data, our penalty terms are only notional. Further studies in the field are needed to examine the effect of time-varying loads on generator efficiency. These studies will lead to more realistic penalty terms that can optimally distribute power between the generator and ESS more accurately.

Author Contributions: Conceptualization, E.C., A.K. and G.O.; methodology, E.C., A.K. and G.O.; software, J.W.L. and E.C.; validation, J.W.L., E.C., A.K. and G.O.; formal analysis, E.C., A.K. and J.W.L.; investigation, J.W.L.; resources, E.C., A.K.; data curation, J.W.L., E.C. and A.K.; writing—original draft preparation, J.W.L.; writing—review and editing, J.W.L., E.C., A.K. and G.O.; visualization, J.W.L.; supervision, E.C., G.O. and A.K.; project administration, A.K. and E.C.; funding acquisition, A.K. All authors have read and agreed to the published version of the manuscript.

Funding: This research was funded by the Energy System Technology Evaluation Program and by the Office of Naval Research.

Institutional Review Board Statement: Not applicable.

Informed Consent Statement: Not applicable.

Data Availability Statement: Source data and code are available upon request from the authors.

Acknowledgments: This research is partially supported by the Naval Postgraduate School. Any opinions or findings of this work are the responsibility of the authors, and do not necessarily reflect the views of the sponsors or collaborators.

Conflicts of Interest: The authors declare no conflict of interest.

Appendix A. FOB Model with Linear Piecewise Penalty

Sets and Indices:

$t \in T = \{0, 10, 20, 30, ..., T\}$	Time periods
$i \in I = \{0, 1, 2, 3, 4\}$	Generator operating region (for linearization purposes)
$h \in H = \{0, 1, 2, 3, 4\}$	Generator fractional change regions (for linearization purposes)

Parameters:

sl^b	Generator base fuel consumption slope (gal/kW)
in^b	Generator base fuel consumption intercept (gal)
$demand_t$	Load that is demanded in time step t (kW)
l_i	Start of fractional change region i (kW)
u_i	End of fractional change region i (kW)
lo_h	Lower boundary of penalty region h (Δ)
up_h	Upper boundary of penalty region h (Δ)
slo_h	Slope of penalty function in region h (gal/Δ)
int_h	Intercept of penalty function in region h (gal)
dt	Time step (hours)
$battcap$	Battery capacity (kWh)
$effd$	Discharge efficiency of battery
$effc$	Charge efficiency of battery
$maxCharge$	Maximum battery rate of charge (kW)
$maxDischarge$	Maximum battery rate of discharge (kW)
$minGen$	Minimum generator load (kW)
$maxGen$	Maximum generator load (kW)
$minSOC$	Minimum battery state of charge (%)
$maxSOC$	Maximum battery state of charge (%)

Decision Variables:

gen_t	Continuous (≥ 0)	Generator power flow in time step t (kW)
$cbatt_t$	Continuous (≥ 0)	Power flow used to charge battery in time step t (kW)
$dbatt_t$	Continuous (≥ 0)	Power flow out of battery in time step t (kW)
SOC_t	Continuous (≥ 0)	Battery state of charge in time step t (%)
abs_chg_t	Continuous (≥ 0)	Absolute difference in generator power flow between time step $t-1$ and t (kW)
$Y_{i,t}$	Binary	1 if gen_t in region i in time step t and 0 otherwise
$P_{i,t}$	Continuous (≥ 0)	Auxiliary variable used for linearization: $P_{i,t} = Y_{i,t} abs\_chg_t$ (kW)
$frac\_chg_t$	Continuous (≥ 0)	Fractional [0-1] change in generator power flow between time step $t-1$ and t
$Y_{i,t}$	Binary	1 if $frac\_chg_t$ in region h in time step t and 0 otherwise
$Q_{h,t}$	Continuous (≥ 0)	Auxiliary variable used for linearization: $Q_{h,t} = W_{h,t} frac\_chg_t$

Objective Function:

$$\min \sum_{t\in T} \left[sl^b gen_t + in^b + \sum_{h\in H} \left(slo_h Q_{h,t} + int_h W_{h,t} \right) \right]$$

Constraints:

$$demand_t = gen_t + effd \cdot dbatt_t - cbatt_t \qquad \forall\, t \in T$$
$$minGen \leq gen_t \leq maxGen \qquad \forall\, t \in T$$
$$SOC_t = SOC_{t-1} - dbatt \cdot \frac{dt}{battcap} + effc \cdot cbatt_t \cdot \frac{dt}{battcap} \qquad \forall\, t > 1$$
$$0 \leq dbatt_t \leq maxDischarge \qquad \forall\, t \in T$$
$$0 \leq cbatt_t \leq maxCharge \qquad \forall\, t \in T$$
$$minSOC \leq SOC_t \leq maxSOC \qquad \forall\, t \in T$$

$$
\begin{array}{ll}
SOC_1 = SOC_T & \\
gen_t - gen_{t-1} \le abs\_chg_t & \forall\, t \in T \\
gen_{t-1} - gen_t \le abs\_chg_t & \forall\, t \in T \\
0 \le P_{i,t} \le Y_{i,t}(maxGen - minGen) & \forall\, i \in I, t \in T \\
abs\_chg_t - (maxGen - minGen)(1 - Y_{i,t}) \le P_{i,t} \le abs\_chg_t & \forall\, i \in I, t \in T \\
\sum_{i\in I} Y_{i,t} l_i \le gen_t \le \sum_{i\in I} Y_{i,t} u_i & \forall\, t \in T \\
\sum_{i\in I} \frac{2P_{i,t}}{l_i + u_i} = frac\_chg_t & \forall\, t \in T \\
\sum_{h\in H} W_{h,t} lo_h \le frac\_chg_t \le \sum_{h\in H} W_{h,t} up_h & \forall\, t \in T \\
0 \le Q_{h,t} \le up_{|H|} W_{h,t} & \forall\, h \in H, t \in T \\
frac\_chg_t - up_{|H|}(1 - W_{h,t}) \le Q_{h,t} \le frac\_chg_t & \forall\, h \in H, t \in T \\
\sum_{i\in I} Y_{i,t} = 1 & \forall\, t \in T \\
\sum_{h\in H} W_{h,t} = 1 & \forall\, t \in T \\
Y_{i,t} \in \{0,1\} & \forall\, i \in I, t \in T \\
W_{i,t} \in \{0,1\} & \forall\, h \in H, t \in T
\end{array}
$$

Table A1. Parameter Values.

Parameter	Value
sl^b	0.0113
in^b	0.0933
dt	$\frac{1}{6}$ h
$battcap$	25 kWh
$effd$	$\sqrt{0.9}\%$
$effc$	$\sqrt{0.9}\%$
$maxCharge$	20 kW
$maxDischarge$	20 kW
$minGen$	15 kW
$maxGen$	60 kW
$minSOC$	20%
$maxSOC$	80%

References

1. Schwartz, M.; Blakely, K.; O'Rourke, R. *Department of Defense Energy initiatives: Background and issues for Congress*; CRS Report R42558; Congressional Research Service: Washington, DC, USA, 2012.
2. Office of the Assistant Secretary of Defense for Energy, Installations, and Environment. New Operational Energy Strategy Released by the Department. 2018. Available online: https://www.acq.osd.mil/eie/OE/OE%20Strategy%20Page.html (accessed on 6 March 2021).
3. Katiraei, F.; Iravani, M.R.; Lehn, P.W. Micro-grid autonomous operation during and subsequent to islanding process. *IEEE Trans. Power Deliv.* **2005**, *20*, 248–257. [CrossRef]
4. Morstyn, T.; Hredzak, B.; Aguilera, R.P.; Agelidis, V.G. Model predictive control for distributed microgrid battery energy storage systems. *IEEE Trans. Control Syst. Technol.* **2018**, *26*, 1107–1114. [CrossRef]
5. Hartono, B.S.; Budiyanto, Y.; Setiabudy, R. Review of microgrid technology. In Proceedings of the 2013 International Conference on QiR, Yogyakarta, Indonesia, 25–28 June 2013; pp. 127–132. [CrossRef]
6. Ton, D.; Reilly, J. Microgrid controller initiatives: An overview of RD by the U.S. Department of Energy. *IEEE Power Energy Mag.* **2017**, *15*, 24–31. [CrossRef]
7. Craparo, E.; Sprague, J. Integrated supply- and demand-side energy management for expeditionary environmental control. *Appl. Energy* **2019**, *233–234*, 352–366. [CrossRef]
8. Sprague, J.G. Optimal Scheduling of Time-Shiftable Electric Loads in Expeditionary Power Grids. Master's Thesis, Naval Postgraduate School, Monterey, CA, USA, 2015.
9. Garcia, K. Optimization of Microgrids at Military Remote Base Camps. Master's Thesis, Naval Postgraduate School, Monterey, CA, USA, 2017.
10. Kiser, E. The Impact of Technologies and Missions on Contingency Base Fuel Consumption. Master's Thesis, Naval Postgraduate School, Monterey, CA, USA, 2018.

11. Anglani, N.; Oriti, G.; Colombini, M. Optimized energy management system to reduce fuel consumption in remote military microgrids. *IEEE Trans. Ind. Appl.* **2017**, *53*, 5777–5785. [CrossRef]
12. Cummins Inc. Advanced Medium Mobile Power Sources (AMMPS). 2017. Available online: http://power.cummins.com/system/files/literature/brochures/Advanced_Medium_Mobile_Power_Sources_Brochure.pdf (accessed on 8 October 2022).
13. Di Cairano, S.; Liang, W.; Kolmanovsky, I.V.; Kuang, M.L.; Phillips, A.M. Power Smoothing Energy Management and Its Application to a Series Hybrid Powertrain. *IEEE Trans. Control Syst. Technol.* **2013**, *21*, 2091–2103. [CrossRef]
14. Ahmed, M.; Vahidnia, A.; Meegahapola, L.; Datta, M. Small signal stability analysis of a hybrid AC/DC microgrid with static and dynamic loads. In Proceedings of the 2017 Australasian Universities Power Engineering Conference (AUPEC), Melbourne, Australia, 19–22 November 2017; pp. 1–6. [CrossRef]
15. Fakhrazari, A.; Vakilzadian, H.; Choobineh, F.F. Optimal energy scheduling for a smart entity. *IEEE Trans. Smart Grid* **2014**, *5*, 2919–2928. [CrossRef]
16. Siritoglou, P.; Oriti, G.; Van Bossuyt, D.L. Distributed Energy-Resource Design Method to Improve Energy Security in Critical Facilities. *Energies* **2021**, *14*, 2773. [CrossRef]
17. Reich, D.; Oriti, G. Rightsizing the Design of a Hybrid Microgrid. *Energies* **2021**, *14*, 4273. [CrossRef]
18. Gamarra, C.; Guerrero, J.M. Computational optimization techniques applied to microgrids planning: A review. *Renew. Sustain. Energy Rev.* **2015**, *48*, 413–424. [CrossRef]
19. Kazem, H.A.; Khatib, T. A novel numerical algorithm for optimal sizing of a photovoltaic/wind/diesel generator/battery microgrid using loss of load probability index. *Int. J. Photoenergy* **2013**, *2013*, 718596. [CrossRef]
20. Katiraei, F.; Abbey, C. Diesel plant sizing and performance analysis of a remote wind-diesel microgrid. In Proceedings of the 2007 IEEE Power Engineering Society General Meeting, Tampa, FL, USA, 24–28 June 2007; pp. 1–8. [CrossRef]
21. Ulmer, N. Optimizing Microgrid Architecture on Department of Defense Installations. Master's Thesis, Naval Postgraduate School, Monterey, CA, USA, 2014.
22. Bahramirad, S.; Reder, W.; Khodaei, A. Reliability-constrained optimal sizing of energy storage system in a microgrid. *IEEE Trans. Smart Grid* **2012**, *3*, 2056–2062. [CrossRef]
23. Chen, S.X.; Gooi, H.B.; Wang, M.Q. Sizing of energy storage for microgrids. *IEEE Trans. Smart Grid* **2012**, *3*, 142–151. [CrossRef]
24. Moradi, H.; De Groff, D.; Abtahi, A. Optimal energy scheduling of a stand-alone multi-sourced microgrid considering environmental aspects. In Proceedings of the 2017 IEEE Power Energy Society Innovative Smart Grid Technologies Conference (ISGT), Washington, DC, USA, 23–26 April 2017; pp. 1–5. [CrossRef]
25. Liu, J.; Huang, X.; Zuyi, L. Multi-time scale optimal power flow strategy for medium-voltage DC power grid considering different operation modes. *J. Mod. Power Syst. Clean Energy* **2020**, *8*, 46–54. [CrossRef]
26. Craparo, E.; Karatas, M.; Singham, D.I. A robust optimization approach to hybrid microgrid operation using ensemble weather forecasts. *Appl. Energy* **2017**, *201*, 135–147. [CrossRef]
27. Bouaicha, H.; Craparo, E.; Dallagi, H.; Nejim, S. Dynamic optimal management of a hybrid microgrid based on weather forecasts. *Turk. J. Electr. Eng. Comput. Sci.* **2020**, *28*, 2060–2076. [CrossRef]
28. del Valle, Y.; Venayagamoorthy, G.K.; Mohagheghi, S.; Hernandez, J.; Harley, R.G. Particle Swarm Optimization: Basic concepts, variants and applications in power systems. *IEEE Trans. Evol. Comput.* **2008**, *12*, 171–195. [CrossRef]
29. Li, H.; Eseye, A.T.; Zhang, J.; Zheng, D. Optimal energy management for industrial microgrids with high-penetration renewables. *Prot. Control Mod. Power Syst.* **2017**, *2*, 12. [CrossRef]
30. Kim, R.K.; Glick, M.B.; Olson, K.R.; Kim, Y.S. MILP-PSO Combined Optimization Algorithm for an Islanded Microgrid Scheduling with Detailed Battery ESS Efficiency Model and Policy Considerations. *Energies* **2020**, *13*, 1898. [CrossRef]
31. Das, A.; Ni, Z. A Computationally Efficient Optimization Approach for Battery Systems in Islanded Microgrid. *IEEE Trans. Smart Grid* **2018**, *9*, 6489–6499. [CrossRef]
32. Zhu, D.; Broadwater, R.; Tam, K.S.; Seguin, R.; Asgeirsson, H. Impact of DG placement on reliability and efficiency with time-varying loads. *IEEE Trans. Power Syst.* **2006**, *21*, 419–427. [CrossRef]
33. Hernandez-Aramburo, C.A.; Green, T.C.; Mugniot, N. Fuel consumption minimization of a microgrid. *IEEE Trans. Ind. Appl.* **2005**, *41*, 673–681. [CrossRef]
34. Bhandari, Y.; Chalise, S.; Sternhagen, J.; Tonkoski, R. Reducing fuel consumption in microgrids using PV, batteries, and generator cycling. In Proceedings of the IEEE International Conference on Electro-Information Technology, EIT 2013, Rapid City, SD, USA, 9–11 May 2013; pp. 1–4. [CrossRef]
35. Xiao, Z.X.; Guan, Y.Z.; Fang, H.W.; Terriche, Y.; Guerrero, J.M. Dynamic and Steady-State Power-Sharing Control of High-Efficiency DC Shipboard Microgrid Supplied by Diesel Generators. *IEEE Syst. J.* **2022**, *16*, 4595–4606. [CrossRef]
36. Kapetanović, M.; Núñez, A.; van Oort, N.; Goverde, R.M. Reducing fuel consumption and related emissions through optimal sizing of energy storage systems for diesel-electric trains. *Appl. Energy* **2021**, *294*, 117018. [CrossRef]
37. Downs, J.; Marinelli, A.; Fowler, K.; Duncan, J.; Hamm, K.; Choi, E.; McCabe, K.; Boyd, P. *Contingency Base Demand Data Collection for Buildings and Facilities*; Technical Report PNNL-26065; Pacific Northwest National Laboratory: Richland, WA, USA, 2016.
38. Singleton, B. (US Army Base Camp Integration Lab, Fort Devens, MA, USA). Email Exchanges and Site Visit, Personal Communication, 2017.

39. Li, K.; Tseng, K.J. Energy efficiency of lithium-ion battery used as energy storage devices in micro-grid. In Proceedings of the IECON 2015—41st Annual Conference of the IEEE Industrial Electronics Society, Yokohama, Japan, 9–12 November 2015; pp. 005235–005240. [CrossRef]
40. Moseley, P.T.; Garche, J. Chapter 4: Applications and markets for grid-connected storage systems. In *Electrochemical Energy Storage for Renewable Sources and Grid Balancing*; Moseley, P.T., Garche, J., Eds.; Elsevier: Amsterdam, The Netherlands, 2015; pp. 440–443. [CrossRef]
41. Majima, M.; Ujiie, S.; Yagasaki, E.; Koyama, K.; Inazawa, S. Development of long life lithium ion battery for power storage. *J. Power Sources* **2001**, *101*, 53–59. [CrossRef]
42. International Business Machines Corporation (IBM). IBM ILOG CPLEX Optimization Studio (CPLEX). 2009. Available online: https://www.ibm.com/products/ilog-cplex-optimization-studio (accessed on 17 October 2022).

energies

MDPI

Article

Exact Feedback Linearization of a Multi-Variable Controller for a Bi-Directional DC-DC Converter as Interface of an Energy Storage System

Gabriel R. Broday [1], Luiz A. C. Lopes [1,*] and Gilney Damm [2]

1 Department of Electrical and Computer Engineering, Concordia University, Montreal, QC H3G 1M8, Canada
2 Department of Components and Systems (COSYS), University Gustave Eiffel, 93162 Paris, France
* Correspondence: lalopes@encs.concordia.ca

Abstract: DC microgrids have shown to be a good approach for better accommodating stochastic renewable energy sources (RES) and for the charging of electric vehicles (EVs) at the distribution level. For this, fast-acting energy storage units (ESSs) are essential. This requires that both the bi-directional power converter topology and the control scheme present the right set of features. The ESS discussed in this paper consists of a new DC-DC converter based on a tapped inductor (TI) for a higher voltage gain at moderate duty cycles. The direction of the current in its intermediate inductor does not need to be reversed for power flow reversal, leading to a faster action. Moreover, it can employ a multi-state and multi-variable modulation scheme that eliminates the right half-plane (RHP) zero, common in boost-type converters. In order to achieve good dynamic performance across a wide range of operating points, a control scheme based on feedback linearization is developed. This paper presents the modeling of the five-switch DC-DC converter operating in the tri-state buck–boost mode. A systematic approach for deriving control laws for the TI current and output voltage based on exact state feedback linearization is discussed. The performance of the proposed control scheme is verified by simulation for a supercapacitor (SC)-based ESS. It is compared to that of a conventional control scheme for a dual-state buck–boost mode with cascaded PI controllers designed based on small-signal models. The results show that both control schemes work similarly well at the operating point that the conventional control scheme was designed for. However, only the proposed scheme allows the SC-based ESS to control the current injected into the DC microgrid with the voltage of the SC varying between the expected range of rated to half-rated.

Keywords: exact feedback linearization; multi-variable controller; bi-directional DC-DC converter; energy storage systems

Citation: Broday, G.R.; Lopes, L.A.C.; Damm, G. Exact Feedback Linearization of a Multi-Variable Controller for a Bi-Directional DC-DC Converter as Interface of an Energy Storage System. *Energies* **2022**, *15*, 7923. https://doi.org/10.3390/en15217923

Academic Editor: Alon Kuperman

Received: 19 September 2022
Accepted: 20 October 2022
Published: 25 October 2022

1. Introduction

The decentralization of power generation has become a topic of high interest for industry and academia. The integration of stochastic renewable energy sources (RESs) at the distribution level is facilitated by incorporating them into a microgrid [1]. DC microgrids are a good option since many RESs (photovoltaic and fuel cells) and energy storage units present DC output characteristics [2]. The efficiency of the DC-DC interfaces tends to be higher than in DC-AC, and issues such as frequency regulation, reactive power control and synchronization are avoided [3,4]. The control of segments of the distribution system as a microgrid also helps with the deployment of new large loads, such as electric vehicles (EVs). However, fast-acting energy storage systems (ESSs) are essential for balancing supply and demand, thus regulating the DC bus voltage [5–7].

The power converters of the ESSs must be bi-directional and can be either isolated or non-isolated. The dual-active bridge (DAB) is an example of the former, allowing higher voltage gains than the latter at the expense of a higher switch count. Among the non-isolated bi-directional DC-DC converters, there are the Class C and the four-switch converter, with two half-bridges and an intermediate inductor. The relatively low voltage gains of these topologies can be increased with topologies that employ tapped inductors (TIs) [8]. In [9], a three-switch converter with a TI is discussed. By employing a tri-state modulation scheme, the right half-plane (RHP) zero that appears in the control-to-output transfer function of the converters can be eliminated, thus increasing the speed of response of the converter [9,10]. It should be noted that for the three topologies mentioned above, the current in the intermediate inductor must be reversed for power flow reversal, which slows down the operation. In [11], a five-switch bi-directional DC-DC converter that does not present this constraint was presented. Moreover, it can also operate with tri-state modulation for an improved dynamic response [12].

Power converters frequently employ linear PI-type controllers [13,14]. They are designed based on converter models linearized around a specific operating point. The performance of PI controllers tends to degrade if the operating point of the converter changes. This might be a significant issue for ESSs that are prone to wide variations in the output current, as well as for the input voltage, when supercapacitors (SCs) are used as storage units. Recall that SCs typically operate with voltages in the range of rated to half-rated. For cases such as this, state feedback linearization techniques are a good and effective alternative. State feedback linearization has been applied to unidirectional buck, boost and buck–boost converters in [15–17], respectively. It was also applied to a bi-directional Class C converter in [18,19], and to a four-switch converter in [20]. One potential issue in these cases is when the inductor current is used as a feedback term in the denominator of the control law. Note that this quantity crosses zero during power flow reversal, leading to a potential singularity. The above converters are all single-input single-output (SISO) systems, but the technique has also been applied to multi-input multi-output (MIMO) systems. In [21], exact feedback linearization was used in a three-port boost converter for interfacing a PV panel to a battery and load. In [22], it was applied to a four-level buck converter with two flying capacitors and an LC output filter. Suitable control laws were developed to generate the duty cycles of the three switches so as to regulate three of the four state variables.

In this paper, state feedback linearization is applied to a new bi-directional DC-DC converter with a TI for an increased voltage gain at moderate duty cycles. A multi-state modulation scheme with multiple modulation signals is used to eliminate the RHP zero typically found in converters with intermediate inductors. Unlike previously reported DC-DC converters, this one allows reversal of the power flow without reversing the intermediate inductor current. This is beneficial since the inductor current, which is often used in the denominator of the control laws, will always be positive, eliminating potential singularities. It is a dual-input dual-output (DIDO) system, which is the simplest form of a MIMO system, where the magnitude of the inductor current presents a certain degree of independence with regard to the output voltage. In principle, it can be adjusted as a trade-off between converter losses and the speed of response of the output voltage.

This paper is organized as follows. In Section 2, the five-switch bi-directional DC-DC converter is introduced. Section 3 presents the modeling of the system for both power flow directions, and how they are combined into a single model. Section 4 presents the derivation of the control law of the converter based on state feedback linearization, while Section 5 presents the system parameters, with considerations for controller design. The performance of the five-switch converter operating in the tri-state buck–boost mode with state feedback by means of simulation is verified in Section 6. It also presents a comparison with the conventional dual-state buck–boost with cascaded inner current outer voltage loops with linear PI controllers. Finally, Section 7 provides the final conclusion of this research work.

2. The Five-Switch Bi-Directional DC-DC Converter

The bi-directional DC-DC converter considered in this paper is shown in Figure 1. It consists of five switches interfacing two DC buses, Bus 1 and Bus 2, through a TI. This is modeled as an ideal transformer with two windings (L_{T1} and L_{T2}) and a magnetizing inductance (L_M) in parallel with the primary winding. The turns ratio of the windings, L_{T1}/L_{T2}, is $n/1$. The converter can present a high voltage gain (V_2/V_1) in the conventional dual-state scheme without an overly high duty cycle, by properly selecting the turns ratio n [23]. Capacitors C_1 and C_2 are low-pass input and output filters. The energy storage unit, in Bus 1, and cabling to the converter, as well as the DC microgrid, are modeled by a voltage source in series with a resistor (V_1–R_1 and V_2–R_2).

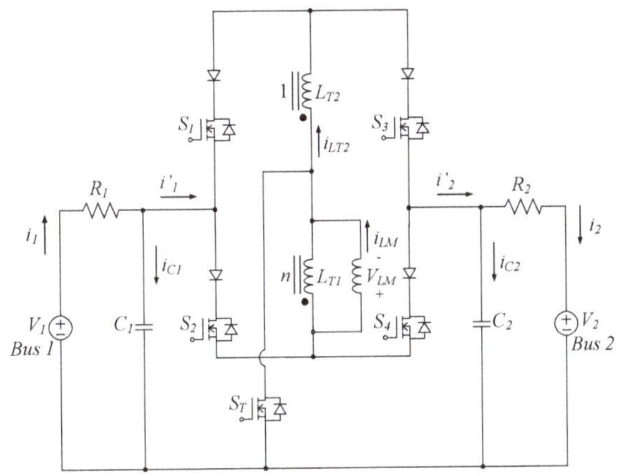

Figure 1. The 5-switch bi-directional DC-DC converter.

For the proper operation of the power converter, switches S_1 to S_4 should be unidirectional in terms of current, such as those used in PWM current source inverters (CSIs): IGBTs with reverse blocking capability or IGBTs/MOSFETs in series with a diode. Conversely, S_T (the switch that taps the inductor to the common node of the DC buses) is a standard MOSFET with an anti-parallel body diode for bi-directional current flow. The converter allows a number of paths for the inductor current, or states of operation, as shown in Table 1. State S_{XY} concerns a moment when switches S_X and S_Y are on/activated together. They can be used in a switching cycle with two or three states [12,23]. Circuit current and voltage waveforms are also presented in [12,23]. Power flow reversal occurs without changing the direction of the current in the magnetizing inductance (i_{LM}), favoring a fast dynamic response.

Table 1. Possible states of operation.

State	Current Path	i'_1	i'_2	v_{LM}	Power Flow
S_{2T}	S_2 and S_T	i_{LM}	0	$v_{C1} > 0$	Bus 1 to inductor
S_{4T}	S_4 and S_T	0	$-\ i_{LM}$	$v_{C2} > 0$	Bus 2 to inductor
S_{3T}	S_3 and $D(S_T)$ *	0	$n\ i_{LM}$	$-n\ v_{C2}$	Inductor to Bus 2
S_{1T}	S_1 and $D(S_T)$ *	$-n\ i_{LM}$	0	$-n\ v_{C1}$	Inductor to Bus 1
S_{23}	S_2 and S_3	$(n\ i_{LM})/(n+1)$	$(n\ i_{LM})/(n+1)$	$(n\ v_{C1} - n\ v_{C2})/(n+1)$	Bus 1 to Bus 2
S_{14}	S_1 and S_4	$(-n\ i_{LM})/(n+1)$	$(-n\ i_{LM})/(n+1)$	$(n\ v_{C2} - n\ v_{C1})/(n+1)$	Bus 2 to Bus 1
S_{34}	S_3 and S_4	0	0	0	Free-wheeling (fw)
S_{12}	S_1 and S_2	0	0	0	Free-wheeling (fw)

* $D(S_T)$ refers to the anti-parallel body diode of S_T.

Based on the operating states and current paths presented in Table 1, Figures 2 and 3 present the current paths, highlighted in red, for all possible operating states for each power flow direction.

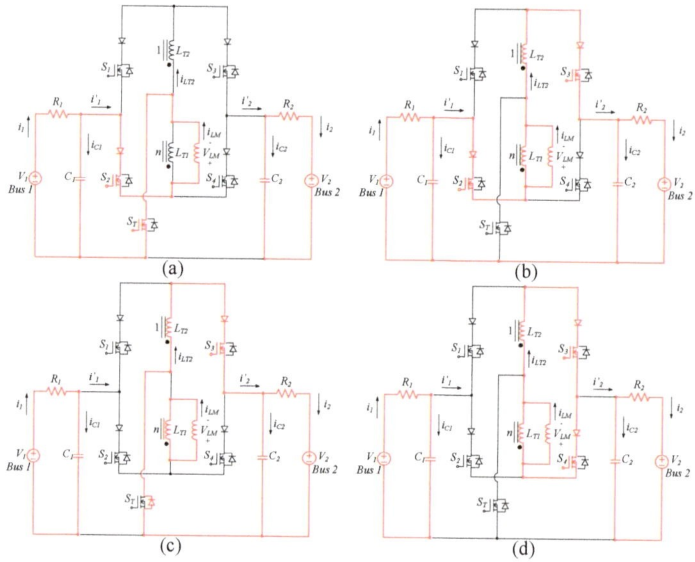

Figure 2. Current paths (in red) for the operating states of the forward mode. (**a**) State S_{2T}; (**b**) State S_{23}; (**c**) State S_{3T}; (**d**) State S_{34}.

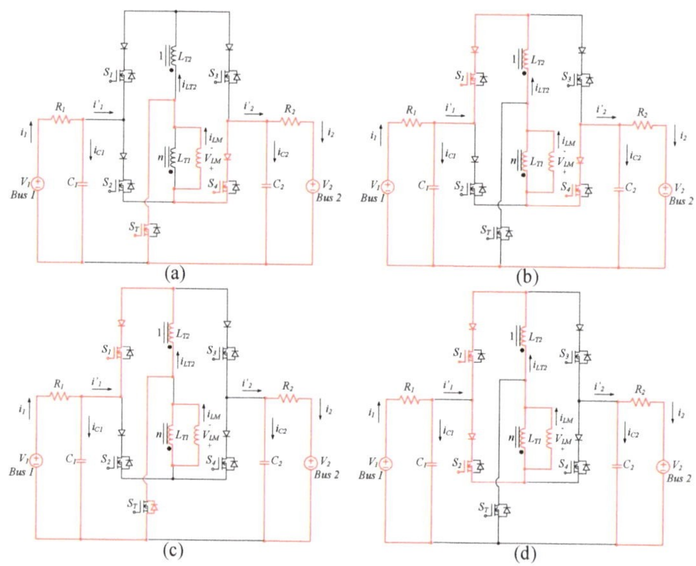

Figure 3. Current paths (in red) for the operating states of the reverse mode. (**a**) State S_{4T}; (**b**) State S_{14}; (**c**) State S_{1T}; (**d**) State S_{12}.

The converter can operate in the dual-state modes [10] and in the tri-state modes [11]. The tri-state buck–boost mode is selected for this paper. In the forward mode, with power flowing from Bus 1 to Bus 2, states S_{2T}, S_{3T} and S_{34} are used. Conversely, in the reverse mode, states S_{4T}, S_{1T} and S_{12} are used. The PWM modulator for this type of operation is shown in Figure 4. It employs two modulation signals, m_1 and m_2, where the magnitude of m_2 is larger than of m_1 and an auxiliary signal (q) that determines the direction of the power flow: $q = 1$ in the forward direction and $q = 0$ in the reverse direction. This modulator is used for obtaining the gating signals for the switches in both forward and reverse modes, shown in Figures 5 and 6.

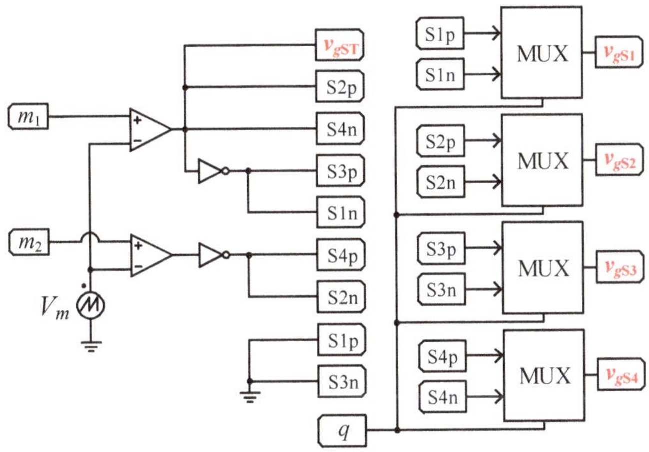

Figure 4. PWM modulator for the tri-state buck–boost mode of operation.

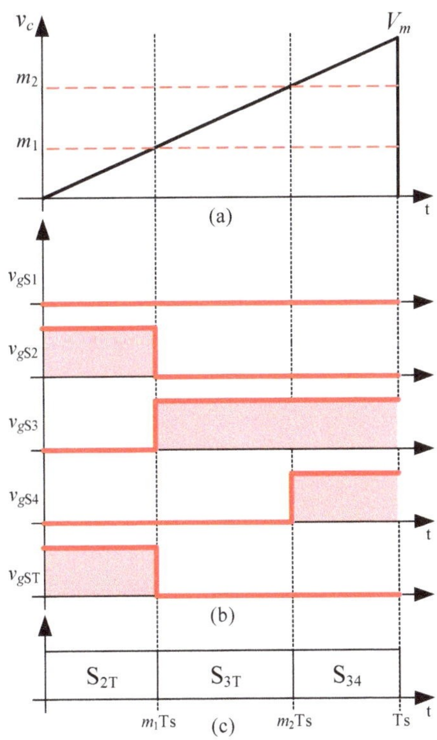

Figure 5. Forward mode: the carrier-based modulation scheme. (**a**) Carrier and modulating signals; (**b**) gating signals of the switches; (**c**) states of operation.

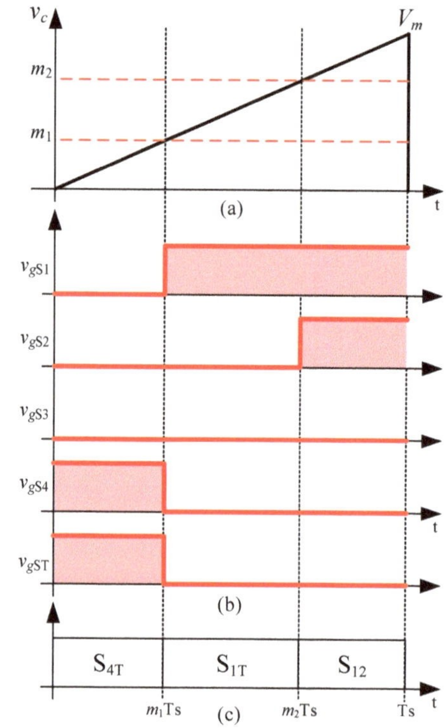

Figure 6. Reverse mode: the carrier-based modulation scheme. (**a**) Carrier and modulating signals; (**b**) gating signals of the switches; (**c**) states of operation.

The impact of the modulation signals (m_1 and m_2) on the gating signals of the switches (v_{gS1}–v_{gST}) and the duration of the states for tri-state buck–boost operation is illustrated in Figures 5 and 6, respectively, for the forward and reverse modes. There, one can see that a switching period consists of three states among those shown in Table 1. Assuming that the peak value of the sawtooth carrier (V_m) is 1 V, and that $m_1 < m_2 \leq 1$, the duty cycle of state $S_{2T} = D_{2T} = m_1$. Likewise, $D_{3T} = (m_2 - m_1)$ and $D_{34} = (1 - m_2)$. In the second segment, m_1T_s to m_2T_s, when only S_3 is on, the inductor current flows through S_3 and the body diode of S_T, sending energy from the inductor to Bus 2. For the reverse mode, $D_{4T} = m_1$, $D_{1T} = (m_2 - m_1)$ and $D_{12} = (1 - m_2)$. According to Table 1, these are equivalent states to those in the forward mode. For instance, in the second segment, m_1T_s to m_2T_s, when only S_1 is on, the inductor current flows through S_1 and the body diode of S_T, sending energy from the inductor to Bus 1.

3. Modeling of the Five-Switch Converter

In this section, the modeling of the converter is performed based on the analysis of the operating states that are generated by the PWM modulator for both the forward and reverse modes. The objective is to find a model that describes the system behavior as a function of the control variables m_1, m_2 and q. It is based on the differential equations of the three state variables of the bidirectional power converter: current in the magnetizing inductance (i_{LM}) and voltages across the filter capacitors (v_{C1} and v_{C2}).

By applying KVL and KCL for the three states of operation and averaging them for the forward mode, one obtains

$$\frac{di_{LM}}{dt} = \frac{v_{C1}}{L_M}(m_1) - \frac{n\,v_{C2}}{L_M}(m_2 - m_1) \tag{1}$$

$$\frac{dv_{C1}}{dt} = \frac{V_1}{R_1C_1} - \frac{v_{C1}}{R_1C_1} - \frac{i_{LM}}{C_1}(m_1) \tag{2}$$

$$\frac{dv_{C2}}{dt} = \frac{V_2}{R_2C_2} - \frac{v_{C2}}{R_2C_2} + \frac{n\,i_{LM}}{C_2}(m_2 - m_1) \tag{3}$$

Repeating the process for the reverse mode, one obtains

$$\frac{di_{LM}}{dt} = -\frac{n\,v_{C1}}{L_M}(m_2 - m_1) + \frac{v_{C2}}{L_M}(m_1) \tag{4}$$

$$\frac{dv_{C1}}{dt} = \frac{V_1}{R_1C_1} - \frac{v_{C1}}{R_1C_1} + \frac{n\,i_{LM}}{C_1}(m_2 - m_1) \tag{5}$$

$$\frac{dv_{C2}}{dt} = \frac{V_2}{R_2C_2} - \frac{v_{C2}}{R_2C_2} - \frac{i_{LM}}{C_2}(m_1) \tag{6}$$

By multiplying (1)–(3) by q and (4)–(6) by $\overline{q}$ and adding them, one obtains a final averaged model, valid for both forward and reverse modes:

$$\frac{di_{LM}}{dt} = \dot{i_{LM}} = \frac{v_{C1}}{L_M}(u_2) - \frac{v_{C2}}{L_M}(u_1) \tag{7}$$

$$\frac{dv_{C1}}{dt} = \dot{v_{C1}} = \frac{V_1}{R_1C_1} - \frac{v_{C1}}{R_1C_1} - \frac{i_{LM}}{C_1}(u_2) \tag{8}$$

$$\frac{dv_{C2}}{dt} = \dot{v_{C2}} = \frac{V_2}{R_2C_2} - \frac{v_{C2}}{R_2C_2} + \frac{i_{LM}}{C_2}(u_1) \tag{9}$$

where the new control variables, u_1 and u_2, are defined as

$$u_1 = (m_2 - m_1)nq - m_1\overline{q} \tag{10}$$

$$u_2 = m_1q - (m_2 - m_1)n\overline{q} \tag{11}$$

4. Exact State Feedback Linearization

The non-linear characteristics of this system can be observed in the model developed in the previous section. It includes cross-products between the state variables and the new control variables u_1 and u_2. Using the conventional linearization approach based on small signal variations around an operating point to design conventional PI controllers, information would be lost and the system would not be fully described by the resulting model.

The main goal of an ESS is usually to control the output current, i_2, which can be achieved indirectly through the output voltage, v_{C2}. Frequently, this is achieved by employing a cascaded inner inductor current and outer output voltage control loop. However, since the developed model presents two control variables, u_1 and u_2, and in order to give more flexibility to the considered non-linear controller, this paper implements a parallel control structure. In this case, the inductor current reference can be chosen according to the system needs, with high values favoring a faster dynamic response and lower ones for improved efficiency.

The first step of the controller design is to rewrite the DIDO system, presented in Figure 7, in an affine linear form to apply the theory of exact state feedback linearization. This is achieved by means of (12) and (13).

$$\dot{x} = f(x) + g(x)u \tag{12}$$

$$y = h(x) \tag{13}$$

where $x \in \mathbb{R}^n$ is the state vector; $u \in \mathbb{R}^m$. relates to the control inputs; $y \in \mathbb{R}^i y$ defines the controlled outputs; f, g and h are differentiable vector fields. Considering the model developed in the last section,

$$x = [i_{LM}, v_{C1}, v_{C2}]^t = [x_1, x_2, x_3]^t \tag{14}$$

$$u = [u_1, u_2]^t \tag{15}$$

$$f(x) = \left[0, \frac{V_1}{R_1C_1} - \frac{x_2}{R_1C_1}, \frac{V_2}{R_2C_2} - \frac{x_3}{R_2C_2}\right]^t \tag{16}$$

$$g(x) = \begin{bmatrix} -\frac{x_3}{L_M} & \frac{x_2}{L_M} \\ 0 & -\frac{x_1}{C_1} \\ \frac{x_1}{C_2} & 0 \end{bmatrix}, g_1(x) = \left[-\frac{x_3}{L_M}, 0, \frac{x_1}{C_2}\right]^t, g_2(x) = \left[\frac{x_2}{L_M}, -\frac{x_1}{C_1}, 0\right]^t \tag{17}$$

$$h_1(x) = x_1 = [1,0,0];\ h_2(x) = x_3 = [0,0,1] \tag{18}$$

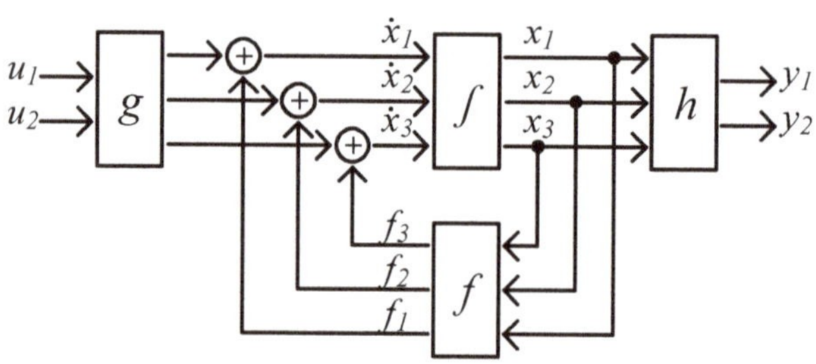

Figure 7. Block diagram representation of the DIDO system.

There are several methods for obtaining the feedback linearization of the DIDO system [24–27]. First, the model is rewritten in a different set of coordinates, using the output y and its successive derivatives. This procedure is performed based on Equation (13), obtaining

$$\dot{y_i} = L_f h_i + \sum_{j=1}^{m} (L_{gj} h_i) u_j \tag{19}$$

where $L_f h$ and $L_g h$ are, respectively, *Lie derivatives* of the smooth scalar vector $h(x)$ as a function of $f(x)$ and $g(x)$. If $L_{gj}h_i(x) = 0$ for all j, then the inputs do not appear, and, as highlighted by [22,24], one needs to differentiate repetitively as

$$y_i^{r_i} = L_f{}^{(r_i)} h_i + \sum_{j=1}^{m} \left(L_{gj} L_f^{r_i - 1} h_i\right) u_j \tag{20}$$

until $L_{gj} L_f^{r_i-1} h_i \neq 0$ for at least one j. The values of r_i are the sub-relative degrees of the DIDO system, and $L_{gj} L_f^{r_i-1} h_i$ is the *Lie derivative* of the i-th output as a function of $f(x)$ and $g(x)$ [22,24]. In such a case, (20) can be rewritten as

$$\begin{bmatrix} y_1^{r_1} \\ \cdots \\ y_i^{r_i} \end{bmatrix} = A(x) + E(x) \begin{bmatrix} u_1 \\ \cdots \\ u_2 \end{bmatrix} \tag{21}$$

where

$$A(x) = \begin{bmatrix} L_f^{r_1} h_1 \\ \cdots \\ L_f^{r_i} h_i \end{bmatrix} \tag{22}$$

and

$$E(x) = \begin{bmatrix} L_{g1} L_f^{r_1 - 1} h_1 & \cdots & L_{gm} L_f^{r_1 - 1} h_1 \\ \cdots & \cdots & \cdots \\ L_{g1} L_f^{r_i - 1} h_i & \ldots & L_{gm} L_f^{r_i - 1} h_i \end{bmatrix} \tag{23}$$

Matrix $E(x)$ is called the decoupling matrix of the DIDO system. If it is nonsingular, the following input transformation can be defined:

$$u = -E^{-1} A(x) + E^{-1}(x) z \tag{24}$$

where z concerns new input variables.

Substituting (24) into (21) results in a linear and decoupled differential relation between the output y and the new input z [19].

$$\begin{bmatrix} y_1^{r_1} \\ \cdots \\ y_i^{i} \end{bmatrix} = \begin{bmatrix} z_1 \\ \cdots \\ z_i \end{bmatrix} \tag{25}$$

It is important to remark that (21), or its closed loop (25), creates a new model for the converter. If this new state vector, represented by the original outputs and their derivatives, does not match the same number of states (n_s) of the original system, it means that this new mathematical model is not describing the dynamics of all states in the converter. As a consequence, any analysis based on this new model would disregard the remaining states. These are called the zero dynamics of this new representation. For a complete analysis of the converter, one must consider these new variables as well as these zero-dynamics.

Taking into account the 5-switch converter operating with the tri-state buck–boost mode with two control variables (u_1 and u_2) and two state variables to be controlled (i_{Lm} and v_{C2}), one can find the sub-relative degrees by considering m = 2 (two inputs) and i = 1 and 2 (two outputs).

By solving the Lie derivatives, one can find the sub-relative degrees of the system as follows.

For $y_1 = h_1 = x_1 = i_{Lm} = [1, 0, 0]$, then

$$\begin{cases} L_{g1} h_1 = [1\ 0\ 0] \left[-\frac{x_3}{L_M}, 0, \frac{x_1}{C_2} \right]^t = -\frac{x_3}{L_M} \\ L_{g2} h_1 = [1\ 0\ 0] \left[\frac{x_2}{L_M}, -\frac{x_1}{C_1}, 0 \right]^t = \frac{x_2}{L_M} \end{cases} \tag{26}$$

For $y_2 = h_2 = x_3 = v_{C2} = [0, 0, 1]$, then

$$\begin{cases} L_{g1} h_2 = [0\ 0\ 1] \left[-\frac{x_3}{L_M}, 0, \frac{x_1}{C_2} \right]^t = \frac{x_1}{C_2} \\ L_{g2} h_2 = [0\ 0\ 1] \left[\frac{x_2}{L_M}, -\frac{x_1}{C_1}, 0 \right]^t = 0 \end{cases} \tag{27}$$

Thus, the sub-relative degrees of the system are r_1 = 1 and r_2 = 1. Knowing these two values, it is possible now to fully define matrix $E(x)$ as

$$E(x) = \begin{bmatrix} L_{g1} L_f^{r_1 - 1} h_1 & L_{g2} L_f^{r_1 - 1} h_1 \\ L_{g1} L_f^{r_2 - 1} h_2 & L_{g2} L_f^{r_2 - 1} h_2 \end{bmatrix} = \begin{bmatrix} L_{g1} L_f^0 h_1 & L_{g2} L_f^0 h_1 \\ L_{g1} L_f^0 h_2 & L_{g2} L_f^0 h_2 \end{bmatrix} \tag{28}$$

Therefore, by replacing (26) and (27) with (28), matrix $E(x)$ is finally given by

$$E(x) = \begin{bmatrix} -\frac{x_3}{L_M} & \frac{x_2}{L_M} \\ \frac{x_1}{C_2} & 0 \end{bmatrix} \tag{29}$$

Next, one computes vector $A(x)$ as

$$A(x) = \begin{bmatrix} L_f^{r_1} h_1 \\ L_f^{r_2} h_2 \end{bmatrix} = \begin{bmatrix} L_f h_1 \\ L_f h_2 \end{bmatrix} \tag{30}$$

where

$$\begin{cases} L_f h_1 = \begin{bmatrix} 1 & 0 & 0 \end{bmatrix} \left[0, \frac{V_1}{R_1 C_1} - \frac{x_2}{R_1 C_1}, \frac{V_2}{R_2 C_2} - \frac{x_3}{R_2 C_2} \right]^t = 0 \\ L_f h_2 = \begin{bmatrix} 0 & 0 & 1 \end{bmatrix} \left[0, \frac{V_1}{R_1 C_1} - \frac{x_2}{R_1 C_1}, \frac{V_2}{R_2 C_2} - \frac{x_3}{R_2 C_2} \right]^t = \frac{V_2}{R_2 C_2} - \frac{x_3}{R_2 C_2} \end{cases} \tag{31}$$

Thus,

$$A(x) = \left[0, \frac{V_2}{R_2 C_2} - \frac{x_3}{R_2 C_2} \right]^t \tag{32}$$

With matrices $E(x)$ and $A(x)$ defined, one obtains the control law of the converter from (24) as

$$u = \begin{bmatrix} \frac{x_3 - V_2}{R_2 x_1} \\ \frac{x_3 (x_3 - V_2)}{R_2 x_1 x_2} \end{bmatrix} + \begin{bmatrix} 0 & \frac{C_2}{x_1} \\ \frac{L_M}{x_2} & \frac{x_3 C_2}{x_1 x_2} \end{bmatrix} z \tag{33}$$

It should be noted that, since, for this particular converter, $i_{LM} = x_1$ does not need to be reversed for power flow reversal and $v_{C1} = x_2 \approx V_1 > 0$ and $v_{C2} = x_3 \approx V_2 > 0$, there are no singularities in the linearized system. Particularizing (25) for this converter, one observes that the equivalent system is linear and decoupled, as shown in (34). It can be represented by a block diagram, as shown in Figure 8.

$$\begin{bmatrix} y_1^1 \\ y_2^1 \end{bmatrix} = \begin{bmatrix} \dot{i}_{Lm} \\ \dot{v}_{C2} \end{bmatrix} = \begin{bmatrix} z_1 \\ z_2 \end{bmatrix} \tag{34}$$

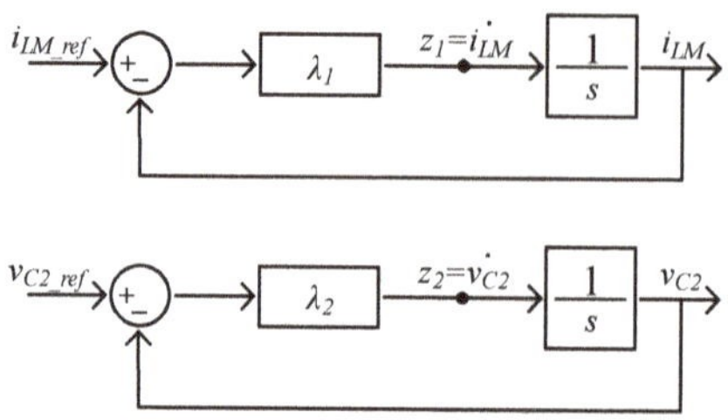

Figure 8. Linear decoupled model of the 5-switch converter.

For tracking reference values for i_{LM} and v_{C2}, given by i_{LM_ref} and v_{C2_ref}, respectively, one can use simple proportional controllers to obtain inputs z_1 and z_2 required in (34). This implies taking

$$\begin{aligned} z_1 &= -\lambda_1 \left(i_{LM} - i_{LM\_ref} \right) \\ z_2 &= -\lambda_2 \left(v_{C2} - v_{C2\_ref} \right) \end{aligned} \tag{35}$$

The overall system, with the original system, state feedback linearization network and simple proportional controllers, is shown in Figure 9. In general, the objective of the "control system" is to provide the modulation signals that will result in the tracking of some reference quantities. In this case, the modulation signals are m_1 and m_2 and the quantities to be tracked are $i_{LM} = x_1$ and $v_{C2} = x_3$. The modulation signals can be obtained from (10) and (11), provided that u_1 and u_2, signals in the input of block *"system"* in Figure 9, are known. According to (33) and as depicted in Figure 9, u_1 and u_2 are functions of the state variables $i_{LM} = x_1$, $v_{C1} = x_2$ and $v_{C2} = x_3$, as well as z_1 and z_2. The latter are the outputs of the proportional controllers of the inductor current and output voltage loops, shown in Figure 8 and (35).

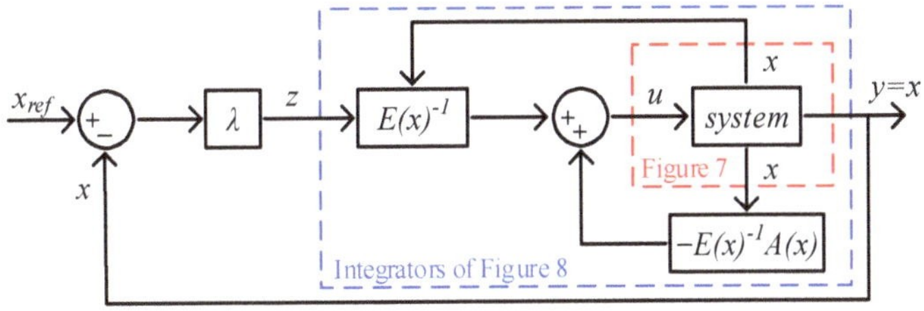

Figure 9. Model of the DIDO system with the state feedback linearization network and simple P-type controllers.

4.1. Remark: Tuning Guidelines

One interesting feature for the proposed control scheme is the fact that it relies on only two parameters, λ_1 and λ_2. These are indeed the desired closed-loop poles imposed by the control algorithm. They impose new dynamics to the closed-loop states represented by i_{LM} and v_{C2}, respectively. Their choice is completely straightforward; they directly impose the new time response of these variables. One can either choose them following practical considerations (desired time constant for these variables) or by a standard pole-placement procedure [28].

4.2. Zero Dynamics Analysis

Now, it is necessary to study the remaining dynamics of the system, the so-called zero dynamics. In the present case, they are trivial and are represented by the remaining state v_{C1}. For this analysis, one considers the dynamics when the controlled ones reach their equilibrium point. Then, the control amounts to (remark that $z = 0$ for this analysis)

$$\overline{u} = \begin{bmatrix} \frac{\overline{v_{C2}} - V_2}{R_2 i_{LM}} \\ \frac{\overline{v_{C2}}(\overline{v_{C2}} - V_2)}{R_2 i_{LM} x_2} \end{bmatrix} \tag{36}$$

and the zero-dynamics become

$$\dot{v_{C1}} = -\frac{v_{C1}}{R_1 C_1} - \frac{\overline{v_{C2}}(\overline{v_{C2}} - V_2)}{C_1 R_2 v_{C1}} + \frac{V_1}{R_1 C_1} \tag{37}$$

This system has two equilibrium points given by

$$v_{C1} = \frac{V_1}{2} \pm \sqrt{\frac{{V_1}^2}{4} - \overline{v_{C2}}(\overline{v_{C2}} - V_2)} \tag{38}$$

Only the larger solution is feasible. This one has as a Jacobian

$$-\frac{1}{R_1 C_1} + \frac{\overline{v_{C2}}(\overline{v_{C2}} - V_2)}{C_1 R_2} \left(\frac{V_1}{2} + \sqrt{\frac{{V_1}^2}{4} - \overline{v_{C2}}(\overline{v_{C2}} - V_2)} \right)^{-2} \tag{39}$$

This is negative inside a large operation region. Hence, the zero dynamics are asymptotically stable in this region, and, consequently, the full system is asymptotically stable.

5. System Parameters

The application scenario for the performance verification of the five-switch converter with the proposed control scheme is as an interface between two DC buses/elements, as depicted in Figure 1. It is assumed that the values of V_2 and R_2 are known. The rated voltages were taken as V_1 = 96 V and V_2 = 380 V and the rated power 1.9 kW. Other converter parameters are as follows: turns ratio n = 2, L_M = 38.8 μH, C_1 = 76.8 μF, C_2 = 76.8 μF, and f_{sw} = 250 kHz. The feeder resistances are R_1 = 0.0625 Ω and R_2 = 0.0625 Ω. These parameters are similar to those used in [12,23], which discuss the power electronic aspects of the five-switch DC-DC converter. The gains (λ_1 and λ_2) of the non-linear controller with state feedback linearization for the converter operating in tri-state buck–boost mode were computed, based on pole placement from linear control theory, as 250 k and 350 k, respectively. These gains are computed taking into account the behavior of a first-order system, where they relate to the time constant (T_s) of such systems. In this case, $T_{s\_\lambda 1}$ = 16 μs and $T_{s\_\lambda 2}$ = 11.4 μs.

In order to demonstrate the feasibility and benefits of the proposed mode of operation, tri-state buck–boost, and state feedback linearization, its performance is compared in the next section with a conventional technique. Usually, a converter such as this operates in dual-state buck–boost mode with a cascaded inner current (i_{LM}) and outer voltage (v_{C2}) control loop, as presented in Figure 10.

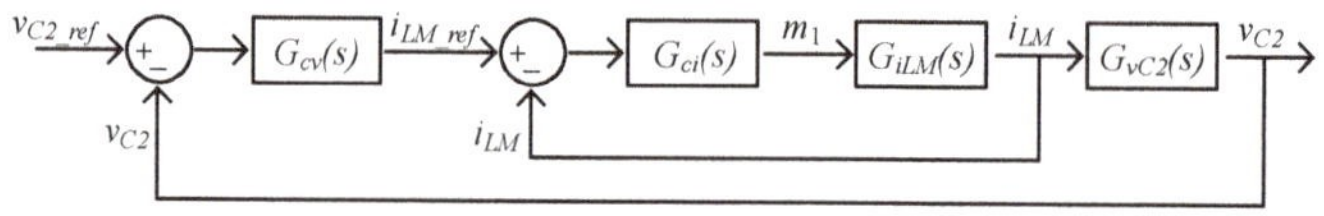

Figure 10. Block diagram of the cascaded control.

The small-signal model is obtained by means of the standard linearization of (7)–(9). For operation in dual-state buck–boost mode, considering the modulation schemes discussed in Section 2, one has to set m_2 to 1 V, to eliminate the free-wheeling states (S_{34} for the forward mode and S_{12} for the reverse mode), and then control the converter with a single input, m_1. The value of q is set at 1, since the forward power flow, boost mode, is the worst case concerning the design of the PI controller. With m_1 as the control variable, m_2 = 1 and q = 1, one substitutes (10)–(11) into (7)–(9) to obtain new state equations as a function of a single control variable: m_1. Then, one employs the conventional linearization technique, based on small-signal variations around a base operating point, to obtain transfer functions ($i_{LM}(s)/m_1(s)$ and $v_{C2}(s)/i_{LM}(s)$). The first can be found to be

$$G_{iLM}(s) = \frac{i_{LM}(s)}{m_1(s)} = \frac{[R_2C_2(V_{C1} + nV_{C2})]s + [n^2R_2(1-m_1)I_{LM} + V_{C1} + nV_{C2}]}{(R_2C_2L_M)s^2 + (L_M)s + n^2R_2(1-m_1)^2} \quad (40)$$

It is used for designing the PI-type controller of the inner current loop for zero error in steady state to step changes. For the design of the outer voltage control loop, considering that its bandwidth is 10–20% of the one for the inner current loop, the latter can be assumed to be a unity gain. Next, one can derive an expression for $v_{C2}(s)/i_{LM}(s)$), applying the current division principle that relates the current in the output capacitor ($i_{C2}(s)$) to $i_{LM}(s)$. This results in

$$i_{C2}(s) = ni_{LM}(s)(1-m_1) - i_2(s) = ni_{LM}(s)(1-m_1) - \frac{v_{C2}(s) - V_2}{R_2} \quad (41)$$

Finally, by assuming that V_2 remains constant, one can compute the transfer function of the output capacitor voltage over the inductor current as

$$G_{vC2}(s) = \frac{v_{C2}(s)}{i_{LM}(s)} = \frac{nR_2(1-m_1)}{(R_2C_2)s + 1} \quad (42)$$

In this particular case, standard type-2 PI controllers can be designed to provide the chosen bandwidths and phase margins for the inner and outer control loops. The parameters of the linear PI controllers were computed as follows. For the inner current loop, the bandwidth is selected as 50 kHz with a phase margin of 60°, resulting in K_{pi} = 0.0427, τ_i = 11.88 µs and f_{pi} = 186.7 kHz. For the outer voltage loop, the bandwidth is chosen as 10 kHz with a phase margin of 60°, leading to K_{pv} = 75.43, τ_v = 12.62 µs and fp_v = 7.93 kHz.

In Figure 11, the Bode diagrams of the linear cascaded control presented in Figure 10 are shown. From this figure, one can see that the compensated systems (orange plots) present the desired bandwidths and phase margin characteristics, validating the design of the type-2 PI controllers employed in the control law.

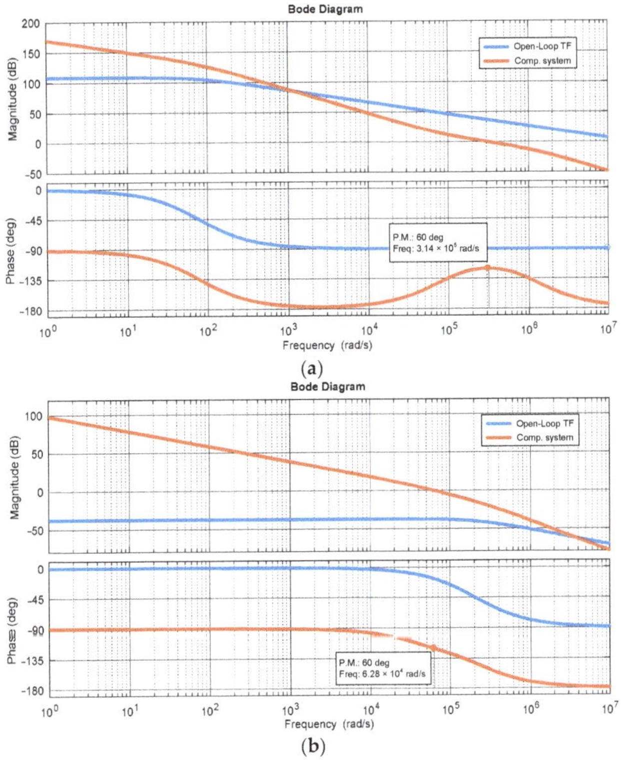

Figure 11. Bode plots of the linear controllers. (**a**) Current control loop, (**b**) voltage control loop.

6. Performance Verification

Two types of tests were conducted, by means of simulations with PSIM, to demonstrate the feasibility and the advantages of the proposed five-switch converter operating in tri-state buck–boost, with a non-linear control scheme based on state feedback linearization. First, it is assumed that the input (V_1) and output (V_2) voltages are kept constant at 96 V and 380 V, respectively. The output current reference signal (i_{2_ref}) is a square waveform of ±5 A and 5 Hz. This can be seen as the case where the five-switch converter (and control law) acts as an interlink converter between two strongly regulated DC buses, meaning that their voltages remain constant regardless of the power injected into, or withdrawn from, the DC buses. This can be considered as the base case. Here, the main goal is to verify whether one can control the output current (i_2) that is injected from one bus to another.

Next, an SC is used in Bus 1 (V_1) to work as a storage unit for a controlled DC microgrid, which is defined in Bus 2 (V_2) for this case study. SCs are devices with high power density and high charge/discharge rates that can be used to provide sudden bursts of power by managing currents with high gradients. Typically, the voltage on an SC is allowed to vary between half-rated and rated values. For the DC microgrid side, to emulate the power variations inherent to a DC microgrid, a sinusoidal ripple of 20 V (peak-to-peak) with a frequency of 120 Hz is added to the voltage source V_2, meaning that instead of staying constant at 380V, this voltage source will vary from 370 V to 390 V. This is a typical condition and concerns the connection of a DC bus to an AC bus (grid or load) through a single-phase DC-AC interface, where such an interface creates DC-side current and voltage harmonics at two times (120 Hz) the grid frequency (60Hz). This case might present a challenge for the performance of the conventional PI controllers, designed for a fixed operating point. Conversely, the non-linear control scheme with state feedback linearization should be able to compensate for these conditions, presenting similar behavior in a wide range of operating points. For this test, the same output current profile from the previous case is used.

Finally, the efficiency of the five-switch DC-DC converter operating in the tri-state buck–boost mode at rated conditions, V_1 = 96 V and V_2 = 380 V, with the output current varying from −5 A (reverse mode) to 5 A (forward mode), is investigated.

6.1. Tests with Fixed Values of V_1 and V_2

Figure 12 shows some key waveforms concerning the operation of the proposed non-linear controller with a fixed voltage at Bus 1 (Figure 12a) and Bus 2 (Figure 12b). In Figure 12c, one sees that the output current (i_2) follows the reference signal (i_{2_ref}). This is achieved indirectly by controlling the output voltage of the converter (v_{C2}), as shown in Figure 12d. The reference (i_{LM_ref}) and inductor current (i_{LM}) are shown in Figure 12e. There, one sees that i_{LM} tracks i_{LM_ref} very well and that power flow reversal, with i_2 changing from −5 A to 5 A, can be obtained without reversing i_{LM}. Moreover, the control of the two state variables, v_{C2} and i_{LM}, is achieved in a decoupled way, as discussed in Section 4. It is worth pointing out that the i_{LM_ref} is kept at 30 A until t = 0.25 s, when it is changed to 40 A, and then changed back to 30 A at t = 0.35 s. This action has two goals. The first is to show the good tracking capability of the inductor current control loop to step changes. A change in i_{2_ref} has a minor effect on i_{LM}. The second is to show the impact of the magnitude of i_{LM} on the values of the modulation signals, as discussed below. The waveforms of control variables (u_1 and u_2), which include the flag "q" with the direction of the power flow, are shown in Figure 12f. Those concerning the modulation signals (m_1 and m_2) used in the PWM modulator can be seen in Figure 12g. They remain within the range of the sawtooth carrier, 0 to 1 V, as expected. There, one can see that as i_{LM_ref} increases, the magnitude of the modulation signals (m_1 and m_2) decreases.

In general, no major difference can be seen between the two control schemes regarding the control of the output current (i_2) in Figures 12c and 13c. Both control schemes present fast and accurate tracking of the output current reference (i_{2_ref}). However, this is achieved with different values for the i_{LM}; see Figures 12e and 13e. For the conventional cascaded control scheme, the inductor current reference i_{LM_ref} varies with the output voltage reference (v_{C2_ref}). When the power flow is reversed, at t = 0.3 s and t = 0.4 s, i_{LM} also changes, but it remains always positive. This is a feature of the five-switch bi-directional DC-DC converter. Thus, i_{LM} and v_{C2} are coupled variables for the cascaded linear controller. Finally, Figure 13f shows that only one modulation signal, m_1, is used for both power flow directions for the converter operating in dual state with the conventional cascaded controller, while the other one, m_2, is kept at 1 V.

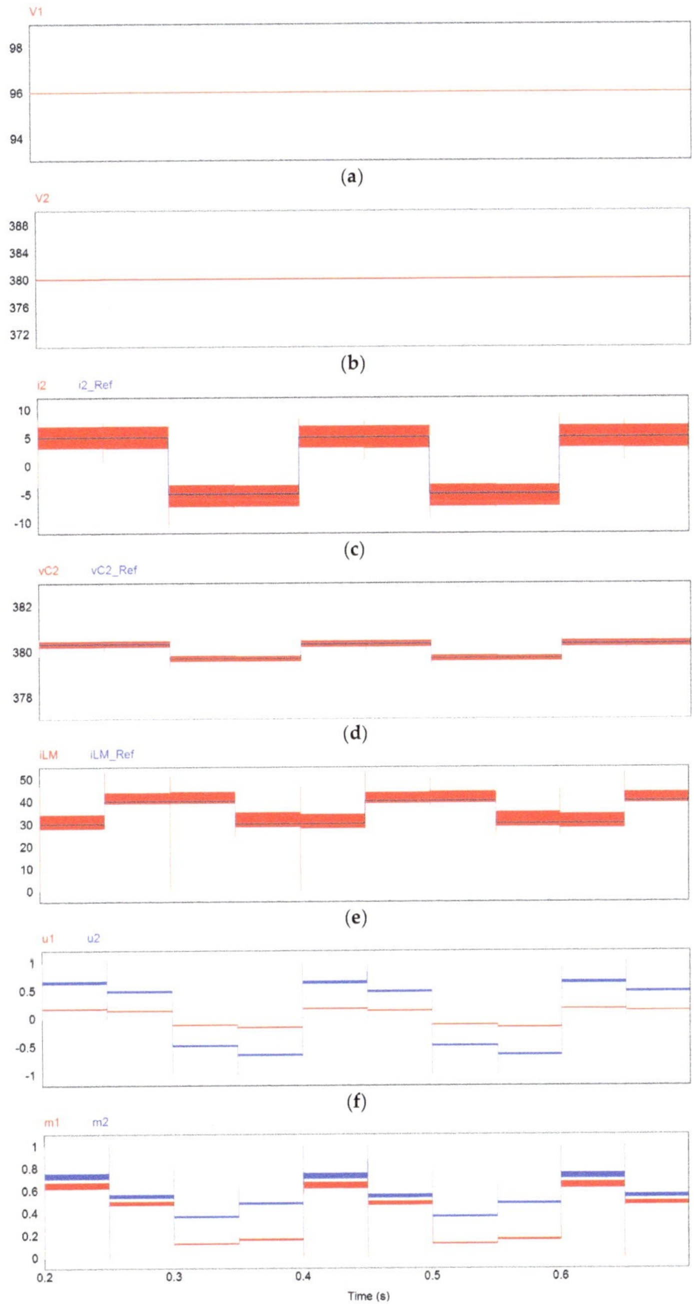

Figure 12. Waveforms for the tri-state buck–boost control with state feedback linearization. (**a**) Voltage at Bus 1 (in V), (**b**) voltage at Bus 2 (in V), (**c**) reference and output current (in A), (**d**) reference and output capacitor voltage (in V), (**e**) reference and inductor current (in A), (**f**) non-linear control variables, (**g**) PWM modulation signals.

Next, Figure 13 presents the results for essentially the same test, but for the converter operating in dual-state mode with the cascaded linear controller. In this case, the output voltage (v_{C2}) is adjusted by varying the inductor current (i_{LM}).

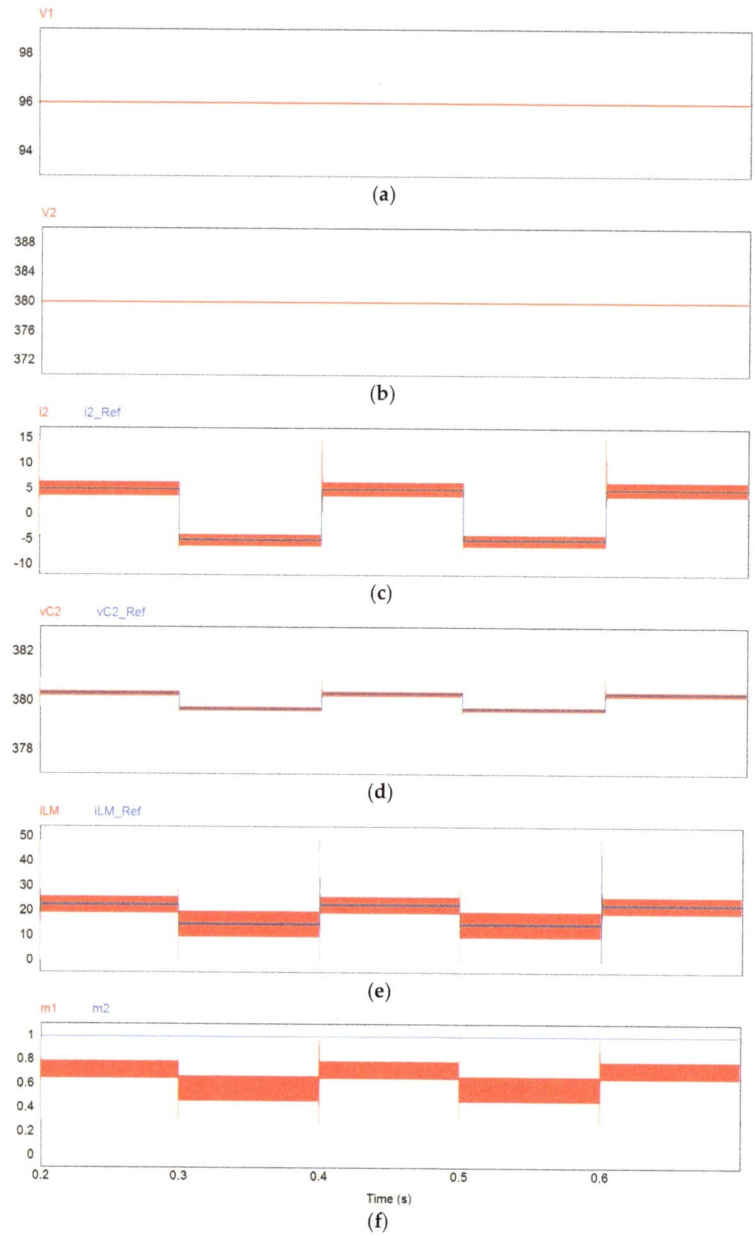

Figure 13. Waveforms for the dual-state buck–boost control with conventional cascaded control loops. (**a**) Voltage at Bus 1 (in V), (**b**) voltage at Bus 2 (in V), (**c**) reference and output current (in A), (**d**) reference and output capacitor voltage (in V), (**e**) reference and inductor current (in A), (**f**) PWM modulation signals.

6.2. Tests with an SC at Bus 1 and Bus 2 with a Sinusoidal Ripple

In this test, the fixed voltages, or "strong buses", previously used in Bus 1 and Bus 2, are replaced, respectively, by an SC (a storage element) and a controlled DC microgrid, which corresponds to another common application for bi-directional DC-DC converters. This concerns a more demanding operating condition where the input (V_1) and output (V_2) voltages of the converter can vary considerably, driving the operating point of the converter further away from the base operating point. There are several papers focused on the sizing of SCs for given applications [29–31], which are typically in the order of 100 s of F. In this particular test, a small supercapacitor (95 mF) was considered for the test so that one can observe the impact of its voltage variation on the response of the inductor current (i_{LM}) and output voltage (v_{C2}) control loops, in a single time frame.

Figure 14 shows some relevant waveforms for this test. In Figure 14a, one sees the DC microgrid voltage at Bus 2 with the 20 V/120 Hz sinusoidal ripple. In Figure 14b, one sees that the output current (i_2) tracks very well the reference signal (i_{2_ref}), a square wave of ±5 A and 5 Hz. The waveform of V_1, the SC voltage at Bus 1, is shown in Figure 14c. It starts at 96 V and decreases due to a positive value for i_2. At t = 0.3 s, it reaches approximately 72 V and starts to increase as the value of i_2 changes to −5 A. Thus, the variation in V_1 and V_2 has no negative impact on the performance of the proposed control scheme. Moreover, v_{C2} in Figure 14d and i_{LM} in Figure 14e track their reference signals accurately. Note that, in this case, the inductor current reference (i_{LM_ref}) was kept constant at 35 A. Good tracking performance was achieved with the waveforms of control variables (u_1 and u_2) shown in Figure 14f and modulation signals (m_1 and m_2) shown in Figure 14g.

The five-switch converter operating with dual-state buck–boost and conventional linear cascaded control loops is subjected to the same test. The results are shown in Figure 15. In Figure 15a, one sees that the DC microgrid voltage waveform at Bus 2 (V_2) is the same as in the test for the non-linear controller. In Figure 15b, one sees that this conventional modulation and control scheme tracks the positive output current reference (i_{2_ref}) well until the moment that the decreasing SC voltage, (V_1 in Figure 15c), reaches a value of approximately 80 V. This is when the modulation signal (m_1) and several waveforms start to oscillate. The control scheme starts to malfunction. At t = 0.3 s, when the power flow reverses and i_{2_ref} = −5 A, the output current (i_2) fails to track it, becoming more negative (lower) than i_{2_ref}. The inductor current (i_{LM}) and inductor current reference (i_{LM_ref}) waveforms, shown in Figure 15e, are limited at the maximum value, 50 A. One sees in Figure 15f that the modulation signal (m_1) decreases following the change of i_{2_ref} to −5 A, but starts to increase, unlike in the previous test for fixed values of V_1 and V_2 (Figure 13f), when its average value remained relatively constant. Overall, as a result of i_2 becoming more negative than i_{2_ref} at t = 0.3 s, V_1 increases more than expected, reaching 140 V at t = 0.4 s. From this point on, the linear controller is able to control the output current, but keeping the input voltage in the range of 120–140 V. In summary, due to the issues in tracking the reference output current/voltage in operating conditions different from the designed ones, the conventional control scheme fails to track the reference output current, which can even lead to damage to the SC and switches, due to overvoltage conditions.

6.3. Results for SC Voltage Falling to Half-Rated Value

For this test, to have the SC voltage decreasing from rated (96 V) to half-rated (48 V) using the same square output current reference (i_{2_ref}) of ±5 A and 5 Hz, the SC used in the previous test is replaced by a smaller one of 55 mF. Figure 16 shows some relevant waveforms for this test. Moreover, since the DC microgrid voltage waveform at Bus 2 (V_2) is the same as in the previous tests, it is omitted in the next figure.

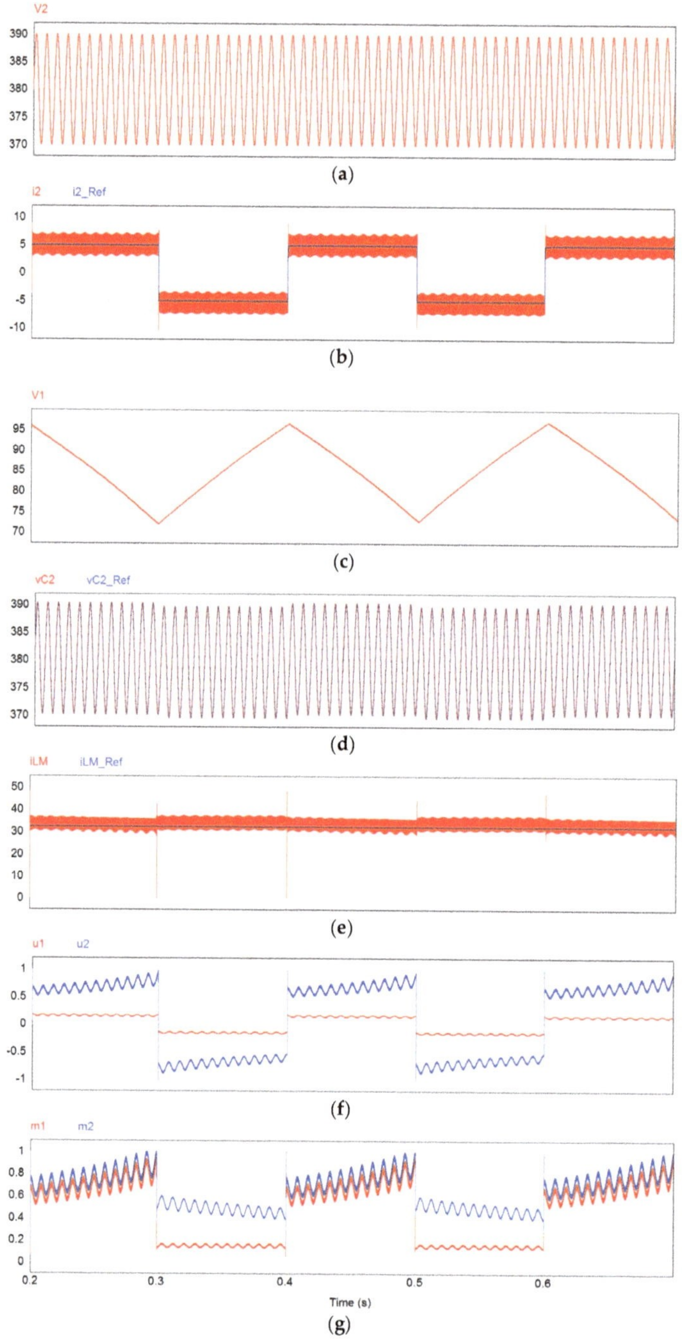

Figure 14. Waveforms for the tri-state buck–boost control with state feedback linearization for test 2. (**a**) Voltage at Bus 2 (in V), (**b**) reference and output current (in A), (**c**) voltage at Bus 1 (in V), (**d**) reference and output capacitor voltage (in V), (**e**) reference and inductor current (in A), (**f**) non-linear control variables, (**g**) PWM modulation signals.

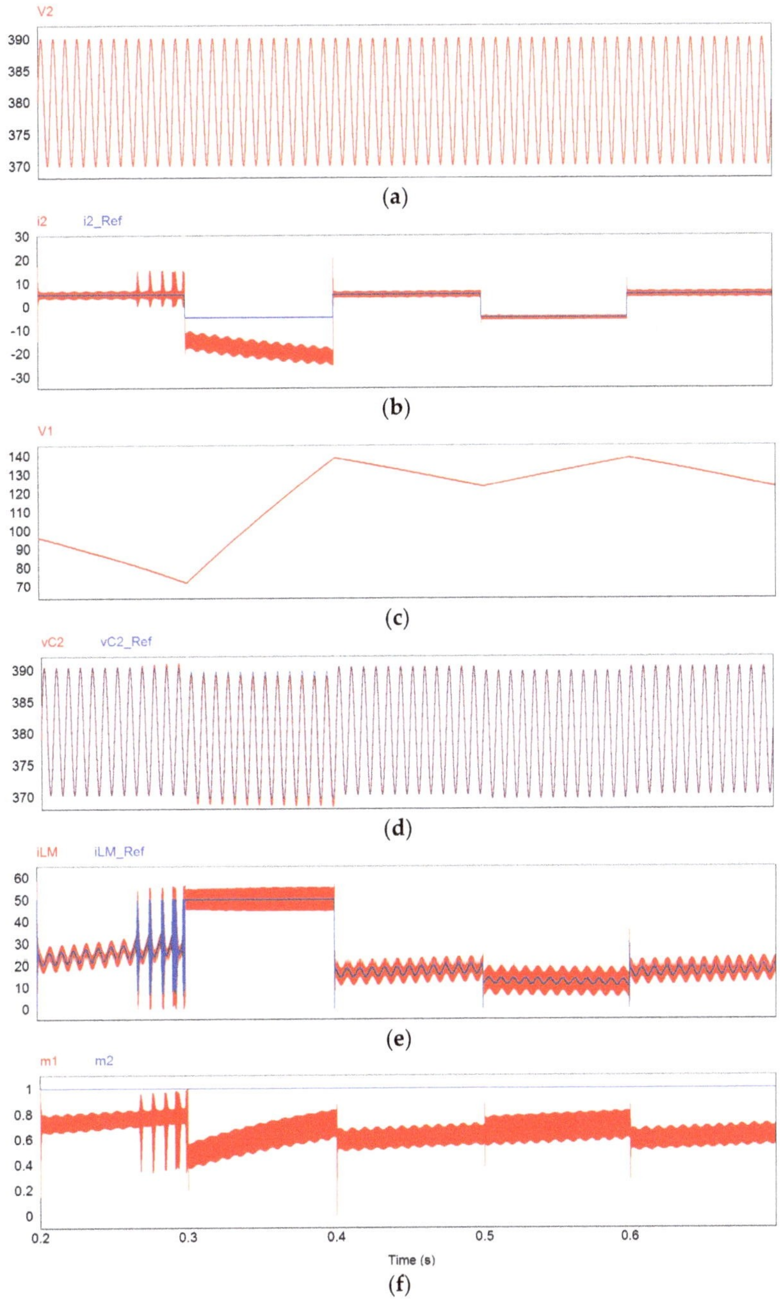

Figure 15. Waveforms for the conventional cascaded control scheme for test 2. (**a**) Voltage at Bus 2 (in V), (**b**) reference and output current (in A), (**c**) voltage at Bus 1 (in V), (**d**) reference and output capacitor voltage (in V), (**e**) reference and inductor current (in A), (**f**) PWM modulation signals.

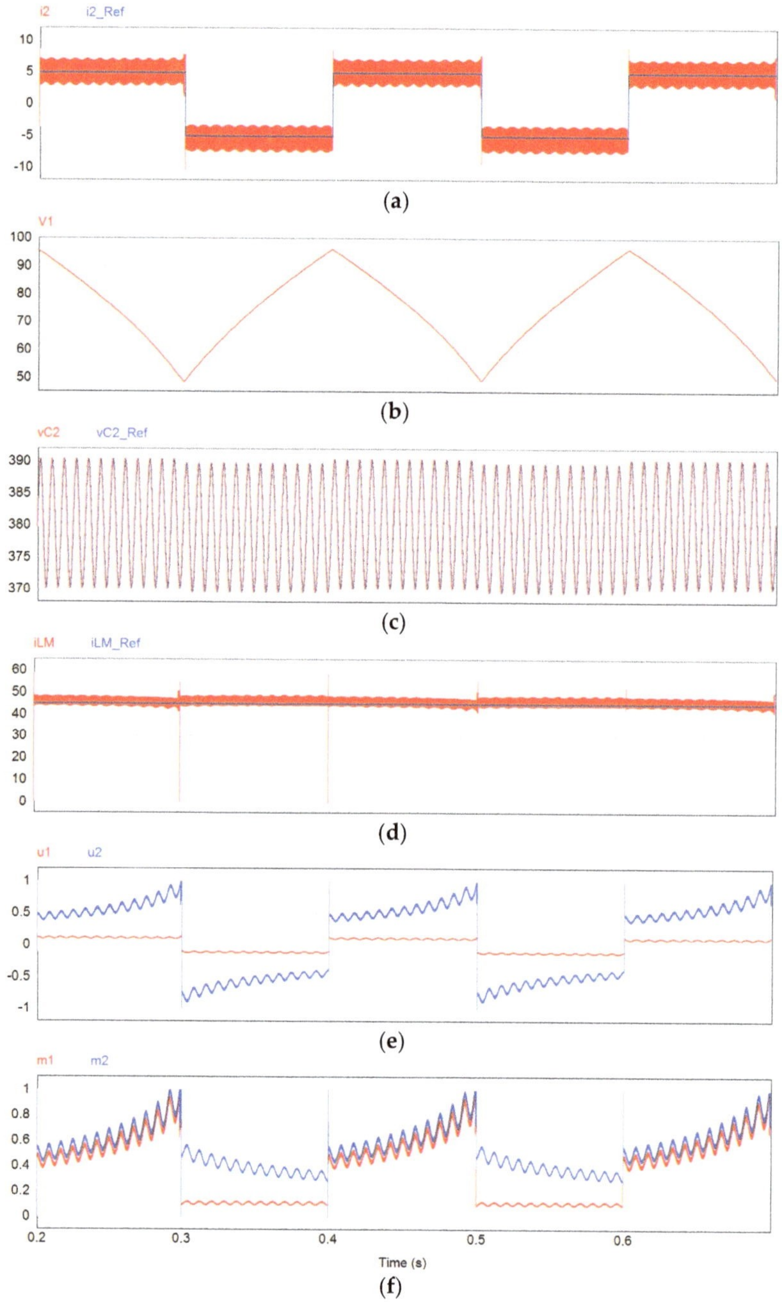

Figure 16. Waveforms for the tri-state buck–boost control with state feedback linearization for test 3. (**a**) Reference and output current (in A), (**b**) voltage at Bus 1 (in V), (**c**) reference and output capacitor voltage (in V), (**d**) reference and inductor current (in A), (**e**) non-linear control variables, (**f**) PWM modulation signals.

In Figure 16a, one sees that the output current waveform (i_2) tracks very well i_{2_ref}. The voltage at the SC (V_1) is shown in Figure 16b. It starts at 96 V and decreases due to the positive value for i_2. At $t = 0.3$ s, it reaches 48 V and starts to increase as the value of i_2 changes to −5 A. Thus, the larger variation in V_1 has no negative impact on the performance of the proposed control scheme. The output capacitor voltage (v_{C2}) in Figure 16c and the inductor current (i_{LM}) in Figure 16d track their reference signals accurately. Note that in this case, the inductor current reference (i_{LM_ref}) was increased to 45 A in order to avoid the saturation of the control variables, which tend to increase as V_1 decreases. Good tracking performance was achieved with the waveforms of control variables (u_1 and u_2) shown in Figure 16e and modulation signals (m_1 and m_2) shown in Figure 16f.

The results for the conventional dual-state modulation and linear cascaded control loops under the same conditions are shown in Figure 17.

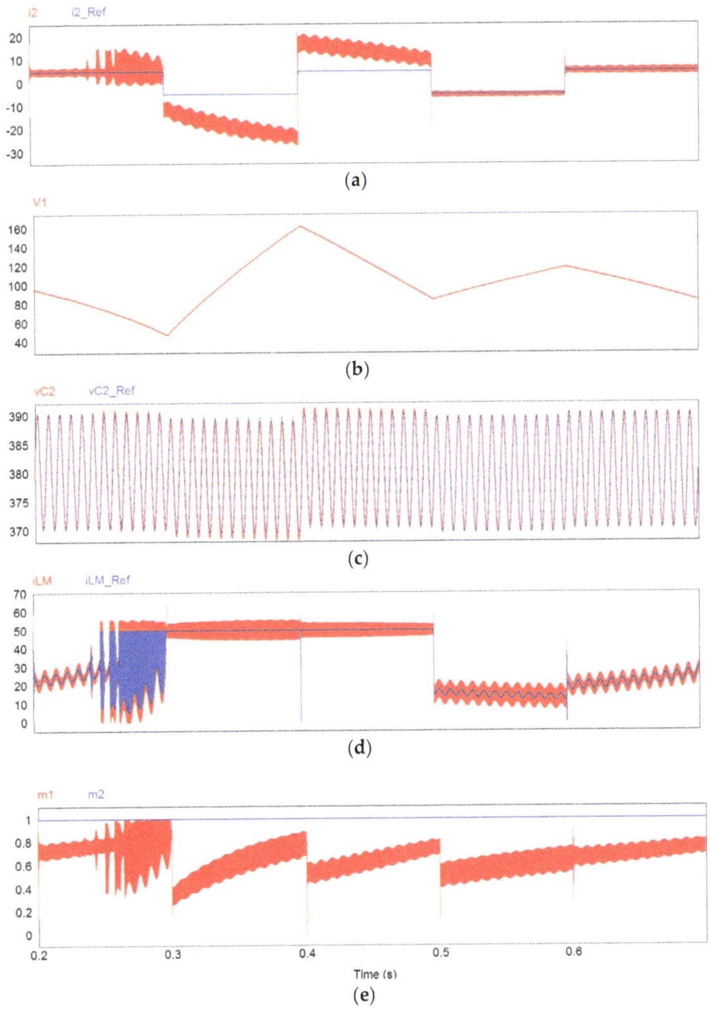

Figure 17. Waveforms for the conventional cascaded control scheme for test 3. (**a**) Reference and output current (in A), (**b**) voltage at Bus 1 (in V), (**c**) reference and output capacitor voltage (in V), (**d**) reference and inductor current (in A), (**e**) PWM modulation signals.

In Figure 17a, one sees that this conventional modulation and control scheme tracks the positive output current reference (i_{2_ref}) well until the moment that the decreasing SC voltage, (V_1 in Figure 17b), reaches a value of approximately 80 V. At this point, similar to the previous test with the slightly larger SC, the modulation signal (m_1) and several waveforms start to oscillate and the control scheme starts to malfunction. At t = 0.3 s, when the power flow reverses and i_{2_ref} = -5 A, the output current (i_2) fails to track it, becoming more negative (lower) than i_{2_ref}. The inductor current (i_{LM}) and inductor current reference (i_{LM_ref}) waveforms, shown in Figure 17d, are limited at the maximum value, 50 A. Overall, as a result of i_2 becoming more negative than i_{2_ref} at t = 0.3 s, V_1 increases more than expected, reaching 160 V at t = 0.4 s. This is when i_{2_ref} becomes positive again but, still, i_2 fails to track it. Nonetheless, V_1 decreases. At t = 0.5 s, when i_{2_ref} becomes -5 A and with V_1 with a value of approximately 85 V, which is close to its rated value (96 V), the conventional control scheme starts to work again, with i_2 tracking i_{2_ref} well, until the end of the test. Again, note that the linear controller is not able to keep the SC voltage within the desired range of 96–48 V.

6.4. Efficiency Estimation

Having shown that the five-switch DC-DC converter operating with tri-state modulation and the proposed non-linear controller behave equally well regardless of the operating point, it is important to assess the impact of such a modulation and control scheme on the power losses. For this, two main sources of losses are considered in this analysis: conduction and switching losses on the semiconductors.

For the conduction losses, recall that switches S_1 to S_4 consist of MOSFETS in series with diodes to realize semiconductor switches that are unidirectional in current, while S_T concerns a MOSFET only. In terms of losses, this means that the conduction losses in such devices (S_1 to S_4) will be defined as the sum of the conduction losses in each semiconductor, while S_T follows the conventional losses calculation for a standard MOSFET. From the power semiconductor theory, one knows that

$$Pcond_{MOSFET} = {i_{S_{RMS}}}^2 R_{DS} \tag{43}$$

$$Pcond_{Diode} = {i_{D_{RMS}}}^2 R_T + i_{D_{AVG}} V_{TO} \tag{44}$$

where R_{DS} and R_T are, respectively, the conduction resistance of the MOSFET and diode, and V_{TO} is the forward voltage of the diode. Subscripts S and D correspond to MOSFET and diode quantities, and RMS and AVG to root-mean-square and average values.

Since the MOSFET and diode are in series, for S_1 to S_4, they have the same values of average and RMS current. Thus, the total conduction losses for S_1 to S_4 can be defined as

$$Pcond_{SD} = {i_{SD_{RMS}}}^2 (R_{DS} + R_T) + i_{SD_{AVG}} (V_{TO}) \tag{45}$$

For the switching losses, the following equations, (46) and (47), are used for the calculations. One important detail for the switching loss estimation is that, if the voltage across the series connection of the MOSFET and the diode is positive, the losses will be all concentrated in the MOSFET; thus, for this case, Psw_{Diode} = 0. Likewise, if the voltage across the series connection of the MOFEST and the diode is negative, the losses will be all concentrated in the diode; thus, for this case, Psw_{MOSFET} = 0.

$$Psw_{MOSFET} = f_s V_{DS} \frac{\left(t_f i_{S_{rise}} + t_r i_{S_{fall}}\right)}{2} + f_s V_{DS} k_{oss} \tag{46}$$

$$Psw_{Diode} = f_s V_D Q_{RR} \tag{47}$$

where f_s is the switching frequency; t_f and t_r are, respectively, the rise and fall times of the MOSFET; k_{oss} is the coefficient related to the losses due to the output capacitance C_{oss} of the

MOSFET; Q_{RR} is the coefficient related to the losses due to the reverse recovery time of the diode.

From the above equations, one sees that it is necessary to obtain the RMS and average currents of the switches/diodes and maximum voltage of the switches (V_{DS}) and the diodes (V_D). This can be derived from the information presented in Table 1 and the analysis of the operating states presented in Figures 2 and 3. Moreover, it is important to select suitable power devices that match the power ratings of the desired application. Considering the voltage and current levels of the case studies discussed previously, Table 2 presents the devices chosen for this loss estimation analysis.

Table 2. Power semiconductors chosen for this study.

Components	Devices	Parameters
S_1 to S_4 and S_T	5× IPW90R120C3 (900 V/36 A)	R_{DS} = 0.12 Ω; t_f = 25 ns; t_r = 20 ns; k_{oss} = 3.9855 × 10^{-8} J/V
D_1 to D_4	4× FFSH30120A (1200 V/30 A)	R_T = 0.013 Ω; V_{TO} = 1.45 V

The final plots considering the estimated efficiency of each mode/power flow direction, as a function of the output current, are presented next. This can be seen in Figures 18 and 19.

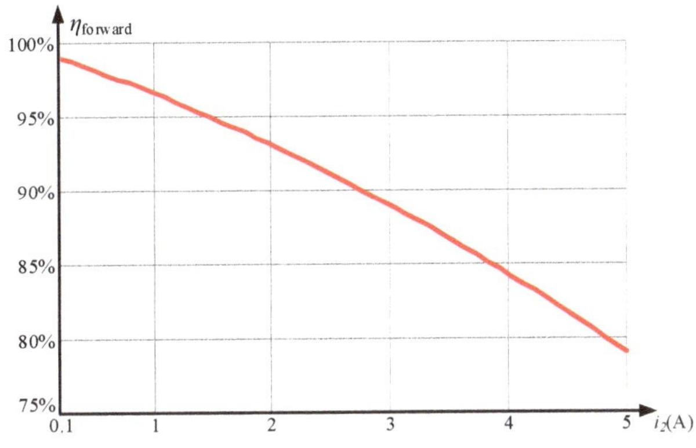

Figure 18. Efficiency estimation for the forward mode.

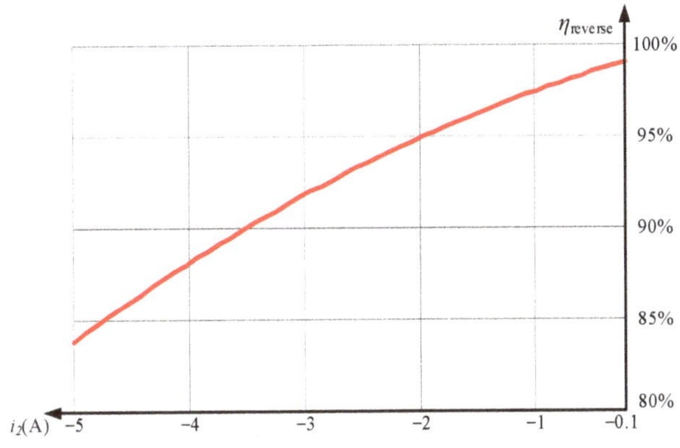

Figure 19. Efficiency estimation for the reverse mode.

As can be seen from Figures 18 and 19, as the output current increases its absolute value, the efficiency of the system tends to drop, which is to be expected, since larger values of output current should increase the RMS and average values of the currents flowing through the semiconductors. Although the efficiency for low output currents is virtually the same for both modes, another aspect to be highlighted is the fact that, as the output current increases its absolute value, the reverse mode presents lower power losses, thus presenting higher efficiency than the forward mode. This can be explained by the fact that, for the reverse mode, the high-voltage side is the one directly connected to the primary winding of the tapped inductor, which means that the lower average current of this DC bus results in a smaller current ripple for the magnetizing current, thus lowering the losses of the overall system.

It should be noted that the power loss/efficiency estimation is performed taking into consideration the power devices presented in Table 2, where one possible option would be to select, depending on the availability, power devices with better characteristics for increased efficiency of operation.

7. Conclusions

The performance of energy storage systems (ESSs), in terms of the dynamic response and tracking of the reference output current, depends on the features of the power converter and control scheme. This paper presented the modeling of a novel five-switch DC-DC power converter, which presents a high voltage gain (V_2/V_1) and does not require the reversal of the intermediate inductor current to achieve power flow reversal. In terms of the control scheme, by operating this converter in a tri-state buck–boost mode, one eliminates the right half-plane (RHP) zero from the output-to-control transfer function, which facilitates the design of the control loop. Moreover, by employing the exact state feedback linearization technique, the converter should be able to operate in a wider range than that with a PI controller designed considering the conventional small-signal model of the converter. A systematic approach for modeling the converter and obtaining a control law with exact feedback linearization was presented. Note that since the inductor current, a state variable, does not need to reverse during power flow reversal in this particular converter, there will be not potential issues of "divide by zero" in the control law.

The performance of the five-switch DC-DC power converter with the proposed modulation (tri-state buck–boost) and control scheme (exact state feedback linearization) was verified by means of simulations. The selected application was as the interface of a supercapacitor (SC)-based ESS to an output DC bus, such as a DC microgrid. It was compared to the converter operating in the conventional dual-state buck–buck mode with linear cascaded current control loops. As expected, both control schemes present similar good behavior at the operating point that the conventional scheme was designed for. When the input voltage (V_1) varies even moderately, which is common in SC-based ESSs, the conventional scheme fails to track the output current reference. Conversely, the proposed modulation and control scheme tracks accurately, with a good dynamic response and zero steady-state error, the output current reference even with the SC voltage varying in a wide range, from rated to half-rated. This is an important feature of the interfaces of SC-based ESSs that is difficult to obtain with conventional power converters and control schemes.

Directions for Further Development

Having validated the mathematical model and proposed a control scheme for the five-switch bi-directional DC-DC converter, the next step is experimental verification. In this way, other aspects, such as the efficiency, power density and electromagnetic compatibility (EMC), of the converter can be assessed. As power semiconductor switches, MOSFETs, with their fast switching speeds and low R_{DS}(ON), are the natural choice. For switches S_1 to S_4 series diodes, to match the switching speeds of MOSFETs, devices with fast switching characteristics and low reverse recovery times, such as Schottky diodes, should be added to the MOSFETs to provide the unidirectional current, or reverse blocking, feature. Regarding the proposed control scheme and PWM modulator, the TMS320F28035 microcontroller from Texas Instruments provides the means for its realization. Note that these are common choices for DC-DC converters used as interfaces of distributed energy resources.

Author Contributions: Conceptualization, G.R.B., L.A.C.L. and G.D.; methodology, L.A.C.L. and G.D.; software, G.R.B. and L.A.C.L.; validation, G.R.B., L.A.C.L. and G.D.; formal analysis, L.A.C.L. and G.D.; investigation, G.R.B., L.A.C.L. and G.D.; resources, L.A.C.L.; data curation, G.R.B.; writing—original draft preparation, G.R.B., L.A.C.L. and G.D.; writing—review and editing, G.R.B., L.A.C.L. and G.D.; visualization, G.R.B., L.A.C.L. and G.D.; supervision, L.A.C.L.; project administration, L.A.C.L.; funding acquisition, L.A.C.L. All authors have read and agreed to the published version of the manuscript.

Funding: This research was funded by the Natural Sciences and Engineering Research Council (NSERC) of Canada: DG-N00863.

Conflicts of Interest: The authors declare no conflict of interest.

References

1. Ghareeb, A.T.; Mohamed, A.A.; Mohammed, O.A. DC Microgrids and Distribution Systems: An Overview. In Proceedings of the IEEE Power & Energy Society General Meeting 2013, Vancouver, BC, Canada, 21–25 July 2013; pp. 1–6.
2. Diaz, E.R.; Savaghebi, M.; Vasquez, J.C.; Guerrero, J.M. An overview of low voltage dc distribution systems for residential applications. In Proceedings of the ICCE-Berlin 2015—5th International Conference on Consumer Electronics, Berlin, Germany, 6–9 September 2015; pp. 318–322.
3. Dragicevic, T.; Lu, X.; Vasquez, J.C.; Guerrero, J.M. DC Microgrids—Part I: A Review of Control Strategies and Stabilization Techniques. *IEEE Trans. Power Electron.* **2016**, *31*, 4876–4891. [CrossRef]
4. Dragicevic, T.; Lu, X.; Vasquez, J.C.; Guerrero, J.M. DC Microgrids—Part II: A Review of Power Architectures, Applications, and Standardization Issues. *IEEE Trans. Power Electron.* **2016**, *31*, 3528–3549. [CrossRef]
5. Jing, W.; Lai, C.H.; Wong, S.H.W.; Wong, M.L.D. Battery-supercapacitor hybrid energy storage system in standalone DC microgrids: A review. *IET Renew. Power Gener.* **2017**, *11*, 461–469. [CrossRef]
6. Ferahtia, S.; Djeroui, A.; Mesbahi, T.; Houari, A.; Zeghlache, S.; Rezk, H.; Paul, T. Optimal Adaptive Gain LQR-Based Energy Management Strategy for Battery–Supercapacitor Hybrid Power System. *Energies* **2021**, *14*, 1660. [CrossRef]
7. Bocklisch, T. Hybrid Energy Storage Systems for Renewable Energy Applications. In Proceedings of the IRES 2015—9th International Renewable Energy Storage Conference, Düsseldorf, Germany, 9–11 March 2015; pp. 103–111.
8. Williams, B.W. Unified synthesis of tapped-inductor DC-to-DC converters. *IEEE Trans. Power Electron.* **2013**, *10*, 5370–5383. [CrossRef]
9. Broday, G.R.; Nascimento, C.B.; Agostini, E.; Lopes, L.A.C. A tri-state bidirectional buck-boost converter for a battery/supercapacitor hybrid energy storage system in electric vehicle applications. In Proceedings of the VPPC 2015—IEEE Conference on Vehicle Power and Propulsion, Montreal, Canada, 19–22 October 2015; pp. 1–6.
10. Viswanathan, K.; Oruganti, R.; Srinivasan, D. A Novel Tri-State Boost Converter with Fast Dynamics. *IEEE Trans. Power Electron.* **2002**, *17*, 677–683. [CrossRef]
11. Broday, G.R.; Lopes, L.A.C. A novel 5-switch tapped-inductor multi-state bidirectional DC-DC converter. In Proceedings of the ICIT 2018—IEEE International Conference on Industrial Technology, Lyon, France, 20–22 February 2018; pp. 596–599.
12. Broday, G.R.; Damm, G.; Pasillas-Lépine, W.; Lopes, L.A.C. Modeling and dynamic feedback linearization of a 5-switch tri-state buck-boost bidirectional DC-DC converter. In Proceedings of the ICIT 2021—IEEE International Conference on Industrial Technology, Valencia, Spain, 10–12 March 2021; pp. 427–432.
13. Georgious, R.; Garcia, J.; Sumner, M.; Saeed, S.; Garcia, P. Fault Ride-Through Power Electronic Topologies for Hybrid Energy Storage Systems. *Energies* **2020**, *13*, 257. [CrossRef]
14. Luo, X.; Barreras, J.V.; Chambon, C.L.; Wu, B.; Batzelis, E. Hybridizing Lead–Acid Batteries with Supercapacitors: A Methodology. *Energies* **2021**, *14*, 507. [CrossRef]
15. Yuan, C.; Huangfu, Y.; Ma, R.; Zhao, B.; Bai, H. Nonlinear PI and Finite-time Control for DC-DC Converter Based on Exact Feedback Linearization. In Proceedings of the IECON 2019—45th Annual Conference of the IEEE Industrial Electronics Society, Lisbon, Portugal, 14–17 October 2019; pp. 6398–6403.
16. Callegaro, L.; Ciobotaru, M.; Pagano, D.; Fletcher, J. Feedback Linearization Control in Photovoltaic Module Integrated Converters. *IEEE Trans. Power Electron.* **2019**, *34*, 6876–6889. [CrossRef]
17. Ding-xin, S. State feedback exact linearization control of Buck-Boost converter. In Proceedings of the PEAC 2014—International Power Electronics and Application Conference and Exposition, Shanghai, China, 5–8 November 2014; pp. 1490–1494.
18. Iovine, A.; Carrizosa, M.J.; Damm, G.; Alou, P. Nonlinear Control for DC Microgrids Enabling Efficient Renewable Power Integration and Ancillary Services for AC grids. *IEEE Trans. Power Electron.* **2018**, *34*, 5136–5146. [CrossRef]
19. Perez, F.; Iovine, A.; Damm, G.; Ribeiro, P. DC Microgrid Voltage Stability by Dynamic Feedback Linearization. In Proceedings of the ICIT 2018—IEEE International Conference on Industrial Technology, Lyon, France, 20–22 February 2018; pp. 129–134.

20. Broday, G.R.; Damm, G.; Pasillas-Lépine, W.; Lopes, L.A.C. A Unified Controller for Multi-State Operation of the Bi-Directional Buck–Boost DC-DC Converter. *Energies* **2021**, *14*, 7921. [CrossRef]
21. Gonzalez, A.; Lopez-Erauskin, R.; Gyselinck, J.; Kei Chau, T.; Ho-Ching Lu, H.; Fernando, T. Nonlinear MIMO control of interleaved three-port boost converter by means of state-feedback linearization. In Proceedings of the PEMC 2018—IEEE 18th International Power Electronics and Motion Control Conference, Budapest, Hungary, 26–30 August 2018; pp. 164–169.
22. Cai, P.; Wu, X.; Sun, R.; Wu, Y. Exact feedback linearization of general four-level buck DC-DC converters. In Proceedings of the CCDC 2017—29th Chinese Control and Decision Conference, Chongqing, China, 28–30 May 2017; pp. 4638–4643.
23. Broday, G.R.; Lopes, L.A.C. A Minimum Power Loss Approach for Selecting the Turns Ratio of a Tapped Inductor and Mode of Operation of a 5-Switch Bidirectional DC-DC Converter. In Proceedings of the ICDCM 2021—IEEE 4th International Conference on DC Microgrids, Arlington, USA, 18–21 July 2021; pp. 1–7.
24. Lee, D.-C.; Lee, G.-M.; Lee, K.-D. DC-bus voltage control of three-phase AC/DC PWM converters using feedback linearization. *IEEE Trans. Ind. Applic.* **2000**, *36*, 826–833.
25. Lu, Z.; Zhang, X.; Wang, Y. Nonlinear Control Strategy of Hybrid Energy Storage Systems Based on Feedback Linearization. In Proceedings of the CVCI 2020—4th CAA International Conference on Vehicular Control and Intelligence, Hangzhou, China, 18–20 December 2020; pp. 674–677.
26. Ming, W.; Liu, J. A new experimental study of input-output feedback linearization based control of Boost type DC/DC converter. In Proceedings of the ICIT 2010—IEEE International Conference on Industrial Technology, Viña del Mar, Chile, 14–17 March 2010; pp. 685–689.
27. Li, X.; Chen, X. A Multi-Index Feedback Linearization Control for a Buck-Boost Converter. *Energies* **2021**, *14*, 1496. [CrossRef]
28. Ogata, K. *Modern Control Engineering*; Prentice Hall: Hoboken, NJ, USA, 2010.
29. Abeywardana, D.B.W.; Hredzak, B.; Agelidis, V.G.; Demetriades, G.D. Supercapacitor Sizing Method for Energy-Controlled Filter-Based Hybrid Energy Storage Systems. *IEEE Trans. Power Electron.* **2017**, *32*, 1626–1637. [CrossRef]
30. Dotelli, G.; Ferrero, R.; Stampino, P.G.; Latorrata, S.; Toscani, S. Supercapacitor Sizing for Fast Power Dips in a Hybrid Supercapacitor-PEM Fuel Cell System. *IEEE Trans. Intrum. and Meas.* **2016**, *32*, 2196–2203. [CrossRef]
31. Arnaudov, D.; Punov, P.; Dimitrov, V. Supercapacitor Sizing for Power Defined Loads. In Proceedings of the ET 2019—IEEE XXVIII International Scientific Conference Electronics, Sozopol, Bulgaria, 12–14 September 2019; pp. 1–3.

energies

MDPI

Article

SOC Balancing and Coordinated Control Based on Adaptive Droop Coefficient Algorithm for Energy Storage Units in DC Microgrid

Guizhen Tian [1,2,*], Yuding Zheng [1], Guangchen Liu [1,2] and Jianwei Zhang [1,2]

[1] College of Electric Power, Inner Mongolia University of Technology, Hohhot 010080, China; zydemailsr@163.com (Y.Z.); liugc@imut.edu.cn (G.L.); zjwzachary@outlook.com (J.Z.)
[2] Inner Mongolia Regional Key Laboratory of Electrical Power Conversion, Transmission and Control, Hohhot 010080, China
* Correspondence: tianguizhen@126.com; Tel.: +86-150-4781-1978

Abstract: In order to achieve a state-of-charge (SOC) balance among multiple energy storage units (MESUs) in an islanded DC microgrid, a SOC balancing and coordinated control strategy based on the adaptive droop coefficient algorithm for MESUs is proposed. When the SOC deviation is significant, the droop coefficient for an energy storage unit (ESU) with a higher (or lower) SOC is set to a minimum value when discharging (or charging). The ESU with the higher (or lower) SOC is controlled to discharge (or charge) with the rated power, while the other ESU compensates for the remaining power when the demanded discharging (or charging) power is greater than the rated power of the individual ESU. Otherwise, when the demanded discharging (or charging) power is lower than the rated power of either ESU, the ESU with the higher (or lower) SOC releases (or absorbs) almost all the required power while the other ESU barely absorbs or releases power, thus quickly realizing SOC balancing. When the SOC deviation is slight, the fuzzy logic algorithm dynamically adjusts the droop coefficient and changes the power distribution relationship to balance the SOC accurately. Furthermore, a bus voltage recovery control scheme is employed to regulate the bus voltage, thus improving the voltage quality. The energy coordinated management strategy is adopted to ensure the power balance and stabilize the bus voltage in the DC microgrid. A simulation model is built in MATLAB/Simulink, and the simulation results demonstrate the effectiveness of the proposed control strategy in achieving fast and accurate SOC balance and regulating the bus voltage.

Keywords: DC microgrid; coordinated control; energy management; SOC balancing; droop control; fuzzy logic algorithm

Citation: Tian, G.; Zheng, Y.; Liu, G.; Zhang, J. SOC Balancing and Coordinated Control Based on Adaptive Droop Coefficient Algorithm for Energy Storage Units in DC Microgrid. *Energies* **2022**, *15*, 2943. https://doi.org/10.3390/en15082943

Academic Editors: Alexander Micallef and Zhaoxia Xiao

Received: 28 February 2022
Accepted: 15 April 2022
Published: 17 April 2022

1. Introduction

Due to the intermittence of renewable energy sources and unpredictable load fluctuations, instantaneous power imbalances are caused. An energy storage system (ESS) is usually used to provide energy support and enhance the system's reliability [1–4]. To meet the power rating of the corresponding converter and ensure the safety and reliability of the ESS, multiple energy storage units (MESUs) are required to be configured into an ESS. When the line impedances and real capacities are different, state-of-charge (SOC) deviation among MESUs is produced [5–7]. Maintaining SOC balance can avoid the overcharging and over-discharging of MESUs and maximize the available energy storage capacity and charging/discharging rates [8–10].

The battery SOC balancing control strategy based on multi-agent is proposed in [11]. The proposed SOC balancing strategy has a plug-and-play capability. Moreover, different capacities of batteries and the decline in the battery capacity after a long-term operation are considered in the control strategy. In [12], the total reflected capacity in the system is estimated by summing the reflected capacity, SOC, and Ah rating of each battery. The total reflected capacity divided by the reference current from the controller gives the reference time for each battery to discharge or charge. In [13], to balance the SOCs of all energy storage units (ESUs), the reference current of each ESU is obtained by a fuzzy controller with SOC and DC-link voltage as inputs.

Droop control is a popular current sharing method in the primary control layer, especially for islanded low-voltage DC microgrids [14–16]. The droop control has been widely applied in DC microgrids [17,18]. In [19], the droop coefficient, inversely proportional to the nth order of SOC, is used to balance the SOCs of ESUs in the DC microgrid. The ESUs with a higher SOC can be controlled to deliver more power, whereas those with a lower SOC deliver less power. In [20], the double-quadrant SOC-based droop control method is used. In the charging/discharging process, the droop coefficient is positively/inversely proportional to the nth order of SOC. In [21], a new decentralized strategy is proposed to balance the SOC of each ESU. The virtual resistance of the droop controller is modified by the fuzzy controller, so the battery with the lowest SOC is charged faster than others. In [22], the droop resistance is adjusted based on the exponential function of the difference between an individual battery's SOC and the average SOC of all the batteries in a DC microgrid. In [23], a highly accurate power-sharing method is presented to balance the SOC. In this method, the objective of the PI controller is to regulate the SOC of each battery equal to the average SOC. The nominal voltage reference of droop control is adjusted by sliding mode control to achieve SOC balancing [24,25]. In [26], a secondary control based on a consensus algorithm has been proposed to regulate the DC-bus voltage reference and balance the SOC of ESUs. In [27], a SOC balancing scheme considering different SOCs, and capacities is achieved by a high-pass-filter-based SOC balancing method. However, under the larger SOC deviation, the above control methods do not consider accelerating SOC balance by controlling ESUs with higher SOC to discharge at maximum power (or to charge at minimum power) or ESUs with lower SOC to charge at maximum power (or to discharge at minimum power).

In order to quickly and accurately balance the SOCs of MESU and ensure stable operation of the DC microgrid, SOC balancing and coordinated control based on an adaptive droop coefficient algorithm are proposed in this paper. The fuzzy logic algorithm does not need the system model. However, it relies on the knowledge formulated by an experienced operator or expert to achieve the control objectives. This paper adopts the fuzzy logic algorithm to dynamically adjust the droop coefficients to realize a more accurate SOC balancing control. The main contributions of this paper are summarized as follows:

(1) The droop coefficient of ESUs with higher/lower SOC under discharge/charge is regulated to a minimum value in the case of a significant SOC deviation. SOCs can be balanced quickly;
(2) The droop coefficient is automatically adjusted by the fuzzy logic algorithm to accurately balance SOC in the case of a slight SOC deviation;
(3) The DC bus voltage recovery control is adopted to eliminate the voltage error caused by the traditional droop control, realizing automatic recovery control of the bus voltage;
(4) To ensure the power balance and stabilize the bus voltage, the energy coordinated management strategy based on SOC balancing of the DC microgrid has been adopted.

2. Analysis of the SOC Unbalance

The studied islanded DC microgrid system is depicted in Figure 1 [28]. The photovoltaic (PV) power generation system provides energy to the DC bus. The two ESUs absorb the excess power from or release power to the bus to achieve the bus voltage stability and supply the load.

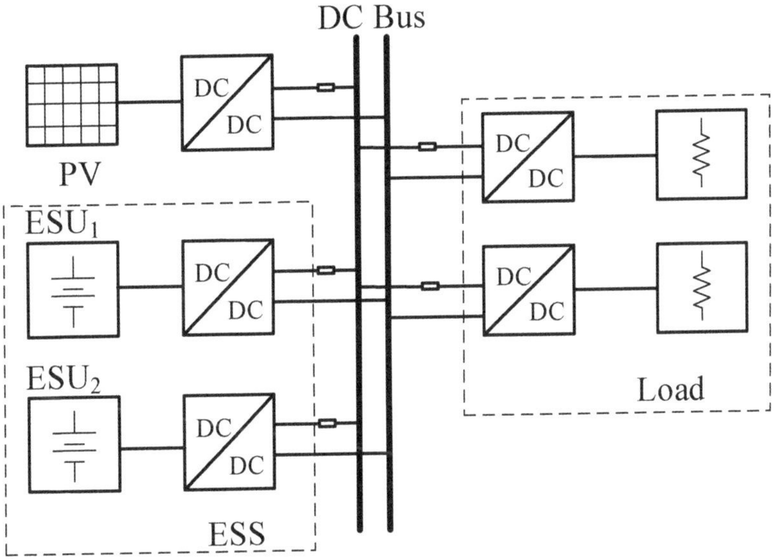

Figure 1. Topology of islanded DC microgrid.

In the DC microgrid, the droop control can be expressed by the following:

$$U_{out} = U_{ref} - kI_{out} \tag{1}$$

where U_{ref} and U_{out} are the reference and actual output voltages of the DC/DC converter; k represents the droop coefficient; I_{out} is the output current of the DC/DC converter.

SOC of the battery is the follows:

$$SOC = SOC_0 - \frac{\eta \int idt}{C_e} \tag{2}$$

where SOC_0 is the SOC initial value; C_e and i represent the capacity and output current; η denotes the battery efficiency.

Considering the line impedance and using (1), current i can be written as follows:

$$i = I_{out} = -\frac{(U_{ref} - U_{out})}{(k + R_{line})} \tag{3}$$

where R_{line} is line impedance.

Combining (2) and (3), the changing rates of SOCs of the two batteries are given by the following:

$$\begin{cases} \dot{SOC_1} = -\frac{\eta(U_{ref1} - U_{out})}{C_{e1} \times (k_1 + R_{line1})} \\ \dot{SOC_2} = -\frac{\eta(U_{ref2} - U_{out})}{C_{e2} \times (k_2 + R_{line2})} \end{cases} \tag{4}$$

where $k_{1,\,2}$, $R_{line1,\,2}$, and $U_{ref1,\,2}$ are the droop coefficient, line impedance, and output voltage reference of ESU_1 and ESU_2.

Further, one can obtain

$$\frac{\dot{SOC_1}}{\dot{SOC_2}} = \frac{C_{e2} \times (k_2 + R_{line2})}{C_{e1} \times (k_1 + R_{line1})} \tag{5}$$

It is known from (5) that SOC deviation would be created when the actual capacity, droop coefficient, or line impedance of these MESU are different.

3. The SOC Balancing Strategy Based on Adaptive Droop Coefficient Algorithm

3.1. The Adaptive Droop Coefficient Algorithm

The average SOC and SOC deviation are shown in (6) and (7).

$$SOC_{avg} = \frac{SOC_1 + SOC_2}{2} \tag{6}$$

$$\Delta SOC_i = SOC_i - SOC_{avg} \tag{7}$$

The SOC balancing strategy based on the adaptive coefficient algorithm is shown in Figure 2, and explanations are described below.

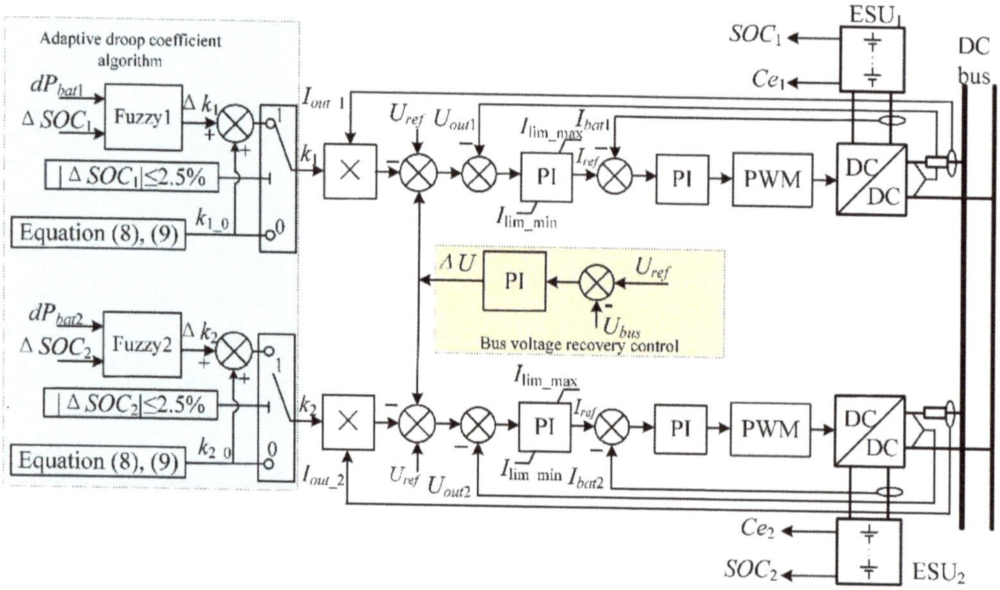

Figure 2. Diagram of the SOC balancing strategy.

(1) $|\Delta SOC_i| > 2.5\%$

When the SOC deviation is significant (i.e., $|\Delta SOC_i| > 2.5\%$), to accelerate SOC equalization, the droop coefficient is set as follows;

In the charging process, as follows:

$$\begin{cases} k_{i\_0} = k_{\min} & \Delta SOC_i \leq -2.5\% \\ k_{i\_0} = \left| \frac{SOC_{avg} - SOC_i}{SOC_{avg} - SOC_{\max}} \right| \cdot k_{\max} & \Delta SOC_i > -2.5\% \end{cases} \tag{8}$$

where k_{i_0} represents the initial coefficient; $i = 1, 2$; $k_{\min}$ and $k_{\max}$ are the minimum and maximum values of the droop coefficient; $SOC_{\max}$ is the maximum value of SOC.

In the discharging process, as follows:

$$\begin{cases} k_{i\_0} = k_{\min} & \Delta SOC_i \geq 2.5\% \\ k_{i\_0} = \left| \frac{SOC_{avg} - SOC_i}{SOC_{avg} - SOC_{\max}} \right| \cdot k_{\max} & \Delta SOC_i < 2.5\% \end{cases} \tag{9}$$

The PI limiter of the outer loop is the following:

$$\begin{cases} I_{\text{lim\_max}} = 0.5C_e \\ I_{\text{lim\_min}} = -0.2C_e \end{cases} \quad (10)$$

The droop coefficients are set according to (8) and (9). In combination with the outer loop limiter, the following two goals can be achieved:

(a) If the power to be released reaches $P_{\text{dmax}} \leq P < 2P_{\text{dmax}}$ or the power to be absorbed reaches $P_{\text{cmax}} \leq P < 2P_{\text{cmax}}$, the battery with the lower (higher) SOC absorbs (releases) the rated power when charging (discharging). The other units compensate for the residual power automatically. Here, P_{dmax} and P_{cmax} represent the rated discharging and charging power;

(b) If the power to be released reaches $P < P_{\text{dmax}}$ or the power to be absorbed reaches $P < P_{\text{cmax}}$, the battery with lower (higher) SOC absorbs (releases) all power when charging (discharging), and the other unit's output power is nearly zero.

(2) $|\Delta SOC_i| \leq 2.5\%$

When the SOC deviation is slight (i.e., $|\Delta SOC_i| \leq 2.5\%$), the droop coefficient is adjusted by the fuzzy control algorithm to balance the SOCs accurately. The ESUs' droop coefficient can be expressed as follows:

$$k_i = k_{i\_0} + \Delta k_i, (i = 1, 2) \quad (11)$$

where Δk_i is the droop coefficient increment and it is determined by the fuzzy control algorithm; k_{i_0} is set as the same as when $|\Delta SOC_i| > 2.5\%$.

The fuzzy control algorithm generates a droop coefficient increment according to the power deviation and the SOC deviation of ESUs to adjust the drooping coefficient and realize SOC balance control.

3.2. Design of the Fuzzy Logic Algorithm

The fuzzy control method has the advantage of not requiring an accurate model and adaptively modifying the key parameters according to the designed fuzzy rules [29–31]. Therefore, the fuzzy control algorithm is used to adjust the droop coefficient to achieve accurate SOC balancing when the SOC deviation is slight.

The variation of the droop coefficient determines the amount of power absorbed or released by each ESU, thus affecting the variation of SOC. Therefore, the deviations in power and SOC between ESUs are selected as the inputs of the fuzzy control algorithm. The corresponding relationship between the physical and fuzzy domains of variables for the fuzzy control algorithm is shown in Table 1.

Table 1. Relationship between physical and fuzzy domain.

Variables	Physical Domain	Quantization Factor	Fuzzy Domain
ΔSOC	[−50,50]	0.02	[−1,1]
dP	Charge: [−10,10]	0.1	[−1,1]
	Discharge: [−25,25]	0.04	[−1,1]
Δk	[−0.2,0.2]	5	[−1,1]

The membership functions of the input variables and output variables of the fuzzy controller are shown in Figure 3. The variables of ΔSOC, dP, and Δk are mapped to three different fuzzy subsets according to the control experience. The fuzzy control rules can be designed according to the experiences shown in Table 2. According to the membership function, and 18 different fuzzy rules in the charging and discharging state, the normalized control surface can be obtained after defuzzification by the centroid method, as shown in Figure 4.

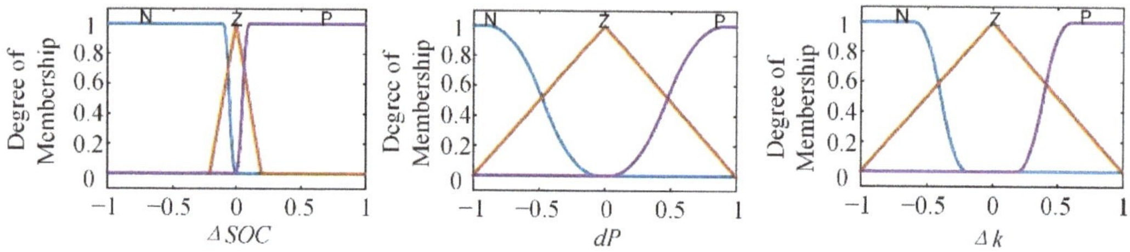

Figure 3. Membership function.

Table 2. Rules of fuzzy controller.

ΔSOC	dP	Δk	
		Charging	**Discharging**
P	P	N	Z
P	Z	P	N
P	N	P	N
Z	P	N	P
Z	Z	Z	Z
Z	N	P	N
N	P	N	P
N	Z	N	P
N	N	Z	N

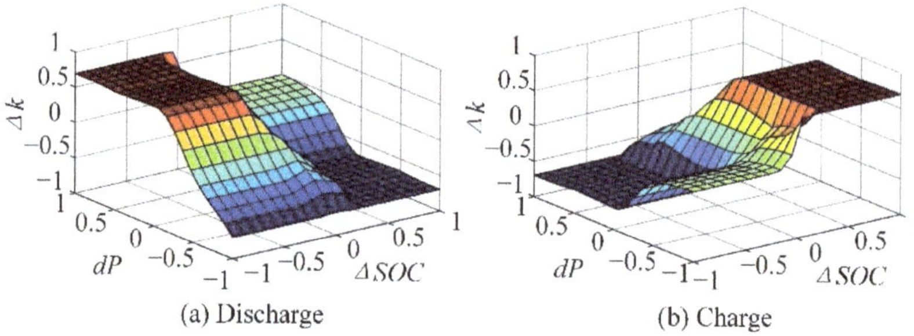

Figure 4. Control surface. (**a**) Discharge; (**b**) Charge.

3.3. Bus Voltage Recovery Control

A voltage deviation equal to kI_{out} is always produced in the traditional droop control. More power transmission leads to more bus voltage drops. Therefore, a bus voltage recovery control is added to overcome the bus voltage drop caused by the traditional droop control and ensure the stability of bus voltage.

Bus voltage can be expressed by the following:

$$\begin{cases} U_{out} = U_{ref} - kI_{out} + \Delta U \\ \Delta U = \left(K_p + \frac{K_i}{s}\right)\left(U_{ref} - U_{bus}\right) \end{cases} \quad (12)$$

where U_{bus} is bus voltage, and K_p and K_i are the proportional and integral coefficients of the PI regulator.

The DC bus voltage (kI_{out}) is compensated by the output voltage (ΔU). When the system reaches a steady state, the bus voltage (U_{bus}) can be controlled to equal the given voltage (U_{ref}). Therefore, the DC bus voltage without control error is realized.

3.4. Simulation Waveforms of SOC Balancing Control

A simulation model is built based on MATLAB/Simulink R2021b from MathWorks (Corporate Headquarters Natick, MA, USA) and parameters are shown in Table 3 [28]. In simulation settings, the solver is set to ode23tb and the step size is 2 µs.

Table 3. Main parameters of the system.

Description	Value
Bus voltage U_{out}	750 V
Capacitance C_{out}	2000 µF
Inductance L	2 mH
PV system	55 kW
Important loads	20 kW
Line impedance of ESU_1 R_{line1}	0.03 Ω
Line impedance of ESU_2 R_{line2}	0.05 Ω
Real capacity of ESU_1 C_{e1}	133 Ah
Real capacity of ESU_2 C_{e2}	130 Ah
Non-important loads	4 × 5 kW

(1) ESU-discharging waveforms

Figure 5 presents the comparative simulation results of the proposed method with the method in [32]. The discharging power of ESS is $P_{dmax} \leq P < 2P_{dmax}$, and PV and load power are 10 kW and 40 kW. To maintain the power balance within the DC microgrid, the ESS needs to release 30 kW, which is more than the maximum discharging power (25 kW) of an ESU. The initial SOCs of the two ESUs are 70% and 50%. As observed in Figure 5a, when the SOC deviation of two ESUs is larger than 2.5% (before 0.45 s), ESU_1 outputs the maximum allowable discharging power of 25 kW since it has a larger SOC. The remaining 5 kW power shortage is provided by ESU_2. The SOC deviation of the two ESUs is decreased to less than 2.5% after 0.45 s. Eventually, the SOC deviation is eliminated.

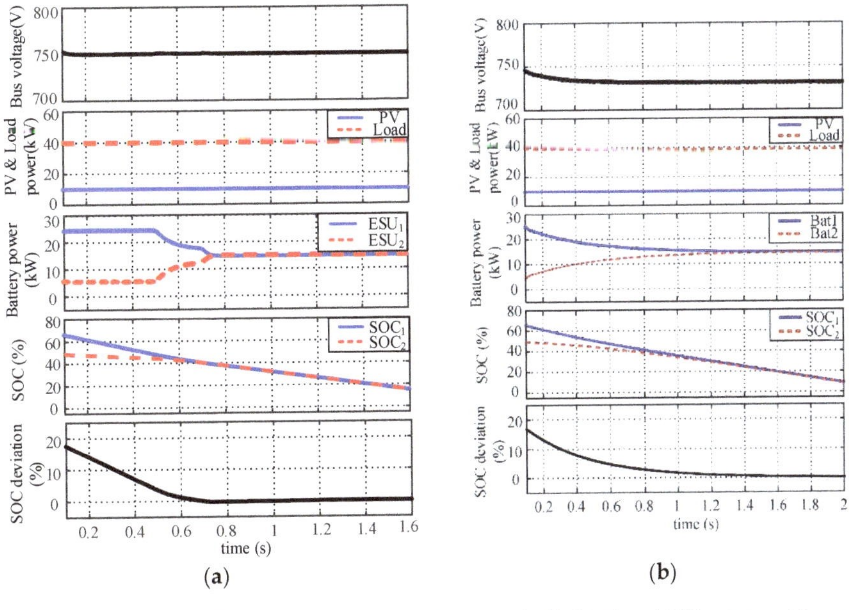

Figure 5. Waveforms of $P_{dmax} \leq P < 2P_{dmax}$ in discharging. (**a**) Waveforms of the proposed method; (**b**) Waveforms of the method in [32].

As seen in Figure 5b, the coefficients are getting close due to the decrease in SOC deviation. Thus, the output powers of the two batteries also get close, and together they output 30 kW all the time. The SOC deviation is eliminated at around 1.8 s, and it is much slower than the proposed method shown in Figure 5a.

When the discharging power ESS is $P \leq P_{dmax}$, the simulation results are shown in Figure 6. In Figure 6, ESU_1 outputs 15 kW of power of the total shortage in the system, and the output power of ESU_2 is nearly 0. The SOC of the two ESUs is balanced at 0.9 s.

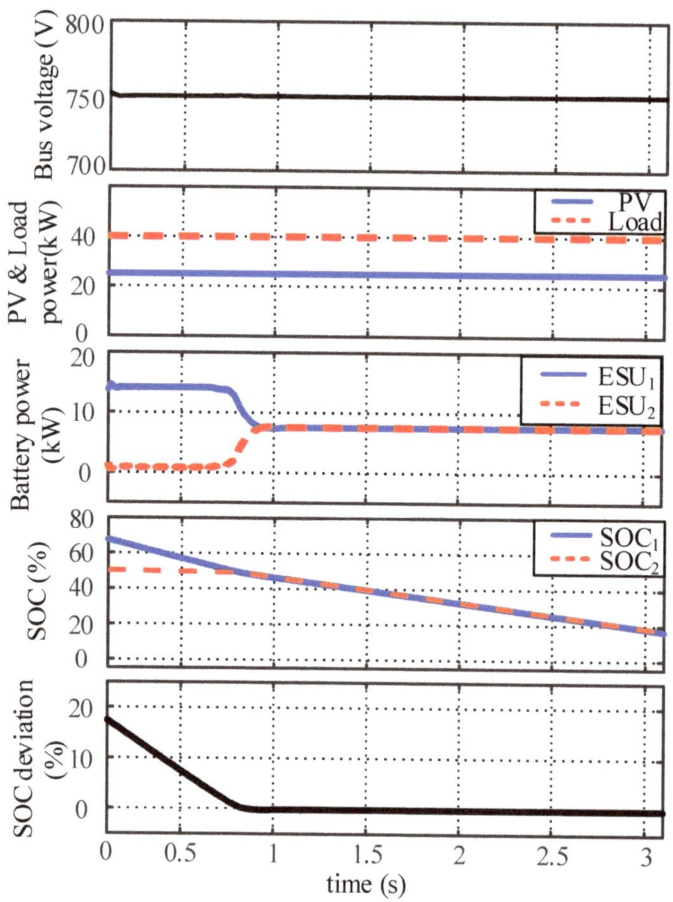

Figure 6. Waveforms of $P \leq P_{dmax}$ in discharging.

As can be seen from Figures 5 and 6, the SOCs can be balanced quickly. The voltage deviation can be eliminated automatically under the voltage recovery control. Finally, the bus voltage is recovered to 750 V, following the defined voltage reference.

(2) ESU-charging waveforms

In Figure 7, PV and load powers are 55 kW and 40 kW, respectively, and there is a 15 kW deficit in the microgrid, larger than the maximum charging power of an EUS. The SOCs of the two ESUs are 50% and 30%. The ESU_2 has a lower SOC, and it is charged with the rated power of 10 kW, while the other 5 kW of power is charged to the ESU_1. At 1.6 s, the controller is switched to the fuzzy controller to approach the droop coefficient. At 2.4 s, the SOCs of the two ESUs are balanced.

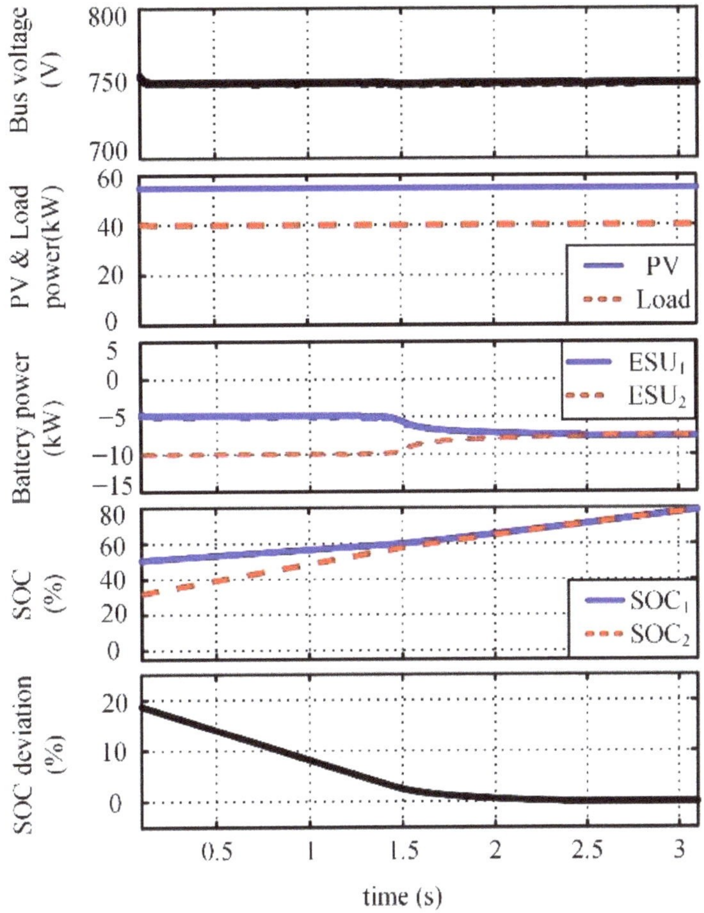

Figure 7. Waveforms of $P_{\mathrm{cmax}} \leq P < 2P_{\mathrm{cmax}}$ in charging.

In Figure 8, the PV outputs 49 kW of power, but the load only absorbs 40 kW. There is 9 kW of power remaining in the system, and it is less than the rated power of an individual ESU. ESU_1 has a higher SOC, so it barely absorbs power. ESU_2 absorbs all 9 kW of power since it has a lower SOC. The SOC balancing of two ESUs is achieved at 1.5 s, as seen in Figure 8.

It can be concluded from the above descriptions that the proposed method has superior performance.

① In terms of significant SOC deviation.

When the demanded discharging or charging power is greater than the rated power of an individual ESU, with the proposed method, one ESU is charged or discharged with the rated power while the other ESU compensates for the remaining power. When the demanded charging or discharging power is smaller than the rated power of an individual ESU, one ESU is charged or discharged with the required power while the other remains on standby to balance the power in the system. Therefore, the proposed method can achieve a fast SOC balance.

② In terms of slight SOC deviation.

The droop coefficient is adjusted dynamically by the fuzzy logic algorithm to achieve accurate SOC balance.

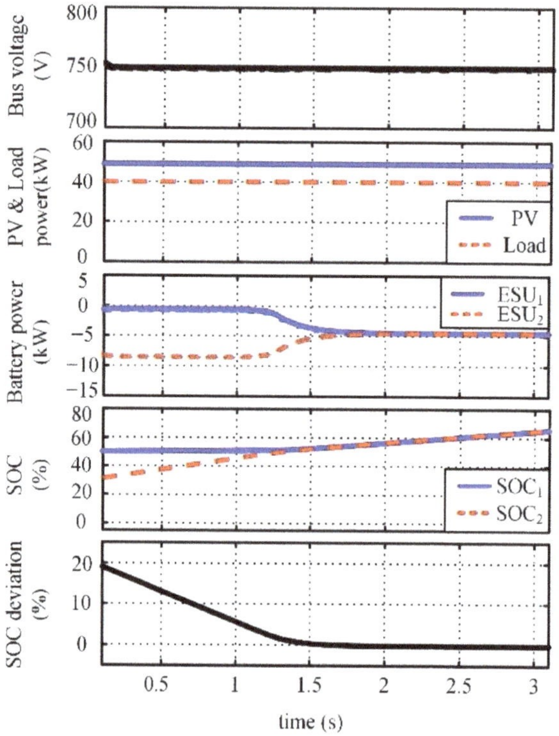

Figure 8. Waveforms of $P \leq P_{cmax}$ in charging.

4. Coordinated Control Based on the Piecewise Adaptive Algorithm

4.1. Coordinated Control Diagram of DC Microgrid

The coordinated control diagram of the DC microgrid is shown in Figure 9.

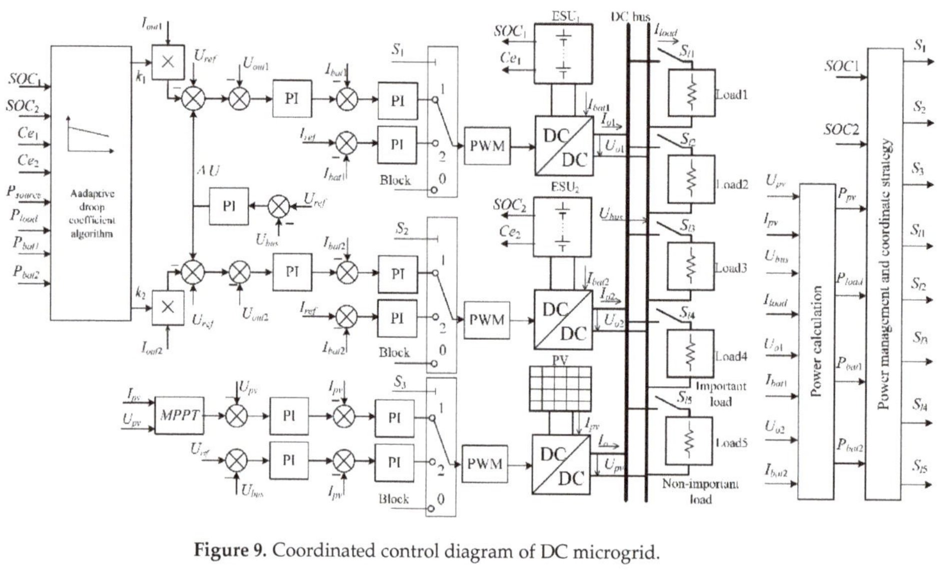

Figure 9. Coordinated control diagram of DC microgrid.

4.2. Power Management and Coordinated Control Strategy

The operation of the DC microgrid can be divided into the following five operation modes according to different conditions of PV, MESU, and loads:

Mode1: The PV is controlled by the constant voltage control (CVC); the ESU is on standby; basic loads are supplied;

Mode2: ① When the residual power is greater than the maximum power of the ESU, the PV is controlled by CVC. The batteries whose SOCs do not reach the limit are controlled to absorb constant power, while those whose SOCs reach the limit are on standby. ② When the residual power is smaller than the maximum power of the ESU, the PV is regulated by the maximum power point tracking (MPPT) control, and the batteries stabilize the bus's voltage or stay on standby;

Mode3: The PV is controlled by the MPPT control; the bus's voltage is maintained by ESU; basic loads are supplied;

Mode4: ① When the shortage of power is greater than the maximum power of the ESU, the PV is controlled by the MPPT control. The batteries whose SOCs do not reach the limit are controlled to stabilize the bus's voltage, while the one whose SOC reaches the limit is on standby; non-important loads are cut off in this scenario. ② When the power shortage is smaller than the maximum power of the ESU, the PV is regulated by the MPPT control and the batteries stabilize the bus voltage or stay on standby, supplying basic loads;

Mode5: The PV is controlled by the MPPT control; ESU is on standby; important loads are cut off gradually according to the PV power.

The correspondence between the DC microgrid and the control mode under each operating condition is shown in Table 4, and the control mode selection flow diagram is shown in Figure 10.

Table 4. Running condition and control mode.

SOC	$P_{pv} > P_{load}$	$P_{pv} < P_{load}$
$SOC_1 > 90\%$ and $SOC_2 > 90\%$	Mode1	Mode3
$SOC_1 > 90\%$ or $SOC_2 > 90\%$	Mode2	Mode3
$10\% \le SOC_1 \le 90\%$ $10\% \le SOC_2 \le 90\%$	Mode3	Mode3
$SOC_1 < 10\%$ or $SOC_2 < 10\%$	Mode3	Mode4
$SOC_1 < 10\%$ and $SOC_2 < 10\%$	Mode3	Mode5

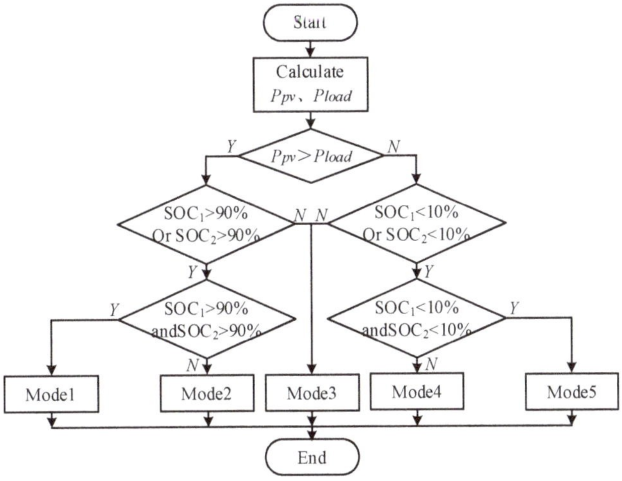

Figure 10. Control mode selection flow diagram.

4.3. Simulation Waveforms of the Coordinated Control

(1) Response to a sudden change of PV power

As shown in Figure 11, the initial SOC values are 85% and 75% for ESU_1 and ESU_2. Irradiance is constant at 900 W/m^2; PV power is 58 kW with MPPT control; the load is constant at 40 kW. During these 0.1~0.42 s, the SOCs of the two ESUs do not reach 90%, and they keep the bus voltage constant. The system works in Mode3. At 0.42 s, SOC_1 reaches 90%, and ESU_1 is on standby. Simultaneously, the maximum PV power is 58 kW, and the load demand is only 40 kW. In order not to exceed the maximum allowable 10 kW charging power of the lithium battery, ESU_2 is controlled to operate with a constant charging current. The PV power outputs 50 kW, and the system switches to control Mode2. During 0.42~0.81 s, ESU_2 is continuously charged, and the bus voltage is controlled by the PV system. At 0.81 s, the SOC_2 reaches 90%, and the charging power of ESU_2 is 0. The PV outputs 40 kW, and the system is switched to Mode1.

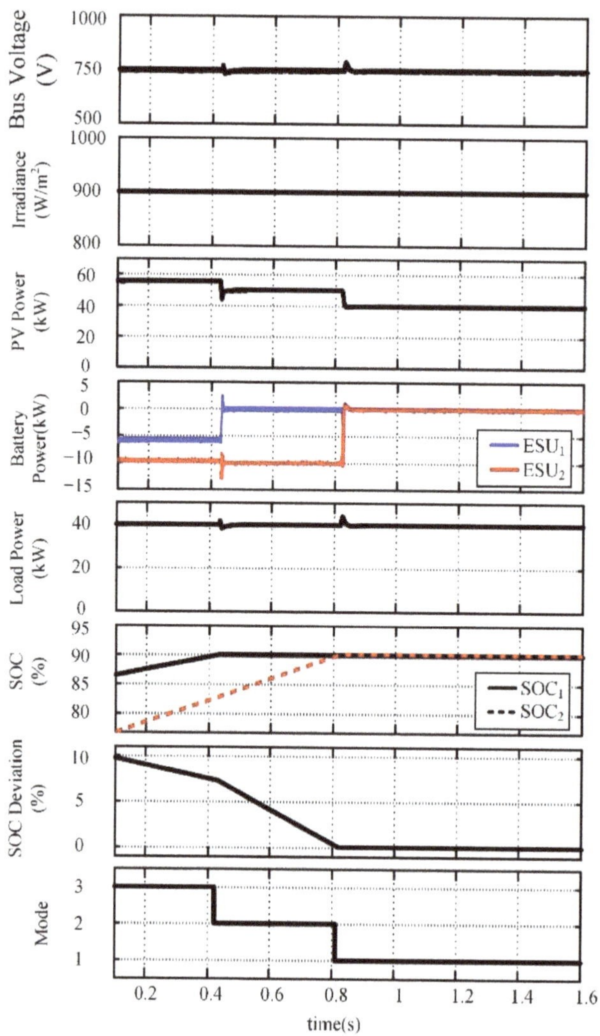

Figure 11. Response to sudden change of PV power.

(2) Simulation results under removal of the load

As shown in Figure 12, irradiance is reduced from 450 W/m^2 to 0 and the corresponding PV power is reduced from 25 kW to 0. The initial SOCs are set at 35% and 15%. Before 0.9 s, the maximum PV power was 27 kW, and the 13 kW of power shortage was allocated by the two ESUs, maintaining the bus voltage. The system works in Mode3 at this time. During these 0.9~1 s, ESU_1 is on standby because SOC_1 reaches the 10% minimum limit. The power shortage in the network is slightly less than 20 kW, which was all released by ESU_2. At this time, the system is working in Mode4. After 1 s, the SOCs of both ESUs reach the lower limit (10%), so they are on standby. However, the maximum power of the PV cannot be maintained to supply the full basic loads. According to the power of PV and load, 20/5/5/5/5 kW load is cut off at 1/1.25/1.8/2.2/2.6 s according to their importance. The remaining power is used to charge the two ESUs. The system is running in Mode5 at this time.

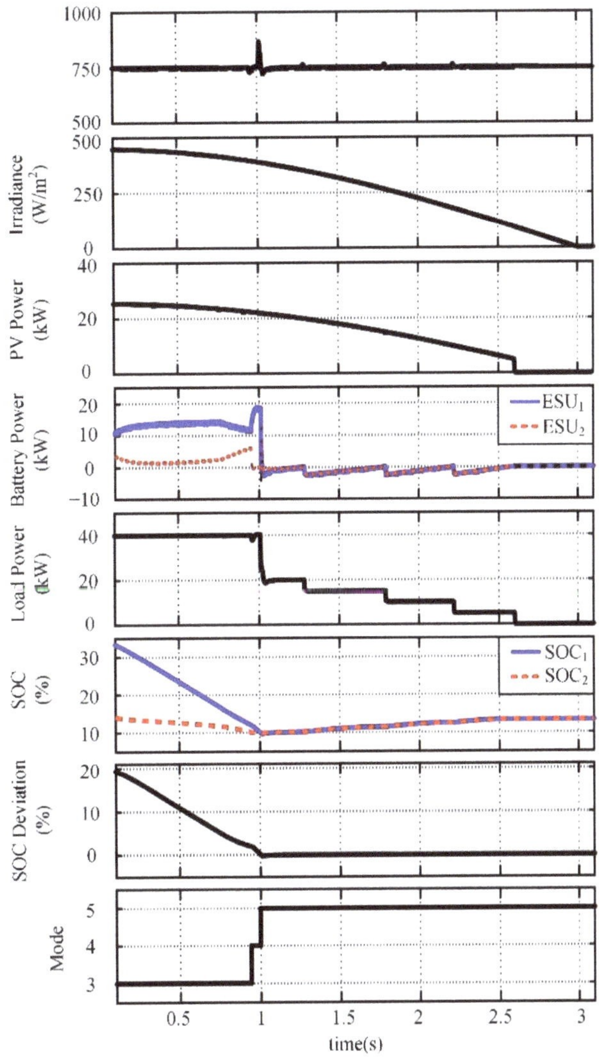

Figure 12. Response to removal of the load.

5. Conclusions

This paper presents the SOC balancing and coordinated control strategy based on the adaptive droop coefficient algorithm for MESU. When the SOC deviation is relatively large, the droop coefficient of an ESU with a lower SOC is set to be the minimum value in the charging process. In the discharging process, the droop coefficient of an ESU with a higher SOC is set to be the minimum value. In the charging process, the ESU with a lower SOC absorbs energy with the rated power or all the charging power, while the one with a higher SOC absorbs the residual charging power or does not absorb any power. Meanwhile, in the discharging process, the ESU with a higher SOC delivers energy with rated power or all discharging power. The ESU with a lower SOC provides the rest of the charging power or does not deliver any power. Hence, rapid SOC balancing can be realized. When the SOC deviation is slight, the droop coefficient is adjusted smoothly and automatically by a fuzzy logic algorithm. The fuzzy logic algorithm adjusts the droop coefficient according to the inputs of the SOC deviation and output power of each ESU. Therefore, SOC deviation caused by the actual capacity and line impedance is eliminated, and accurate SOC balancing is achieved. The coordinated control of the DC microgrid is being studied to ensure the stable operation of the DC microgrid and the stability of bus voltage. The simulation results show that the proposed control strategy can realize the rapid SOC balance of multiple ESUs and maintain the bus voltage stability.

Author Contributions: Data curation, J.Z.; Formal analysis, Y.Z.; Investigation, G.T.; Methodology, Y.Z. and G.T.; Project administration, G.L.; Software, J.Z.; Supervision, G.T.; Validation, G.T.; Visualization, Y.Z.; Writing—original draft, Y.Z.; Writing—review and editing, G.T. All authors have read and agreed to the published version of the manuscript.

Funding: This work was funded by the National Natural Science Foundation of China (NO. 51767019, 51867020), the S&T Major Project of Inner Mongolia Autonomous Region under Grant (No. 2021ZD0040), the Inner Mongolia Science & Technology Plan under Grant (Research on key technology of multi-source coordinated control in islanded operation DC microgrid) and the Inner Mongolia Autonomous Region Natural Science Foundation under Grant (No. 2021MS05003, 2020BS05027).

Institutional Review Board Statement: Not applicable.

Informed Consent Statement: Not applicable.

Conflicts of Interest: The authors declare no conflict of interest.

References

1. Li, L.; Dragičević, T.; Guerrero, J.M.; Vasquez, J.C. Supervisory control of an adaptive-droop regulated dc microgrid with battery management capability. *IEEE Trans. Power Electron.* **2014**, *29*, 695–706.
2. Gao, F.; Kang, R.; Cao, J.; Yang, T. Primary and secondary control in dc microgrids: A review. *Mod. Power Syst. Clean Energy* **2019**, *7*, 227–242. [CrossRef]
3. Han, Y.; Ning, X.; Yang, P.; Xu, L. Review of power sharing, voltage restoration and stabilization techniques in hierarchical controlled dc microgrids. *IEEE Access* **2019**, *7*, 149202–149223. [CrossRef]
4. Yan, H.W.; Narang, A.; Tafti, H.D.; Farivar, G.G.; Ceballos, S.; Pou, J. Minimizing energy storage utilization in a stand-alone dc microgrid using photovoltaic flexible power control. *IEEE Trans. Smart Grid* **2021**, *12*, 3755–3764. [CrossRef]
5. Morstyn, T.; Hredzak, B.; Agelidis, V.G. Control strategies for microgrids with distributed energy storage systems: An overview. *IEEE Trans. Smart Grid* **2018**, *9*, 3652–3666. [CrossRef]
6. Madadi, M.; Bhattacharya, S. Adaptive nonlinear droop control with dynamic state-of-charge balancing capability for batteries in dc microgrids. In Proceedings of the IEEE Applied Power Electronics Conference and Exposition, Phoenix, AZ, USA, 14–17 June 2021; pp. 55–61.
7. Yang, Y.; Tan, S.; Hui, S.Y.R. Mitigating distribution power loss of dc microgrids with dc electric springs. *IEEE Trans. Smart Grid* **2018**, *9*, 5897–5906. [CrossRef]
8. Meng, T.; Lin, Z.; Wan, Y.; Shamash, Y.A. State-of-charge balancing for battery energy storage systems in dc microgrids by distributed adaptive power distribution. *IEEE Control Syst. Lett.* **2021**, *6*, 512–517. [CrossRef]
9. Maharjan, L.; Inoue, S.; Akagi, H.; Asakura, J. State-of-charge (SOC)-balancing control of a battery energy storage system based on a. cascade PWM converter. *IEEE Trans. Power Electron.* **2009**, *24*, 1628–1636. [CrossRef]
10. Huang, W.; Qahouq, J.A.A. Energy sharing control scheme for state-of-charge balancing of distributed battery energy storage system. *IEEE Trans. Ind. Electron.* **2015**, *62*, 2764–2776. [CrossRef]

11. Hong, Y.; Xu, D.; Yang, W.; Jiang, B.; Yan, X.G. A novel multi-agent model-free control for state-of-charge balancing between distributed battery energy storage systems. *IEEE Trans. Emerg. Top. Comput. Intell.* **2021**, *5*, 679–688. [CrossRef]
12. Bhosale, R.; Gupta, R.; Agarwal, V. A novel control strategy to achieve soc balancing for batteries in a dc microgrid without droop control. *IEEE Trans. Ind. Appl.* **2021**, *57*, 4196–4206. [CrossRef]
13. Fagundes, T.A.; Fuzato, G.H.F.; Ferreira, P.G.B.; Biczkowski, M.; Machado, R.Q. Fuzzy controller for energy management and soc equalization in dc microgrids powered by fuel cell and energy storage units. *IEEE J. Emerg. Sel. Top. Ind. Electron.* **2022**, *3*, 90–100. [CrossRef]
14. Liu, C.; Zhao, J.; Wang, S.; Lu, W.; Qu, K. Active identification method for line resistance in dc microgrid based on single pulse injection. *IEEE Trans. Power Electron.* **2018**, *33*, 5561–5564. [CrossRef]
15. Zhi, N.; Ding, K.; Du, L.; Zhang, H. An SOC-based virtual dc machine control for distributed storage systems in dc microgrids. *IEEE Trans. on Energy Convers.* **2020**, *35*, 1411–1419. [CrossRef]
16. Baharizadeh, M.; Golsorkhi, M.S.; Shahparasti, M.; Savaghebi, M. A two-layer control scheme based on *P-V* droop characteristic for accurate power sharing and voltage regulation in dc microgrids. *IEEE Trans. Smart Grid* **2021**, *12*, 2776–2787. [CrossRef]
17. Braitor, A.C.; Konstantopoulos, G.C.; Kadirkamanathan, V. Current-limiting droop control design and stability analysis for paralleled boost converters in dc microgrids. *IEEE Trans. Control Syst. Technol.* **2021**, *29*, 385–394. [CrossRef]
18. Zhang, Y.Y.; Qu, X.H.; Tang, M.D.; Yao, R.Y.; Wu, C. Design of nonlinear droop control in dc microgrid for desired voltage regulation and current sharing accuracy. *IEEE J. Emerg. Sel. Top. Circuits Syst.* **2021**, *11*, 168–175. [CrossRef]
19. Lu, X.; Sun, K.; Guerrero, J.M.; Vasquez, J.C.; Huang, L. State-of-charge balance using adaptive droop control for distributed energy storage systems in dc microgrid applications. *IEEE Trans. Ind. Electron.* **2014**, *61*, 2804–2814. [CrossRef]
20. Lu, X.; Sun, K.; Guerrero, J.M.; Vasquez, J.C.; Huang, L. Double-quadrant state-of-charge-based droop control method for distributed energy storage systems in autonomous dc microgrids. *IEEE Trans. Smart Grid* **2015**, *6*, 147–157. [CrossRef]
21. Diaz, N.L.; Dragičević, T.; Vasquez, J.C.; Guerrero, J.M. Intelligent distributed generation and storage units for dc microgrids—a new concept on cooperative control without communications beyond droop control. *IEEE Trans. Smart Grid* **2014**, *5*, 2476–2484. [CrossRef]
22. Oliveira, T.R.; Silva, W.W.A.G.; Donoso-Garcia, P.F. Distributed secondary level control for energy storage management in dc microgrids. *IEEE Trans. Smart Grid* **2017**, *8*, 2597–2607. [CrossRef]
23. Hoang, K.D.; Lee, H. Accurate power sharing with balanced battery state of charge in distributed DC microgrid. *IEEE Trans. Ind. Electron.* **2019**, *66*, 1883–1893. [CrossRef]
24. Morstyn, T.; Savkin, A.V.; Hredzak, B.; Agelidis, V.G. Multi-agent sliding mode control for state of charge balancing between battery energy storage systems distributed in a DC microgrid. *IEEE Trans. Smart Grid* **2018**, *9*, 4735–4743. [CrossRef]
25. Xu, D.; Zhang, W.; Jiang, B.; Shi, P.; Wang, S. Directed-graph-observer-based model-free cooperative sliding mode control for distributed energy storage systems in dc microgrid. *IEEE Trans. Ind. Inform.* **2020**, *16*, 1224–1234. [CrossRef]
26. Chen, X.; Shi, M.; Sun, H.; Li, Y.; He, H. Distributed cooperative control and stability analysis of multiple dc electric springs in a dc microgrid. *IEEE Trans. Ind. Electron.* **2018**, *65*, 5611–5622. [CrossRef]
27. Lin, X.; Zamora, R.; Baguley, C.A. "A fully filter-based decentralized control with state of charge balancing strategy for battery energy storage systems in autonomous dc microgrid applications. *IEEE Access* **2021**, *9*, 15028–15040. [CrossRef]
28. Zheng, Y.; Tian, G.; Zhang, J. SOC balancing control strategy based on piecewise adaptive droop coefficient algorithm for multienergy storage units in dc microgrid. In Proceedings of the IEEE 4th International Conference on Electronics Technology, Chengdu, China, 7–10 May 2021; pp. 432–436.
29. Long, B.; Liao, Y.; Chong, K.T.; Rodríguez, J.; Guerrero, J.M. Enhancement of frequency regulation in ac microgrid: A fuzzy-MPC controlled virtual synchronous generator. *IEEE Trans. Smart Grid* **2021**, *12*, 3138–3149. [CrossRef]
30. Coelho, V.N. A self-adaptive evolutionary fuzzy model for load forecasting problems on smart grid environment. *Appl. Energy* **2016**, *169*, 567–584. [CrossRef]
31. Wang, Y.; Jin, Q.; Zhang, R. Improved fuzzy PID controller design using predictive functional control structure. *ISA Trans.* **2017**, *71*, 354–363. [CrossRef]
32. Ghanbari, N.; Mobarrez, M.; Bhattacharya, S. A review and modeling of different droop control based methods for battery state of the charge balancing in dc microgrids. In Proceedings of the 44th Annual Conference of the IEEE Industrial Electronics Society, Washington, DC, USA, 21–23 October 2018; pp. 1625–1632.

MDPI
St. Alban-Anlage 66
4052 Basel
Switzerland
www.mdpi.com

Energies Editorial Office
E-mail: energies@mdpi.com
www.mdpi.com/journal/energies

www.ingramcontent.com/pod-product-compliance
Lightning Source LLC
Chambersburg PA
CBHW061445210926
51850CB00008B/1620